T0376079

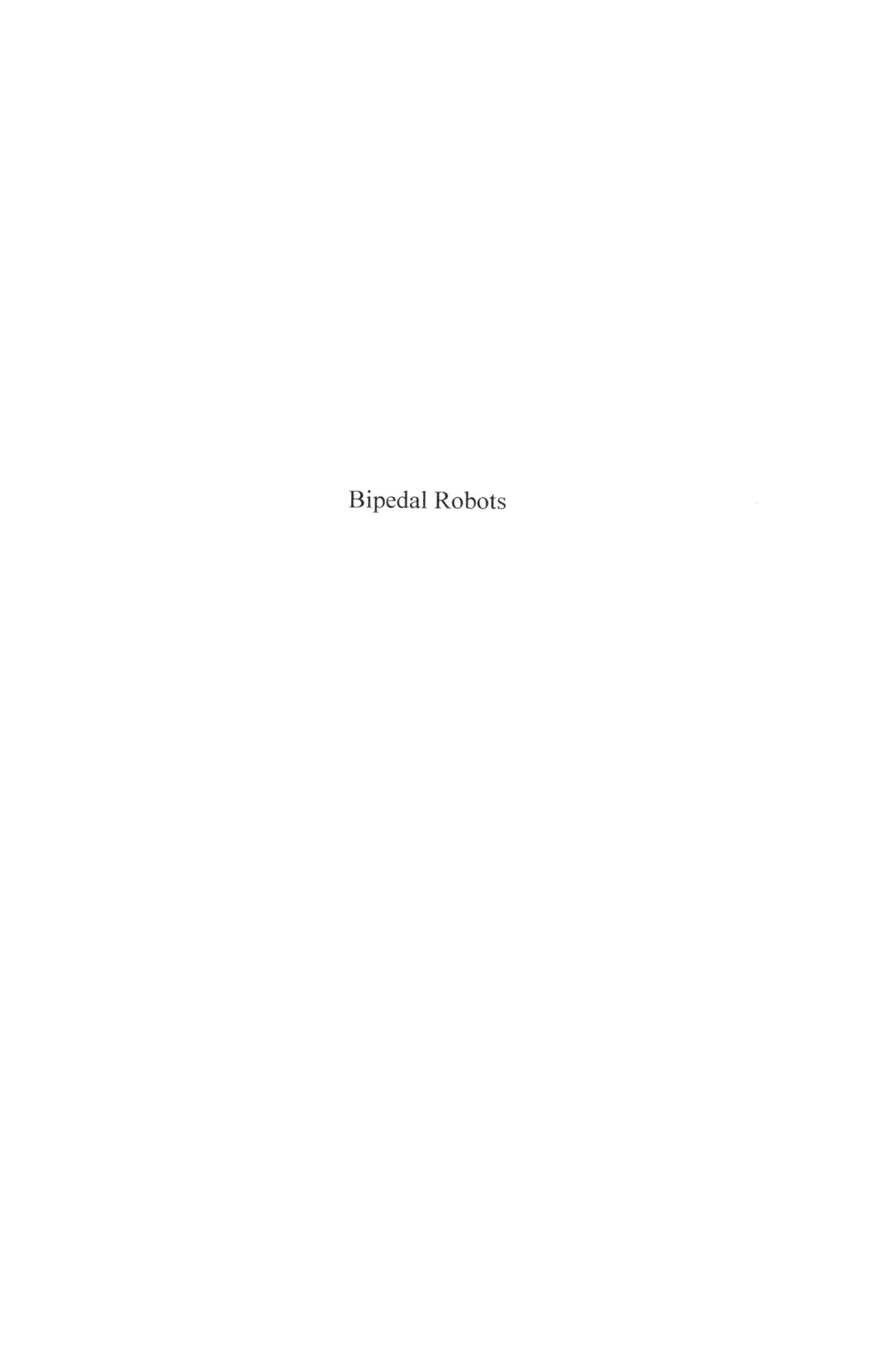

Bipedal Robots

Bipedal Robots

Modeling, Design and Walking Synthesis

Edited by
Christine Chevallereau
Guy Bessonnet
Gabriel Abba
Yannick Aoustin

First published in France in 2007 by Hermes Science/Lavoisier entitles: *Les robots marcheurs bipèdes : modélisation, conception, synthèse de la marche* © LAVOISIER, 2007
First published in Great Britain and the United States in 2009 by ISTE Ltd and John Wiley & Sons, Inc.

ISTE Ltd
27-37 St George's Road
London SW19 4EU
UK

www.iste.co.uk

John Wiley & Sons, Inc.
111 River Street
Hoboken, NJ 07030
USA

www.wiley.com

Library of Congress Cataloging-in-Publication Data

[Robots marcheurs bipèdes. English]
Bipedal robots : modeling, design and walking synthesis / edited by Christine Chevallereau ... [et al.]
p. cm.
Includes bibliographical references and index.
ISBN 978-1-84821-076-9
1. Robots--Motion. 2. Walking. I. Chevallereau, Christine. II. Title.
TJ211.4.R6313 2008
629.8'932--dc22
2008035231

British Library Cataloguing-in-Publication Data
A CIP record for this book is available from the British Library
ISBN: 978-1-84821-076-9

Table of Contents

Chapter 1. Bipedal Robots and Walking

Chapter 2. Kinematic and Dynamic Models for Walking

Chapter 4. Walking Pattern Generators

Chapter 5. Control

Chapter 1

Bipedal Robots and Walking

1.1. Introduction

Man has always been interested in the relationship between himself and the living world and, more particularly, in understanding how he is different from other animals. Since Antiquity and Aristotle, then during the Renaissance when the first studies were carried out in medicine and physiology, scientists have tried to understand the influence of bipedia on human evolution. In his 1680 publication, *De Motu Animalium*, Borelli [BOR 80] compared different bipedal species, analyzed the importance of pendulous movement and introduced spring-mass models in order to understand walking and running in humans and other bipedal animals.

In the 18th century, Doyon [DOY 66] built a whole ensemble of automats. More recent studies have tried to make parallels between passive human walking and walking executed by prototype robots such as Collins' *Walker* [COL 05] or McGeer's *Straight-legged biped* [MCG 90]. These studies can help us to have a better understanding of the laws needed for commands, stability or the generations of trajectories for future humanoid robots.

Research carried out in biomechanics has enabled us to interpret in more detail the principles of kinetic and potential energy transfer which contribute to defining walking. Tendons and muscles are particularly used during walking and running, acting as actuators but also as shock absorbers. Studies have shown that they temporarily store kinetic energy which is redistributed in the propulsion phases. The simple model of an inverted pendulum mounted on a compression spring placed in the leg can be used to describe running. These studies also show that one of the most influential parameters of stability during running is the start angle between the equivalent virtual leg and the ground [SEY 02]. This angle even seems to be an auto-stabilizing running parameter. These behavioral models have been confirmed by tests carried out at MIT on a one-legged jumping robot [RAI 86].

Biomechanics has also enabled us to design kinematic and dynamic walking and running models from biological systems used during simulations with interactions with the ground, and to determine internal forces. These models have enabled us to develop equivalent methods in the context of robotics.

The number of international research teams working on the subject is steadily rising. The disciplines interested in the theme of bipedal robots are: robotics, but also automatics, mechanics, information technology, biomechanics, medicine (with work on improving orthopedic prostheses and medical rehabilitation) and cinematography (e.g. computer animation). Studies in the field of sport have been concentrated on optimizing training and detecting the limiting constraints of articulations and tendons.

Section 1.2 will present a non-exhaustive overview of the biological and biomechanical approaches to this question. The notions of similarity, the characteristics of energy consumption and mobility are particularly focused on.

Section 1.3 concerns human walking. The structure of lower limbs and their muscles are given in detail to illustrate the specificity of human bipeds. Experimental results obtained from the capture of a walking movement will conclude this section.

Section 1.4 offers a historic overview of the different bipedal robots that have been created worldwide in the past.

Finally, the present day and future applications of robot bipedia will be presented in section 1.5.

1.2. Biomechanical approach

1.2.1. *Biomechanical system: a source of inspiration*

Among the ensemble of living things, only humans and birds use bipedia as a form of locomotion. Certain insects (cockroaches in flight for example), reptiles (e.g. lizards, geckos and monitor lizards) and mammals (e.g. primates, bears, mongooses and rats) use bipedia in exceptional cases. In the history of species, bipedia appeared during the Mesozoic periods among a large number of dinosaur reptiles. The anatomic structure of the legs of these animals is very similar, consisting of a thigh, a leg and a foot. Each lower limb has three main articulations: the hip, the knee and the ankle. The hip is a spherical link with three degrees of mobility. The knee is a link with one or two degrees of freedom (DoF). The ankle brings together two main mobilities. An important difference is the relative length of the foot and its average position during walking which can perceptibly modify the extreme values of joint angles.

Physiologists and anatomists have developed the laws of similarity from measurements taken from animals and humans. In the 1980s, the comprehensive work of Alexander's team [ALE 77, ALE 00, ALE 04, ALE 05] enabled them to formulate the laws of geometric and energetic similarity for a large number of living animals. Figure 1.1 depicts, for example, the results obtained from the measurements of the length and diameter of femurs and tibias. For the femurs of primates, for example, these laws of similarity can be expressed in the following formula:

$$l = 100\, m^{0.34} \qquad [1.1]$$

$$d = 6\, m^{0.39} \qquad [1.2]$$

where l and d represent the length and diameter of the femur (mm), respectively, and m the mass of the body (kg). It has been convened that the laws of similarity can be defined by normalizing the sizes in question. The preceding relations therefore become:

$$l^* = (m^*)^{0.34} \qquad [1.3]$$

$$d^* = (m^*)^{0.39} \qquad [1.4]$$

where l^*, d^* and m^* are the normalized length, diameter and mass, respectively. For the same species, studies have shown that they are satisfactory for a large number of bones in the skeleton.

In the same way, the respective durations of the single and double support phases and the relations giving the length and frequency of steps during walking or running have been established by researchers. Figure 1.2 show the measured values for the lengths of steps for different living bipeds The dotted vertical lines on the right-hand side of Figure 1.2 indicate the transition zone between walking and running for the bipedal animals under consideration [HIL 67].

We notice that this zone is small which implies that all bipeds modify their gaits for the same relative speed (as defined by Froude's number) even when their size and morphology are different.

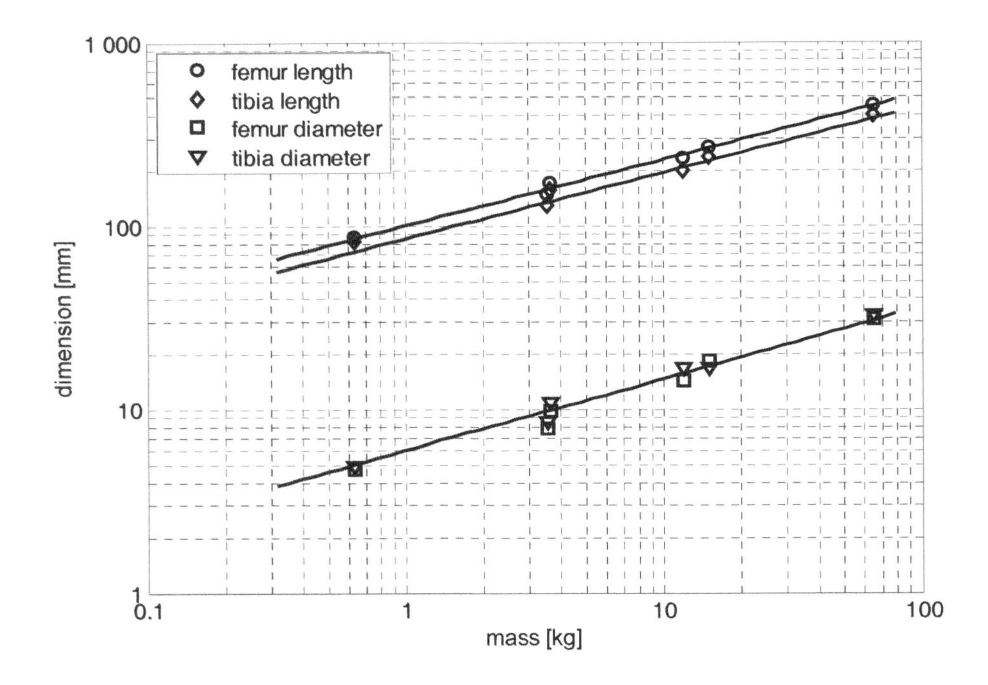

Figure 1.1. *The law of similarity which defines the length and diameter of the femurs and tibias of primates according to Alexander et al. [ALE 79]*

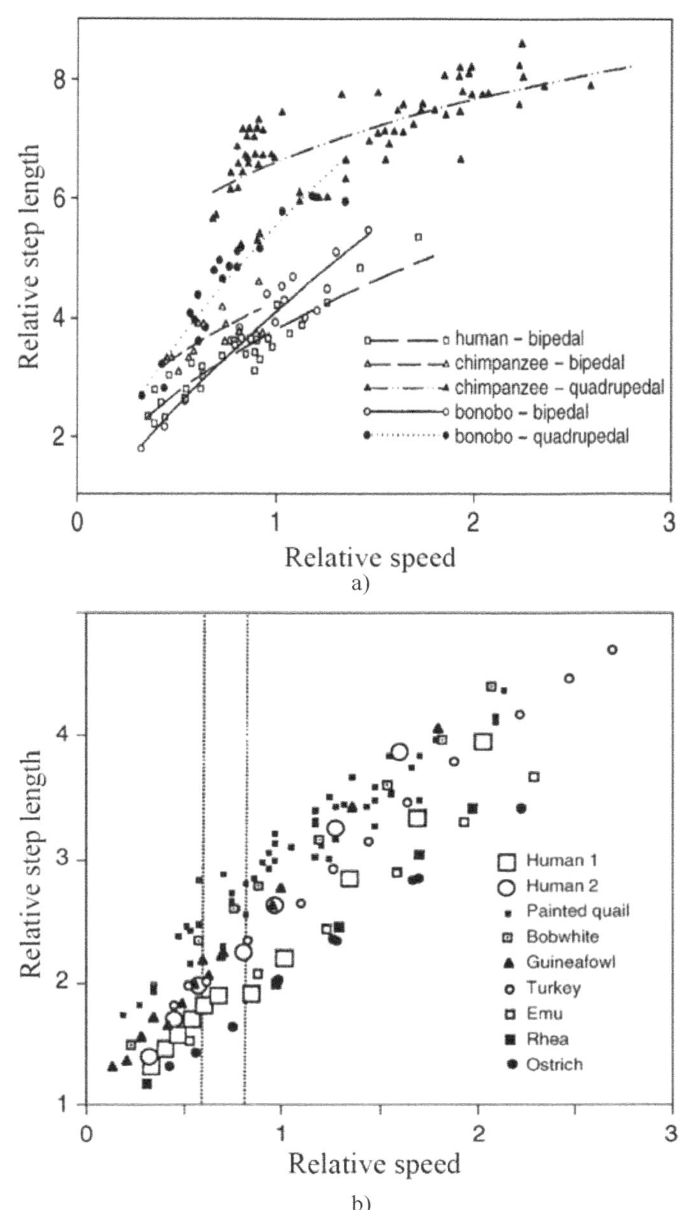

Figure 1.2. *Relative length of walking and running steps for humans and different bipedal animals dependent on the relative speed of progression (according to [ALE 04]). The relative speed is the ratio of the forward speed divided by* $\sqrt{gl}$ *(where g is acceleration due to gravity and l is distance from the hip to ground in the upright position)*

Energy consumption is a very important criterion in the design and study of bipedal robots. The study of the measurements of the energetic consumption of living things is also of evident interest. It is not easy to take measurements of consumed energy per living thing unit of time during walking or running or during any other physical activity. The measurement of the consumption of oxygen per unit of time seems to be the most representative of the total amount of consumption linked to physical activity. The studies of [MAN 80, THY 01] have provided a lot of information on this subject.

Simulation models have also been established from physiological data and have enabled us to plot the energy consumed per human during walking for different speeds and per unit of distance traveled.

Figure 1.3 shows the results obtained by Sellers *et al.* [SEL 04] which measure the energy consumption of a walker during 1 hour and traveling over a given distance (see graph legend) depending on the speed of walking (the tested subject remains immobile when he has traveled the given distance in less than an hour).

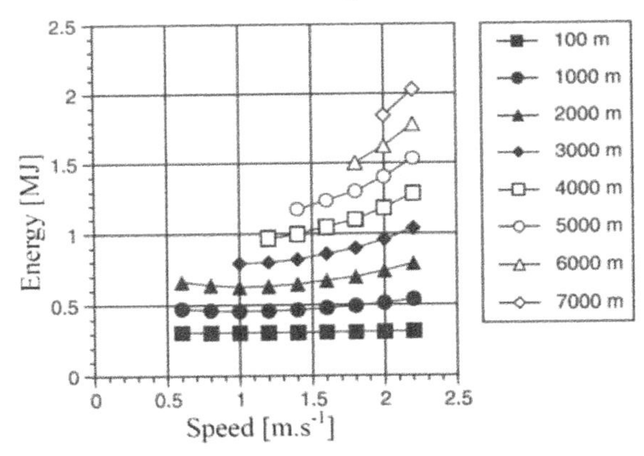

Figure 1.3. *Energy consumed per human during walking depending on the average speed of progression and the distance traveled (according to [SEL 04])*

Data recently published by Marden and Bejan [BEJ 06, MAR 05] show that the actuators used in nature and electromechanical or thermal actuators follow identical laws of similarity.

Marden has identified two family groups of actuator: the first produces linear movements (myosin, DNA polymerase muscle, linear electric motor) and develops a maximum force given by relation [1.5], whereas the second family group corresponds to rotation or beat movements [SCH 04] (insect or bird flight, swimming fish, electric or thermal rotating motors) which deliver a force given by relation [1.6]:

$$F_{max} = 891 m^{0.67} \quad [1.5]$$

$$F_{max} = 55 m \quad [1.6]$$

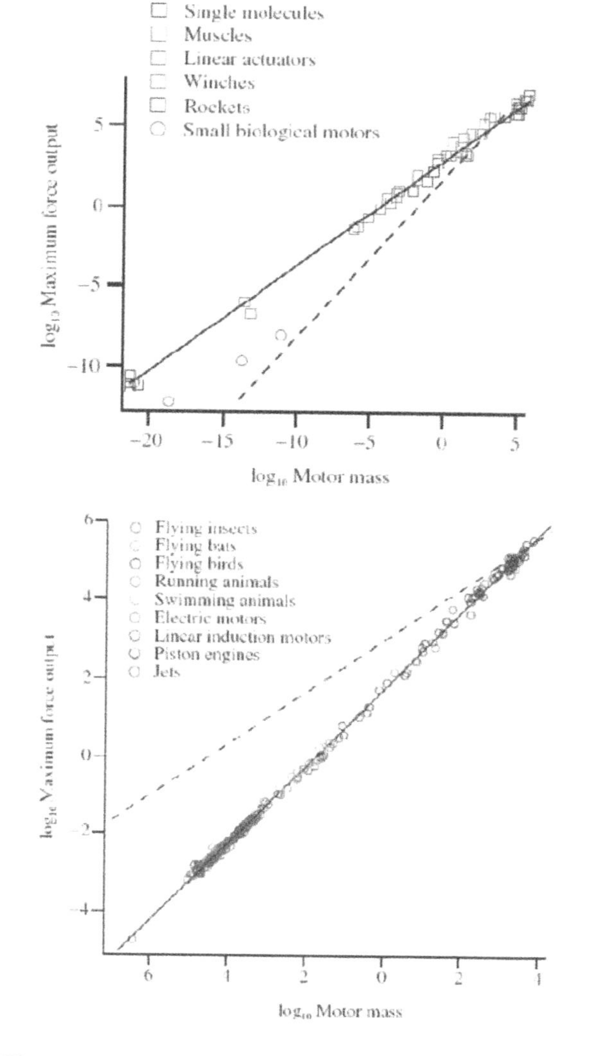

Figure 1.4. *Mass-force relation for different types of actuator (according to [MAR 05])*

Figure 1.4 shows the force produced by an actuator depending on its mass for a set of very different actuators with respect to the origin of the energy that they convert. The corpus of the different types of actuator can therefore be described by the curves in Figure 1.4. Relation [1.5] is represented by the continuous lined curve in the top graph, and the dotted lined curve in the bottom graph. Relation [1.6] is represented by the continuous lined curve in the bottom graph, and the dotted line in the top graph.

In short, biologists have put to the forefront the similarities between the locomotive behavior of animals regardless of their sizes and weights. Following measurements taken from animals of similar anatomic types but of differing sizes, Hill [HIL 50] arrived at the conclusion that for animals of similar morphology, speed is independent of sizes to within a scale factor. In this way, an animal makes movements of amplitude which are proportional to its size and at a frequency which is inversely proportional to this same size. During the study of the effects of scale on the structural adaptation of animals, three main types of similarity were brought to light.

The first is of geometric similarity, where two structures are geometrically similar if one can be obtained from the other by a uniform change in scale factor.

The second is the elastic similarity which allows deformations of the spinal cord without risking lesions [MCM 75].

The third is the dynamic similarity which includes morphological variations of animals of different sizes, the evolution of the movements and their delivered efforts. According to Alexander [ALE 83, ALE 84], two movements are dynamically similar if one can become the other by the uniform change of one or more of the three scale factors: length, time and force.

It has been demonstrated that when the forces of gravity and inertia are preponderate, two movements are dynamically similar only if they have the same Froude number:

$$F = \frac{v^2}{gl} \qquad [1.7]$$

where g is the acceleration of gravity, v is the average horizontal speed and l is the height of the hip from the ground in the upright position. Man moves from walking to running at a Froude number of 0.6 [BRU 98].

1.2.2. *Skeletal structure and musculature*

Bipedia has two main gaits: walking and running. When walking, there is always one of the two locomotive limbs in contact with the ground.

When running, a grounded monopodal impulse propels the body up and forwards in the form of a leap, followed by a new grounding with a shock to the other locomotive limb.

The kinetic energy of this ballistic phase is partially recovered at renewal support, thanks to the elasticity of the articulation, muscles and tendons.

The energy necessary for the leap is partially recovered when new contact is made with the ground. There is a left to right alternate weight transfer between the lower limbs.

Monkeys only use bipedia occasionally. They are not actual quadrupeds either because their upper and lower limbs end in hands with opposable thumbs, which are adapted for brachiation. They walk in this way on four hands which, in addition, have no pedal arches.

Bears also use bipedia very occasionally. Nevertheless, they are real plantigrade quadrupeds; they do not have opposable thumbs on any of their legs and when walking, their pedal arches are fully in contact with the ground.

The locomotive system of kangaroos is made up of lower pelvic members and a tail.

Birds, whose origins can be traced back to the theropod dinosaurs of the Jurassic period [DER 70], use a terrestrial system of locomotion where only the pelvic limbs remain, dissociated from the caudal region from their distant Jurassic ancestors.

The feet of birds are characterized by a tri-segmented Z structure: the femur, the tibiotarsus and the tarsometatarsus [MED 06]. The spinal cord of birds is fixed from the pelvis to the nape. Its central skeleton can therefore be considered as an undeformable solid.

The human spinal cord has a great number of individual mobilities. The trunk is also articulated in this way. Its movements have an influence on its locomotion. The spinal cord has a system of muscular tensors which take into account the constraints due to the alternating weight transfers from the lower right to the left limb during walking or running.

The human locomotive system stores and gives back energy (see section 1.2.2) due to the elasticity of the foot's arch, its muscles, its ankles and its spinal discs.

Nordez's [NOR 06] work on this matter gives a very detailed characterization of the passive stretch of the musco-articular complex. It was noticed that a static stretch brings about a mainly transitory increase in the length of muscles, whereas the dissipative properties and stiffness are only modified after cyclic stretches.

Nevertheless, to maintain an adequate level of mechanical energy for locomotion, the muscles transform chemical energy into mechanical energy, resulting in a global energetic demand.

To carry out the mechanical modeling of anthropomorphic structures, some authors have suggested models for the distribution of mass, the choice of segmented joint-links (and the number of segments) and their geometry [HAN 64].

To illustrate this, Figure 1.5 shows the skeletons of two bipedal animals (a bird and iguanodon) and a human in the upright position.

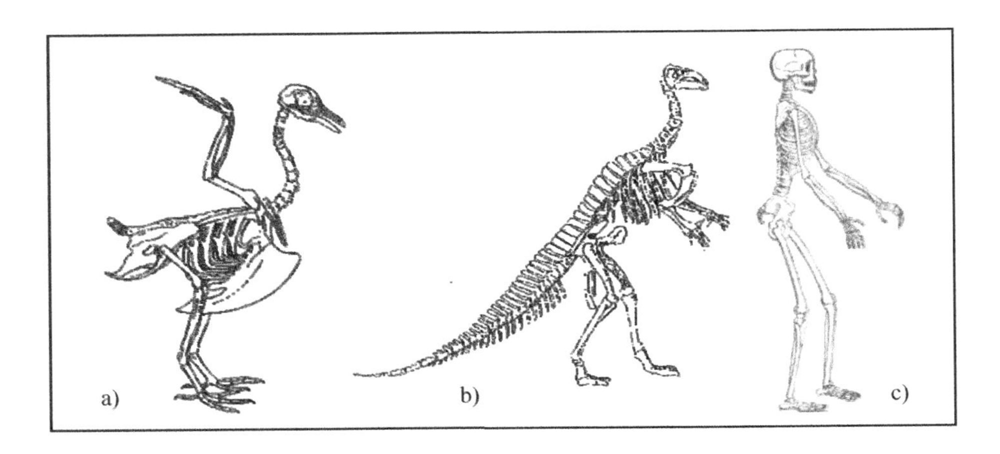

Figure 1.5. *Skeletons of three bipeds: (a) a bird; (b) an iguanodon; (c) a human*

1.3. Human walking

1.3.1. *Architecture*

The architecture of a human's lower limbs is very complex (see Figure 1.6). Its bone structure regroups 44 bones, of which the femur, tibia and fibula are the main bones. Apart from the knee-cap, the remaining bones make up the foot, which can be considered as a deformable composite corpus.

Each member therefore has an ensemble of three main corpuses (thigh, leg and foot) linked together by articulations which have many degrees of mobility.

In this way, the hip articulation has 3 revolutionary degrees of mobility. The articulations of the knees and the ankles each have two DoF.

The musculature of a human's lower limbs is made up of 46 skeletal muscles. The muscles which intervene in the propulsion movement are longer muscles. In this way, we notice that many muscles intervene simultaneously to assist in the motion of an articulation. The muscles make an effort of traction and they always work in conjunctive pairs with another opposite muscle.

We can also separate muscles into those which only act on a single articulation (e.g. the iliacus muscle on the hip) and others, which act on separate body-links, which are separated by two articulations (e.g. the rectus femoris muscle).

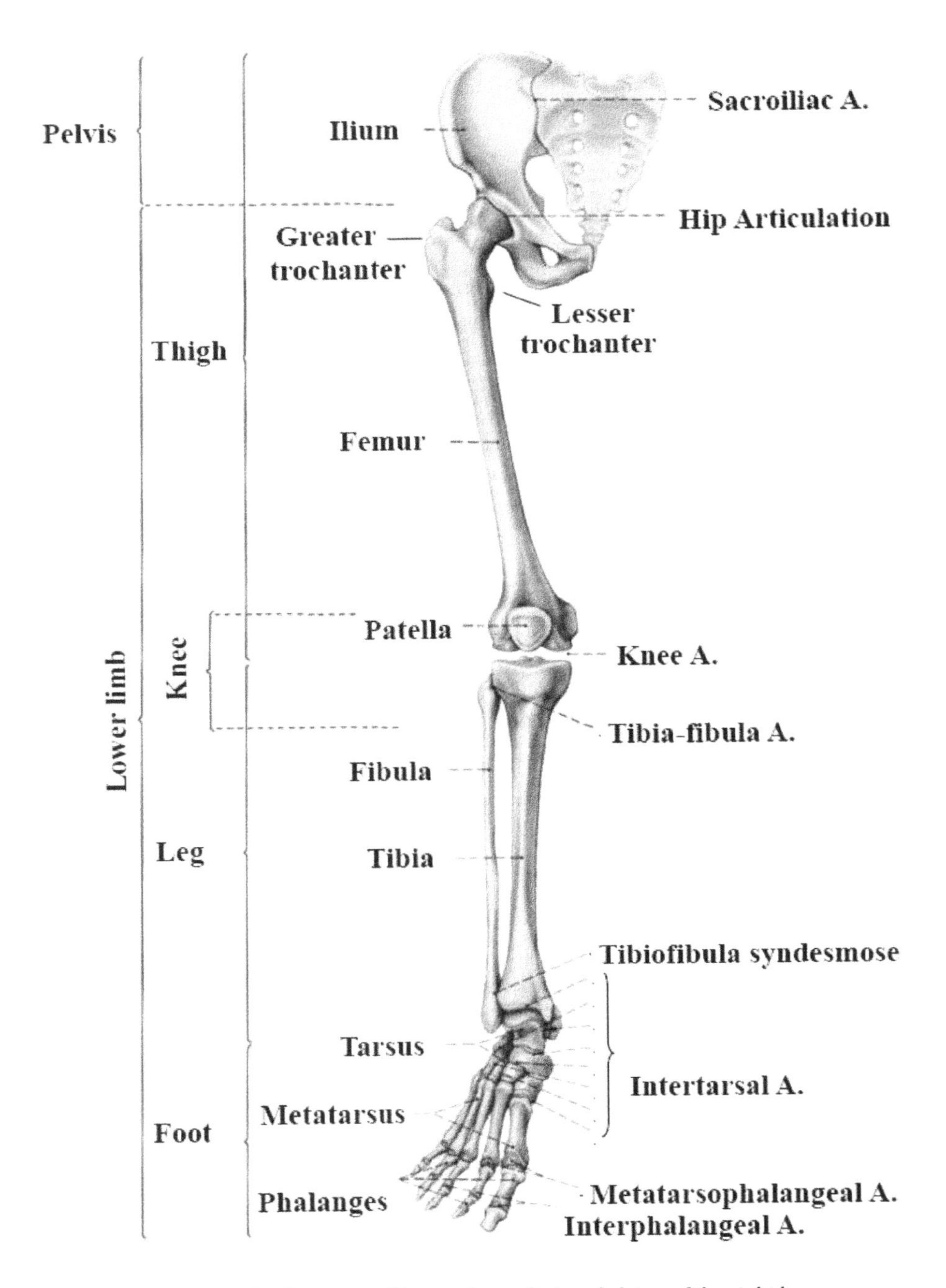

Figure 1.6a. *Structure of human lower limbs: skeleton of the right leg*

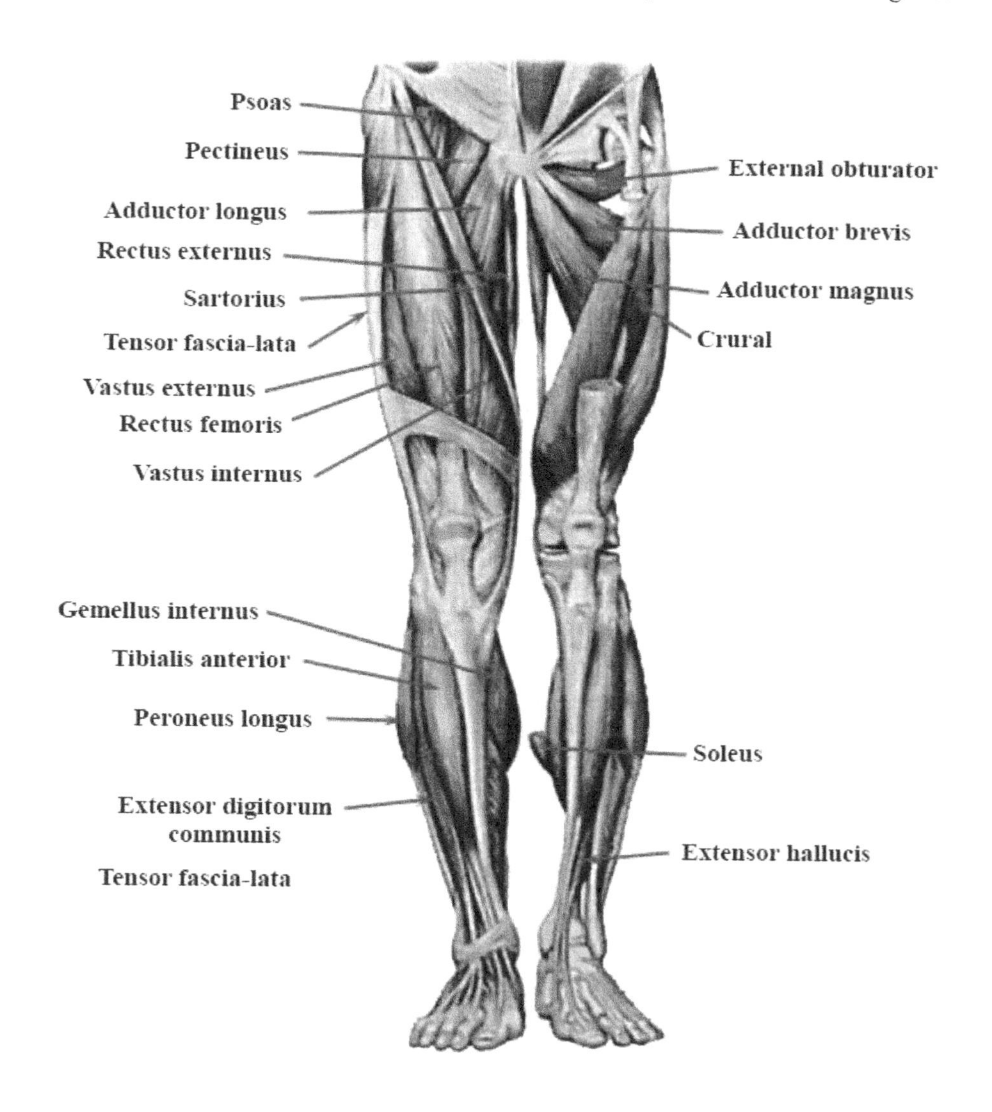

Figure 1.6b. *Structure of human lower limbs: ventral frontal view of the muscular groups*

1.3.2. *Walking and running trajectory data*

In a healthy subject, locomotion can be reduced to a cycle [GAG 90, LAA 92, NIL 89], where the articulate sequence is repeated as long as the speed is constant.

Walking, for example, is a succession of grounding phases and balance. It has been convened that the walking cycle starts when contact is made with the right heel, followed by two steps, left then right, where the walker maintains at least one grounding. By dividing up this cycle into percentages of its duration [BEA 03, VAU 84], the contact of the right heel with the ground is considered as instantaneous (0% of the duration of the cycle) (Figure 1.7).

Both feet are grounded for 0–15% of the cycle. This is a phase of double support which corresponds to the grounding reception of the right leg, and the propulsion of the left leg. In a progressive way, the left foot leaves the ground, until the big toe has left the ground. For 15–50% of the cycle, only the right foot is grounded in a monopodal phase. The left leg is in a balancing phase. When the left heel impacts with the ground and the left foot grounds, there is a second phase of double support for 50–65% of the cycle. This phase is completed by the big toe of the right foot leaving the ground. For 65–100% of the cycle, we are in the monopodal phase on the left foot. The right leg is in the balancing phase. The cycle terminates by a new grounding of the right heel.

Human walking is characterized by a phase of double support which disappears during speeds faster than 2.1 m s^{-1}. This disappearance corresponds to the theoretical transition of walking with an instantaneous double support to running which represents 50% of the cycle (see Figure 1.7).

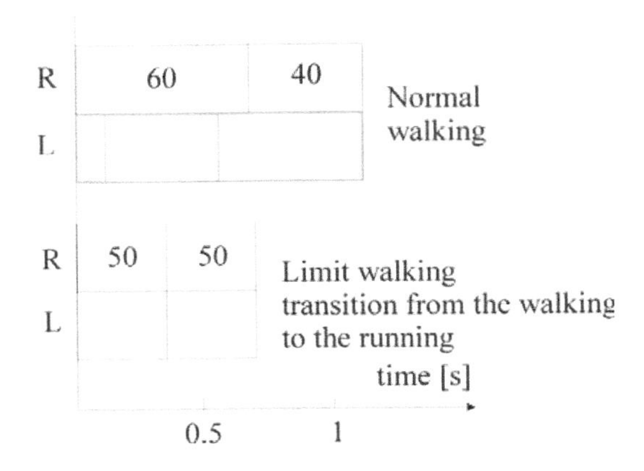

Figure 1.7. *Grounding phases (gray areas) expressed in percentages of the walking cycle duration*

According to various studies [AND 77, LAR 80], distance *D* which is covered during a cycle increases linearly with speed, until it reaches a maximum. Van Emmerik and Wagenaar [VAN 96] have shown a variation of 0.6–1.4 m for speeds of 0.3–1.3 m s^{-1}. A hysteresis phenomenon has also been shown [VAN 96]. For a given velocity, *D* is noticeably longer when the speed is increasing than when it is decreasing.

During walking, the frequency *F* of a step significantly increases with the speed [CAV 86]. Thus, *F* varies by around 0.5–1.0 Hz for speeds from 0.3 m s^{-1} to 1.3 m s^{-1}. A phenomenon of hysteresis has also been observed: for a given speed, *F* reaches a higher value at increasing speed than at decreasing speed.

The main articulations associated with human locomotion are those of the hips, the knees, the ankles and the metatarsal articulations. The movement of the hip combined with the rotation of the pelvis enables humans to lengthen their step [NOV 98]. During a walking cycle, the movement for the hip in the sagittal plane is essentially sinusoidal. In this way, the thigh moves from back to front and vice versa. The articulation of the knee allows for movements of flexion and extension in the leg during locomotion. As for the ankle, the movements of flexion occur when the heel is re-grounded. There is a second flexion during the balancing phase.

The results shown in Figures 1.8–1.11 are relative to measurements[1] of kinematic variables for the movement of a walking male of a height of 1.7 m. The optoelectronic acquisition system used to carry out these measurements gives a three-dimensional analysis of the movements of a polyarticulated system such as the human body, equipped with reflective anatomic frames. Seven infrared cameras (which are synchronized and set to 60 Hz) give to the acquisition software [BEA 03] an image and the coordinates of each frame. Figure 1.8 shows the movements of flexion and extension according to the time measurement of the posterior surface of the left iliac spine in the sagittal plane during one of the cycles of an established walk. The sinusoidal feature of the movement of the hip articulation is easy to identify, even if there isn't perfect symmetry between the two extensions. Figure 1.9 shows the flexion and extension movements according to the time measurement of the left knee in the sagittal plane during the same cycle. For the left leg, the indications given on the duration of the grounding and balancing phases are given in Figures 1.10 and 1.11, which show the vertical sides of the heel and big toe of the left foot.

We notice that the heel is in contact with the ground from 0 to 0.4 s and after 0.9 s. The big toe is grounded from approximately 0.35–0.6 s. There is therefore a balancing phase for the left leg for a time span of 0.6–0.9 s.

---

1. These measurements were communicated to us by Franck Multon, who is part of the Physiology and Muscular Exercise Laboratory of the University of Rennes 2 and the ENS of Cachan, France.

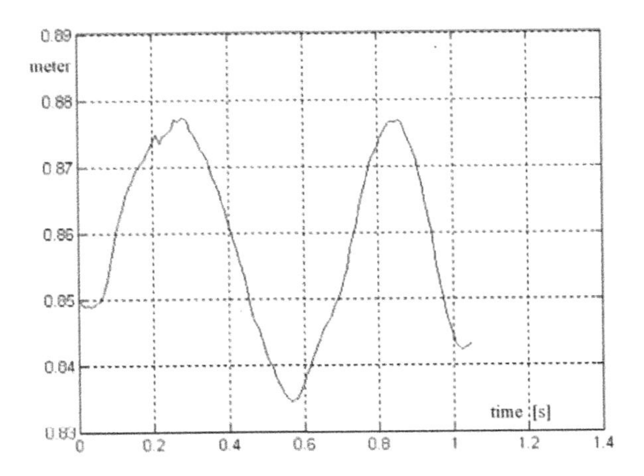

Figure 1.8. *Flexion-extension of the posterior surface of the left iliac spine during a walking cycle*

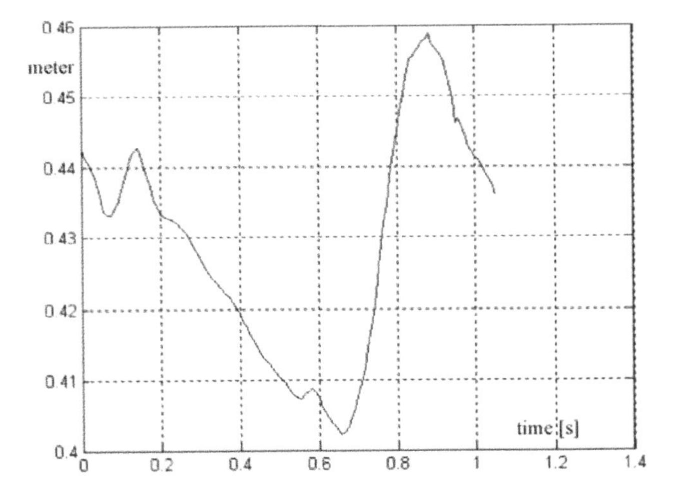

Figure 1.9. *Flexion-extension of the left knee in the sagittal plane during a walking cycle*

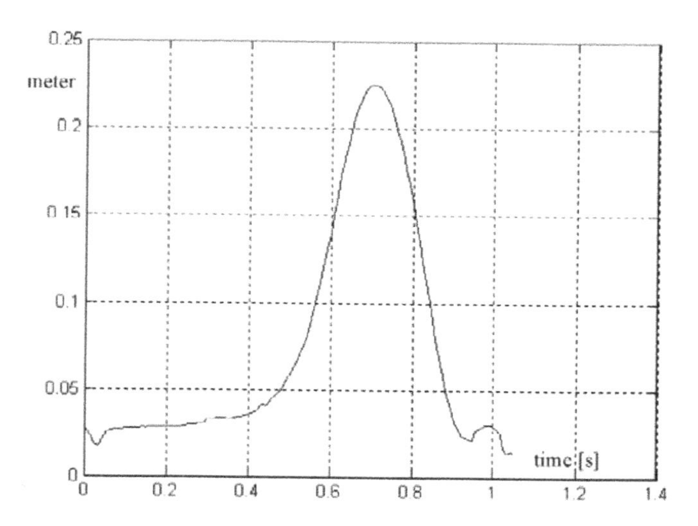

Figure 1.10. *Evolution of the altitude of the left heel during a walking cycle*

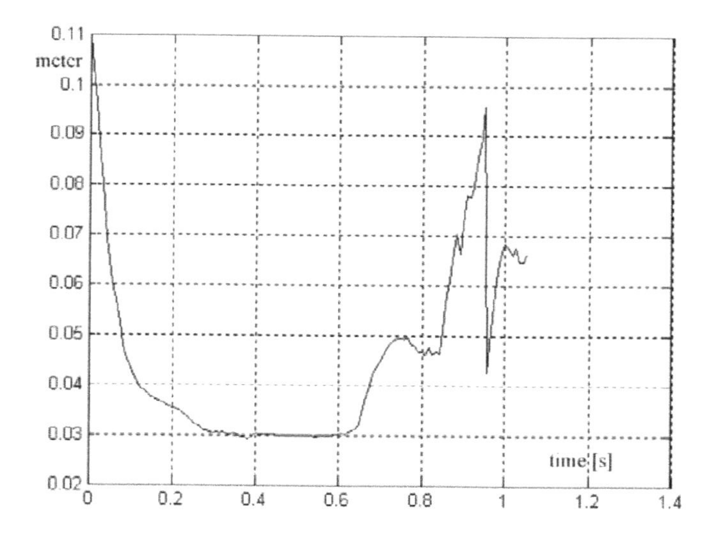

Figure 1.11. *Evolution of the altitude of the left big toe during a walking cycle*

A human's center of gravity is situated at about 55% of its height from the ground. When walking, the human center of gravity oscillates both vertically, by following a path of approximate cycloidal amplitude equal to about 75 mm and horizontally, by adopting a sinusoidal amplitude equal to about 30 mm. Researchers [CAI 94] have shown that the actions of rotating the pelvis around a vertical axis, tilting the pelvis to the side of the non-carrying leg, flexing the knee when the heel grounds, moving the foot and the heel, coordinating movements of the knee and ankle and the lateral shift of the pelvis all help to stabilize walking by playing on the position of the center of gravity.

When walking, the potential and kinetic energies of the human body are in opposition phases. The level of energy is maintained due to the transfer between the potential and kinetic energies. The potential and kinetic energies reach both a maximum and minimum value during the phases of double support. The potential energy reaches its maximum value when the mass center point goes over the grounded foot [BIA 98, NOV 98]. This corresponds to the position of the highest mass center point. When in a running gait, the potential and kinetic energies are synchronized and they reach their minimum and maximum values in the middle of the grounded phase and in the middle of the airborne phase. The transfers of kinetic/potential energy are no longer possible. The level of energy is maintained due to the storage of potential energy as a result of the elastic deformation of the body parts (e.g. the pedal arch) and due to the transfer of energy of a body segment to another through muscular articulations [NOV 98]. The muscles must therefore accomplish more mechanical work during running than during walking.

1.3.3. *Study cases*

The design of bipedal robots and especially humanoid robots is naturally inspired from the functional mobilities of the human body. Nevertheless, the complex nature of the skeletal structure as well as the human muscular system cannot be reproduced in robotics. The number of internal mobilities is therefore limited to the essential and the actuating system must be simplified. A bipedal robot or a humanoid robot therefore has fewer DoF than a human body. The choice of the number of DoF for each articulation is very important. The approach consists of analyzing the structure of the robot from three main planes: the sagittal, frontal and transversal planes. The movement of walking mainly takes place in the sagittal plane; all bipeds have the largest number of important articulations in this plane.

Figure 1.12 shows the typical configurations of bipedal robots in the sagittal plane. The segmented bodies are usually modeled by a punctual mass placed at their center of gravity. In Figure 1.12, the articulations are located by the black circles.

The structure in Figure 1.12a corresponds to the simplest robot, which only has two articulations at the hip. Structure 1.12b has two additional articulations at the knees.

The last figure, Figure 1.12c, is more complete with two additional articulations at the ankles. This is the configuration that is most often used for the construction of bipedal robot prototypes.

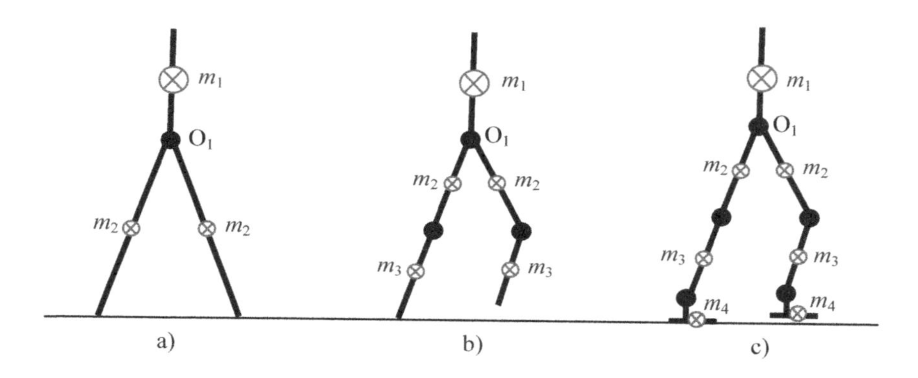

Figure 1.12. *Studies of configurations for bipedal robots in the sagittal plane*

The analysis of articulations in the frontal plane shows their importance for the lateral stability of walking.

Figure 1.13 shows the given solutions obtained by the designers. The stability in the frontal plane depends on the position of the center of gravity in relation to the contact point of the stance leg with the ground. In this way, there are two possible structures for moving trunk mass in the frontal plane.

The structure in Figure 1.13a allows for the displacement of the trunk's mass above the grounded leg with a lateral flexion of the trunk at the level of the pelvis-trunk articulation.

The structure of 1.13b enables it to laterally move the trunk mass due to the articulation of the hips. An independent movement of the balancing leg is equally possible due to the independent articulation of the second hip.

The structure of Figure 1.13c is more complete; it enables it to perform a lateral transfer of the trunk by combining a double movement of adduction-abduction at the hips and ankles. In addition, keeping the feet flat is guaranteed by a better adherence to the ground.

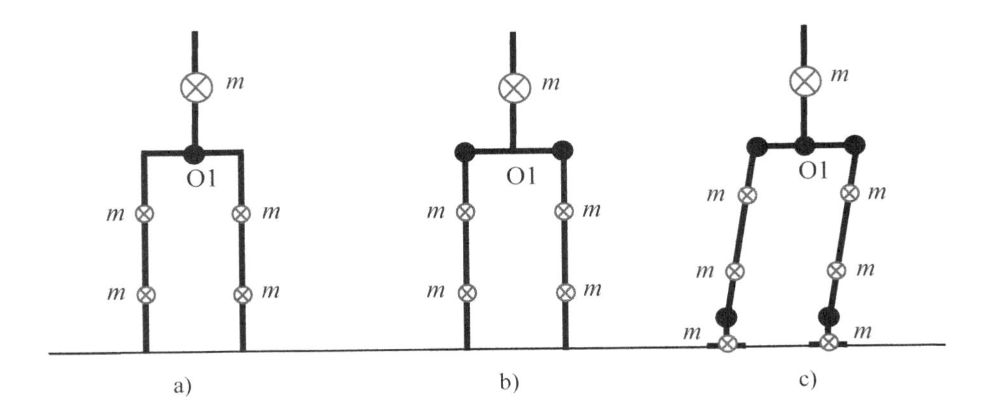

Figure 1.13. *Bipedal robot study cases in the frontal plane*

The articulations in the transversal plane only exist in the most complex bipeds. These robots have one to five articulations in the transversal plane. Figure 1.14 shows the possible configurations.

The structure of Figure 1.14a, with its movement of trunk rotation, enables it to compensate for the coupled reactions in the transversal plane due to the movement of the balancing leg.

The structure of Figure 1.14b enables the robot to pivot its balancing leg by a movement of internal-external rotation, which enables it to make changes in direction when walking.

Finally, the structure of Figure 1.14c is the most complete and, in addition, enables it to orientate its foot when it is about to touch the ground or to pivot on the grounded foot.

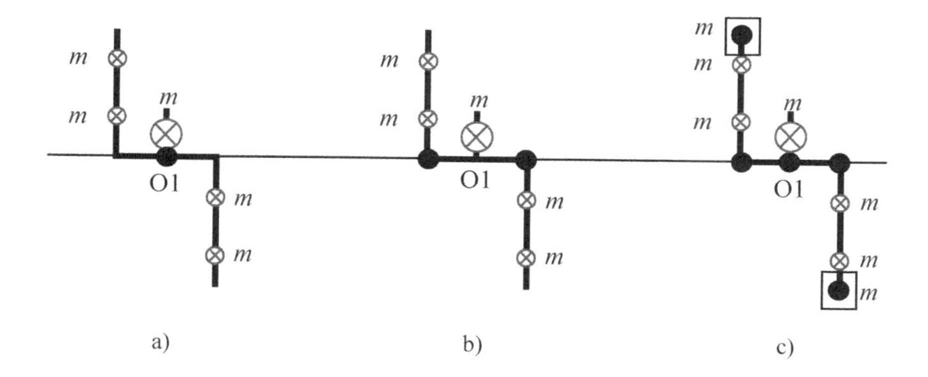

Figure 1.14. *Configuration studies of bipedal robots in the transversal plane*

1.4. Bipedal walking robots: state of the art

1.4.1. *A brief history*

Man has always been fascinated by making systems in his own image. Leonardo de Vinci is probably the first man to have drawn (and perhaps even built) a humanoid mechanism (Figure 1.15).

The 18th century was a fertile period, with the creation of many automats able to reproduce human movements when placed in specific contexts of tasks (e.g. writing or playing music).

The 19th century was a period of construction when the *Steam Man* (moved by steam-engine) was built by John Brainerd and the *Electric Man* was built by Frank Reade Junior (Figure 1.16).

At the beginning of the 20th century, the Westinghouse society made the *Elektro* humanoid. It was during the 1960s and 1970s that legged robots really started to appear, especially in Japan (section 1.4.2). In Russia, the University of Lomonosov in Moscow and the University of Saint Petersburg built legged robots very early on (*Mascha*, *Rikscha* and *OstRover* robots) [GRI 94, GUR 81]. In the USA, the first legged robot creations controlled by computers were quadrupeds and hexapods [BRO 89, OZG 84]. The hexapod CMU, built during the period 1980–1983, reached a maximum speed of 0.11 m s^{-1}.

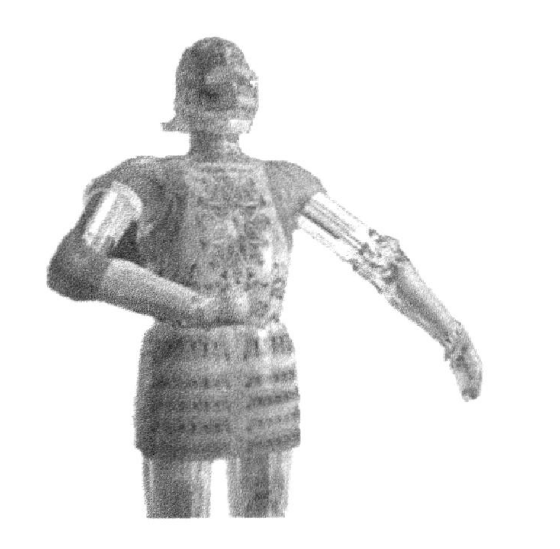

Figure 1.15. *Leonardo de Vinci's Humanoid*

Figure 1.16. *Frank Reade Junior's Electric Man*

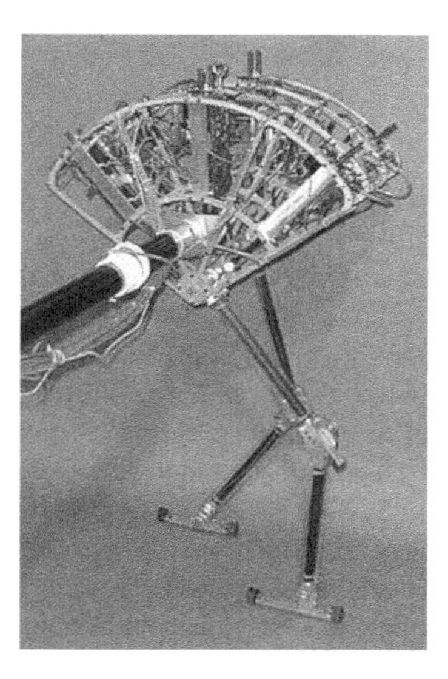

Figure 1.17. *Spring Flamingo from Gill Pratt's MIT laboratory*

Figure 1.18. *3D bipedal robot from Marc Raiberts MIT laboratory*

We should also mention the work carried out at MIT (Massachusetts Institute of Technology) in the 1980s [MUR 84, RAI 83, RAI 86] on jumping robots. The bipedal robots called *Biped Planar*, *Spring Flamingo*, *Spring Turkey*, *Uniroo* and *3D Biped* (Figures 1.17 and 1.18) were among the first to perform walking and running movements in a dynamic and stable gait. Their dynamic performances and the variety of their tested moving gaits were remarkable.

1.4.2. *Japanese studies and creations*

In the 20th century, the first studies concerning bipedal robots were carried out in Japan, where certain researchers had been interested in the subject since the 1970s. The robotic team of Waseda University must be mentioned in this context as they developed a whole family of WL (*Waseda Legged*) robots. Figure 1.19 shows one of the latest developments of these robots, the *WL-10R* [TAK 84] which has 10 articulations motorized by electrical servomotors and body parts made of plastic which are reinforced with carbon fibers. This robot was able to walk both forwards and backwards and turn around, which was a real achievement in 1983. These developments were brought about by Professor Kato's team as part of the Wabot project, who also constructed the first anthropomorphic robot which was entirely controlled by hydraulic actuators.

The Waseda university team then developed a whole variety of bipedal and humanoid robots; the most recent have 41 motorized joints. The *WABIAN-2R* robot [OGU 06] is one of the most accomplished examples (see Figure 1.20). It is 1.53 m in height and weighs a total of 64.5 kg (4.5 kg for the battery). Each of its legs has 7 articulations, 4 articulations placed at the pelvis and trunk and the remaining 23 in its arms and neck. The majority of the joints are activated by the Harmonic Drive gearbox, coupled with DC motors. Each foot is equipped with a force sensor with 6 components and the control is based on the ZMP feedforward drive (see Chapter 5). Its average walking speed is 0.36 m s^{-1} with a period of 0.96 s per step.

Among the most interesting of Japanese research, the work of Sano and Furusho's team is of particular interest [SAN 90, SAN 91]. From 1984 to 1988 they worked on the *BLR-G2* robot which had 9 DoF and was controlled by DC motors. This robot's maximum speed of progression was 0.35 m s^{-1}. In the same way, Kajita and Tani [KAJ 96] built the *MELTRAN II* robot in the 1990s, which had passive articulations at the ankles. One of its laws of control was a function which depends on the angle of the equivalent virtual leg (see Chapter 5).

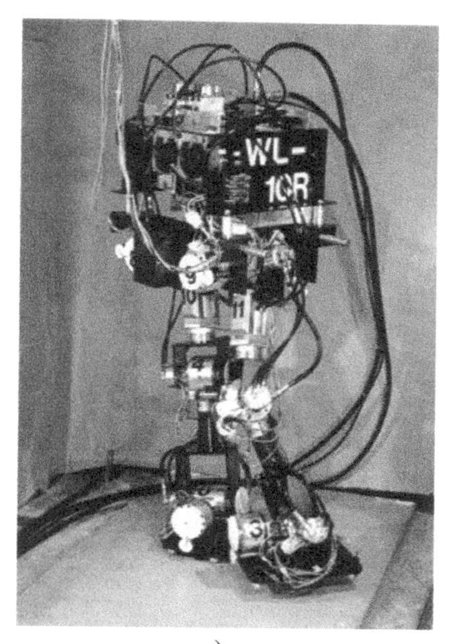

a)

b)

Figure 1.19. *(a) WL-10R Robot, 1983; (b) Wabot-1 robot, Waseda University, Tokyo, Japan, 1973*

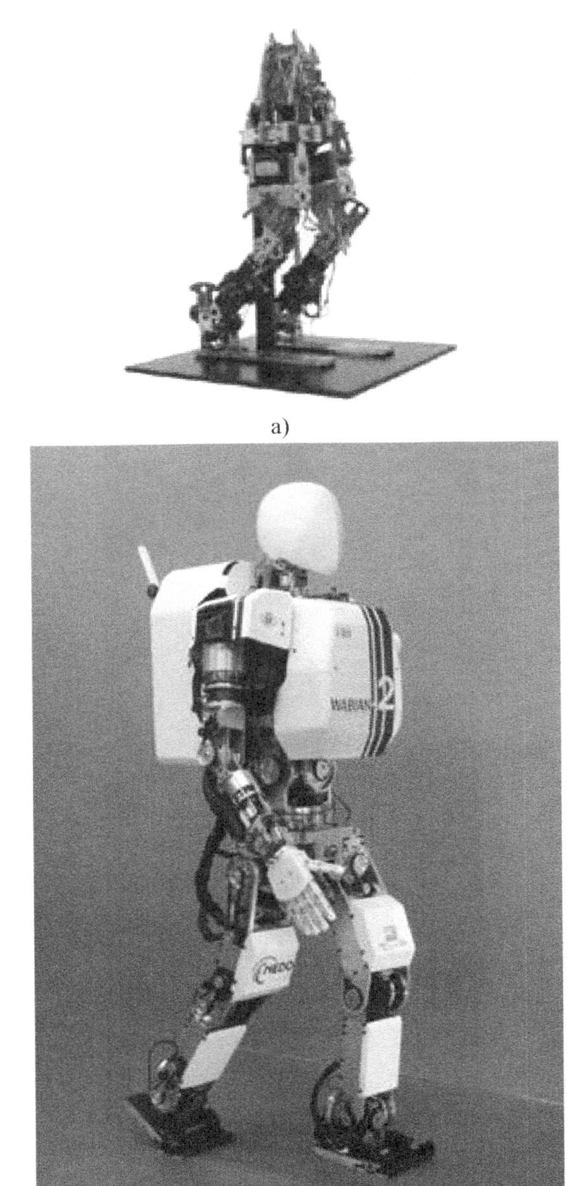

b)

Figure 1.20. *(a) BIPER-4 robot, Tokyo University, 1984; (b) WABIAN-2R, Waseda University, Tokyo, Japan, 2004*

In the 1980s, Miura and Shimoyama [MIU 84] developed the bipedal robot family called *BIPER* which was statically unstable but which had a dynamically stable walk. The *BIPER-4* robot (Figure 1.20) for example had non-motorized articulation at the ankles, very big feet and no articulation at the knees. The analogy of an inverted pendulum's movement was used to define its gait.

In Japan, the industrial companies developed bipedal and humanoid robots very early. Honda, in particular, built a whole range of bipedal robots from 1986 onwards. First there was *E0* to *E6*, then humanoid robots called *P1* to *P3* and finally, the most complete, *ASIMO* (se Figure 1.21). The *ASIMO* robot is 1.4 m high, has 26 DoF and is moved by 26 electric motors.

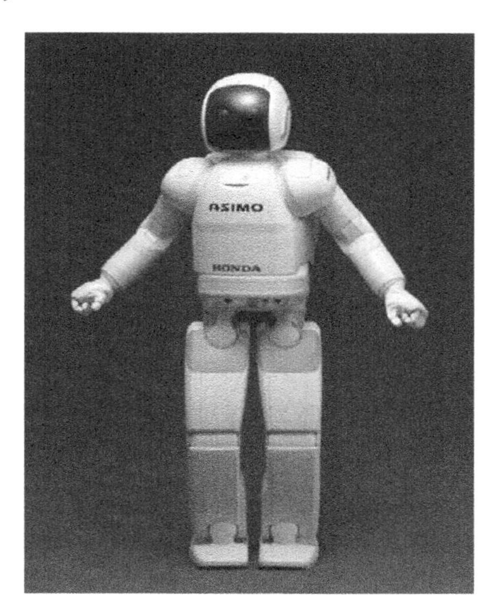

Figure 1.21. *ASIMO humanoid robot built by HONDA, Japan*

1.4.3. *The situation in France*

The first studies of legged robots were carried out at Strasbourg University in the LSIT laboratory, which culminated in the creation of a bipedal robot made up of six body parts but with no trunk (just legs and feet) [CHA 93]. We can also mention the work on monopodal jumping robots [FRA 98].

Another dimension was reached with the design and construction (two models were made) of the *BIP2000* biped, which was jointly made by the INRIA Rhône-Alpes and the Poitiers LMS (Figure 1.22). This 3D bipedal robot is 1.8 m high and weighs 105 kg. Its locomotive system has 12 basic mobilities which enables it to perform walking gaits similar to that of a human. It also has a pelvis-trunk articulation with three DoF. It is made up of eight main body-parts and seven articulations, which adds up to a total of 15 degrees of internal mobility. Statically stable trajectories were obtained for a walking speed equal to 0.36 km h^{-1}.

The *Rabbit* project, which started in 1998 with CNRS backing, enabled the French bipedal robotic community to further their studies on stability and the control of underactuated bipedal robots. The aim of this project was to obtain walking and running gaits which were dynamically stable, based on a simple mechanical anthropomorphic structure which only had a few DoF.

The project culminated in a five-bodied biped with four motors, two on the hip and two at the top of the thighs (see Figure 1.22). Each of the four gearboxes can give a maximum nominal couple of 150 Nm. This value is necessary for running gaits. The biped is maintained in the vertical position by a pivoting horizontal beam, and moves along a circular trajectory in this way. This secure layout enabled them to perform successful walking and running trials for significant distances.

The LIRIS Laboratory at the University of Versailles made an experimental anthropomorphic biped called *ROBIAN* (Figure 1.23) [MOH 04] (with backing from the French Ministry of Education, Research and Technology).

This biped has a three-dimensional kinematic architecture and has 16 degrees of motorized freedom. It weighs 29 kg and is 1.30 m high. The foot is made up of an articulated forefoot along a transversal axis moved with a compliant link. The trunk is made up of a mechanism which activates three mobile masses so that it can make weight transfers in three directions. The hip module is made up of a parallel kinematic system which enables the activation of its three DoF. Two linear actuators create movements of abduction-adduction and internal-external rotation. A rotating actuator enables flexion-extension.

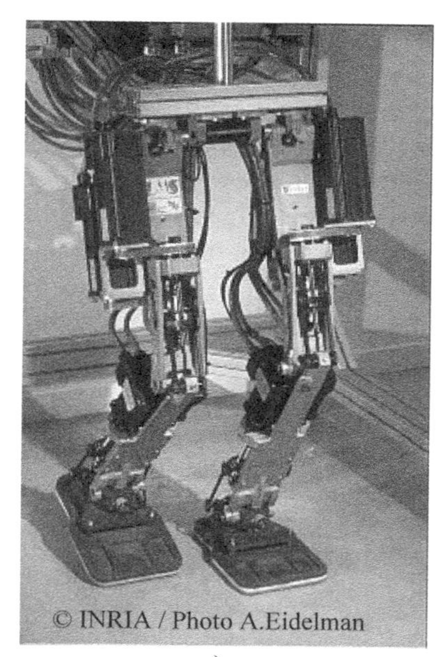

a)

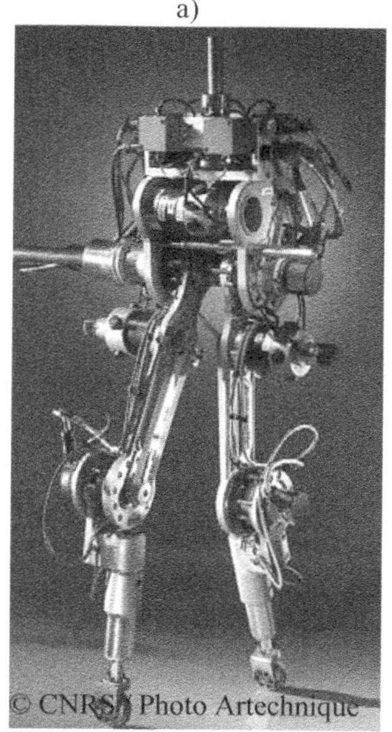

b)

Figure 1.22. *(a) BIP robot (INRIA Rhône-Alpes, LMS Poitiers); (b) Rabbit robot (CNRS Grenoble)*

In Japan, the *Humanoid Robotics Project* came about under the instigation of the Ministry of Economics, Commerce and Industry (METI). The project came into being with the creation of a simulation platform (*OpenHRP*) and the creation of a humanoid. Its second version is called *HRP-2* (Figure 1.24). This humanoid was designed by the National Institute of Science and Advanced Industrial Technologies (AIST) and built by Kawada Industries. There are 15 *HRP-2* models in the world. 14 are to be found in different laboratories in Japan and one is at the Analysis and Architecture of Systems Laboratory (LAAS) at the CNRS in Toulouse. The purchase of this humanoid was made within the context of the Franco-Japanese Joint Robotics Laboratory (JRL) [KHE 07]. The *HRP-2* robot is 1.5 m high, weighs 58 kg and has 30 DoF. It can move at a speed of 2.5 km h^{-1}. It has vision cameras and force and attitude sensors so that it can control its own balance, as well as plan and control its tasks.

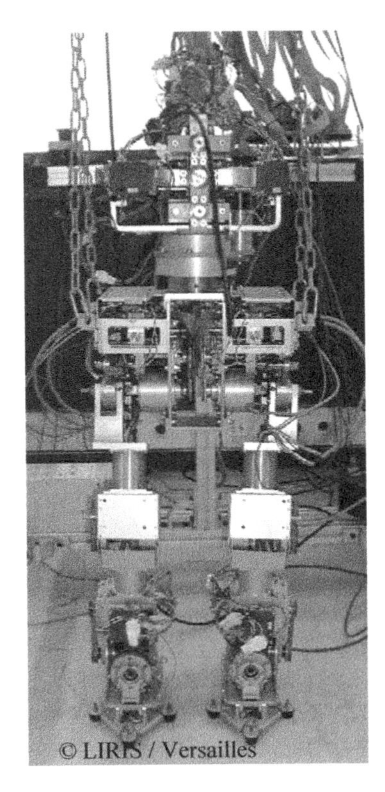

Figure 1.23. *Robian robot made by LIRIS*

Figure 1.24. *HRP2 Humanoid made by Kawada*

1.4.4. *General evolution tendencies*

In the following years, the European Network of Excellence EURON carried out a very interesting study on the development of robotics. The 2005 report [EUR 05] summed up the previsions for the different types of mobile robots and detailed the future applications of service robots in the workplace and at home. The study estimated that a total of 2 million service robots of all kinds would be used in the world by the end of 2004, either for professional or domestic use. This highly increasing number could quadruple before the end of the present decade. The estimated rise is higher still for all robots destined for domestic use (robots to help with housework or robot toys). Within the category of robots for professional use, robots which help with surgery and defense robots (surveillance, de-mining and warfare) are on the highest increase. However, this report gives us no indication of the part to be played by bipedal or humanoid robots within this market.

Important projects were launched in Japan and in the USA to develop robotics and application domains. The Japanese Ministry for Economics, Commerce and Industry (METI) invested more than 17 million dollars of its 2007 budget in order to back the development of perfectly autonomous intelligent robots which could make their own decisions in the workplace. The final aim is to introduce intelligent robots to the market by 2015. The types of intelligent robots envisaged by METI included security and cleaning robots. We could simply indicate a type of task to carry out, and it would be executed in an intelligent manner.

In the USA, among the many projects underway in the different universities, the projects of the DARPA (Department of Defense Agency) are of particular interest. Indeed, many of their projects are centered on the development of autonomous machines (*unmanned autonomous vehicles*) to help improve transportation and exploration in unknown territories.

Legged vehicles are therefore particularly adapted for accompanying ground troops and for traveling in difficult terrain with obstacles or escarpment. Many exoskeleton projects are also underway. We should also note the very ambitious NASA project to create an astronaut robot. Indeed, the *Robonaut* project consists of building and testing an autonomous humanoid robot capable of helping other astronauts, taking down information and manipulating objects in a space capsule or when exploring planets.

1.5. Different applications

Robotics is a priority in technologically advanced countries. With the ageing population, the needs in terms of service and innovating technological advance create a particularly propitious context for robotics. Indeed, there are many domains of application for robotics. The 2005 report by the EURON [EUR 05] network mentions two main robot categories for application purposes: service robots for professional use and service robots for domestic use.

For professional use robotics, the applications most often used are: agricultural and environmental robots (e.g. harvesting robots, milking robots or lumberjack robots), mining company robots, hostile territory exploration (e.g. planets or underwater), environmental surveillance (forest fires), cleaning robots (e.g. floors, walls, exterior walls or pipes), inspection robots, service robots (e.g. surveillance, security, handling or logistics), building robots for contractors and demolishers, medical robots (e.g. surgery and therapy), military robots, robot for welcome desks and as guides (e.g. hotels or museums) and graphic animation for the cinema industry in particular.

In robotics for domestic use, there are service robots (e.g. carers, housework or leisure activities), robots for company, interfaces for man-machine interactions and robots for helping the handicapped. In the following section, we will present a few examples of applications that we have already mentioned.

1.5.1. *Service robotics*

Service robotics for companies and individuals will be on the increase in the next few years. This domain is currently dominated by wheeled robots, but some of the bigger Japanese companies have been working on many different humanoid robot projects with an end to using them as service robots for daily tasks (e.g. maneuvering or distribution), help with housework and hospital servicing work.

The most well-known projects are those of Honda and Sony as well as the *HRP-2* robot which has already been mentioned. These companies clearly advertise their objectives, which are to market robots to do the housework or to help the elderly.

Bipedal robots will be improved and developed within this context, as they will be better adapted to an environment that is usually destined for humans. Moreover, their anthropomorphic aspect will ensure that they are accepted psychologically in a more spontaneous way. A robot companion may be better accepted by a person it is caring for as it cannot make a moral judgment about this.

In service robotics, the most advanced bipedal robot project concerns the elderly. Researchers at the University of Waseda have come up with a mobile chair mounted on legs (see Figure 1.25).

The KAIST research center at the HUBO Lab has created a bipedal armchair which is in complete working order (Figure 1.26). These legged armchairs have been designed to help improve the mobility of paraplegics. In 2005, Toyota presented the *i-foot* prototype to help people of reduced mobility.

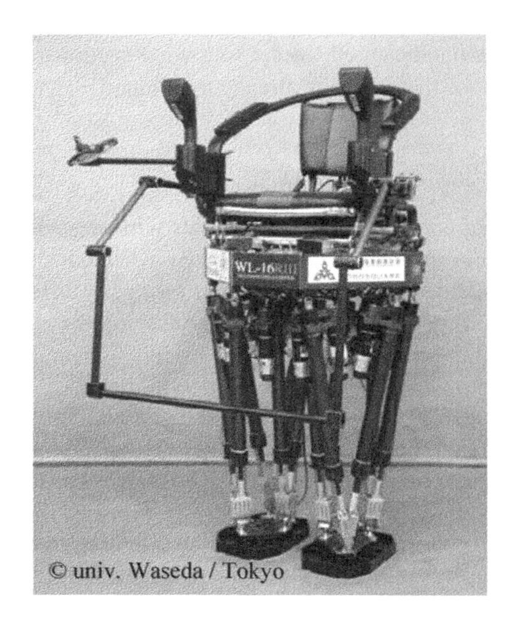

Figure 1.25. *Robot to help the elderly, Waseda University, Japan*

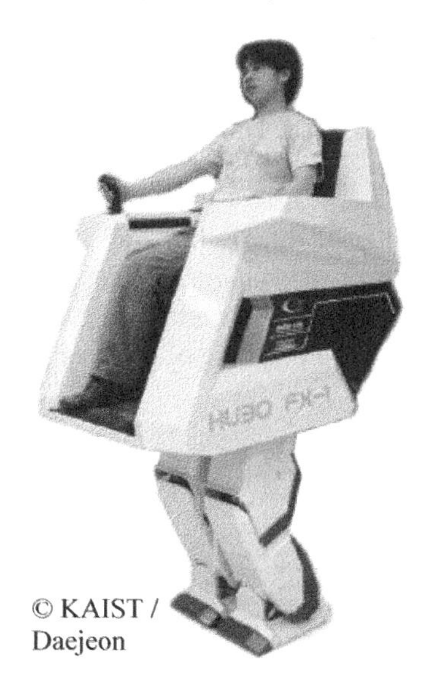

Figure 1.26. *Hubo FX-1 robotic armchair, HUBO Lab, KAIST, South Korea*

1.5.2. *Robotics and dangerous terrains*

Robotics for dangerous environments will also be further developed. Applications in the nuclear domain [GEL 90] or for interventions in dangerous situations (accidents) are being studied by many international teams.

To offer assistance during fires, researchers first designed wheeled and tracked-type robots, or snake-like robots with a water-hose nozzle. An important project to build a robot to help firemen is currently under way in Tokyo [MIY 02]. Buildings or sites that are mainly dedicated to human activity need mobile machines which are capable of moving around rapidly [AMA 02]. Bipedal robots have the potential mobile capacity to be able to intervene in this type of environment. In the case of catastrophes, such as an earthquake, robots can also intervene to explore destroyed buildings or locate and save victims. In the case of nuclear catastrophes, the robotic machines are also capable of exploring and transmitting collected data from contaminated zones and, if necessary, getting even closer to the source of contamination [BRI 98]. Numerous projects are underway in various countries for intervention tasks or for dismantling in the nuclear industry. However, robotic interventions of this kind are currently still carried out by wheeled or tracked-type machines as their operation is sufficiently mastered and secure.

1.5.3. *Toy robots and computer animation in cinema*

It is possible to buy kits of little humanoid robots such as those sold by *Hitec Robotics* (who commercialized the *Robonova I* robot, Figure 1.27). It is 0.4 m high and weighs 1.6 kg. In the same vein, Sony developed the robot dog *AIBO* (*Artificial Intelligence roBot*) and marketed it for the general public. It can interact with the environment and has sound, voice and facial recognition sensors. Mini football tournaments were organized using the AIBO robots as players. Sony has recently developed a humanoid called *QRIO* (Figure 1.28) which is not out on the market yet.

Other robots are dedicated to computer animation for cinema and are sometimes referred to as the growing branch of robotics called "animatronics". With an increase in the number of scenarios including special effects with prehistoric, legendary or science fiction animals, the cinema industry is constantly looking for materials and software to help them realistically reproduce the movements of these real or long-lost animals.

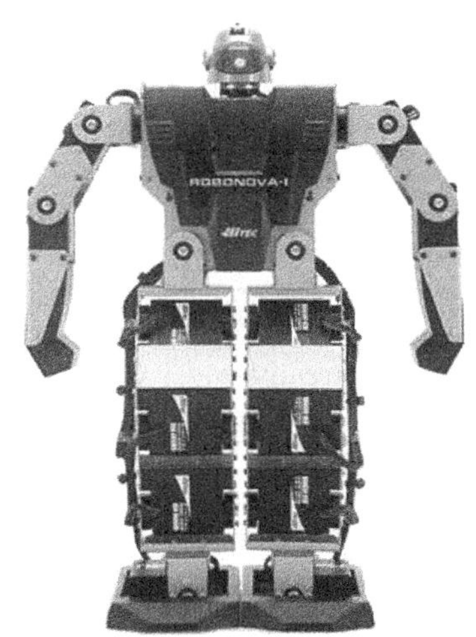

© Hitec

Figure 1.27. *Robonova-I robot made by Hitec Robotics, Japan*

Figure 1.28. *QRIO robot made by Sony, Japan*

1.5.4. *Defense robotics*

In the USA, the Pentagon has predicted that robots will make up a major part of the American military force in the next ten years. A budget of 127 billion dollars has been given to the project called *Future Combat Systems* which represents the biggest military contract in the history of the USA. The first generation of robotic systems will take the form of a remote-controlled family of rolling machines. With developments in technology, a second generation could evolve towards more diverse conformations and morphologies, including human forms, (although certain specialists think that this will not be possible for another 20–30 years). By becoming increasingly "intelligent", these robots will also be increasingly autonomous. To military minds, these machines will be able to distinguish friend from foe and soldier from innocent citizen.

In France, the General Direction for Armament (DGA) has put together the *SYRANO* prototype (robotic acquisition system for neutralizing objectives). The Robosoft company also presented an all-terrain robot called *Roburoc 6.* This machine (1.60 m in length, weighing 160 kg) can carry loads of 100 kg and is destined for military use such as reconnaissance, surveillance and mine detection. It can be equipped with numerous sensors and cameras.

iRobot is the only company to offer military robotic applications as well as personal robots for the general public. *PackBot* is one of its best-selling products. It is a tracked-type robotic machine prototype and can help soldiers in reconnaissance missions. This military robot was developed with the help of the American Defense Advanced Research Projects Agency (DARPA). It can survive a fall of 3 m and land on a concrete floor.

Another US army project is the UGV (*Unmanned Ground Vehicles)* project which aims to develop a Mule robot to help ground troops by acting as a logistics trailer, a light de-mining machine or as a robotic ammunitions stand. Similarly, Vecna Technologies have presented a project for a semi-humanoid robot called *BEAR* to help wounded soldiers on the battlefield. It can also be converted for domestic use for helping the elderly. A functional prototype may be ready within the next 2–3 years. The bipedal robot *LandWalker* made by Sakakibara Kikai in Japan is no doubt the first armed bipedal robot which fires sponge balls. It has an imposing stature and a relatively small advancing speed (1.5 km h^{-1}) (Figure 1.29).

Figure 1.29. *Land Walker robot made by Sakakibara Kikai, Japan*

Figure 1.30. *BLEEX Exoskeleton California University, Berkeley*

Exoskeleton robots are robotic systems which will enable the military, firemen and life savers to transport heavy loads in the future without having to bear the weight of these heavy loads themselves. The *BLEEX* project [MAI 05, ZOS 05] at the University of California, Berkeley, in conjunction with Schilling Robotics, is one of the exoskeleton developments. The *BLEEX* is made up of a sort of armature similar to that of a rucksack, with reinforced mechanisms linked to the feet, legs and hips of the carrier. A person who is equipped with a *BLEEX* can carry a rucksack weighing 45 kg, but with the impression that he or she is carrying a load of only 2.5 kg (Figure 1.30).

1.5.5. *Medical prostheses*

The use of robotics for medical assistance is on the increase. Research in bipedal robots has led to the development of active or passive leg prostheses. Robots can also be used to provide functional and re-educational assistance, to help strengthen weakened limbs or assist with training sessions. The work on passive prostheses has been developing for a number of years now, and has culminated in interesting performances.

The projects recently developed have led to leg prostheses where the pendulous phase movement is controlled by a microchip (the *C-Leg* prosthesis made by Otto Bock for example; see Figure 1.31) and helps to improve stability and make walking more comfortable. The balancing movement is controlled by pneumatic or hydraulic settings. Another application of this domain is the creation of devices for functional rehabilitation of the lower limbs. A robotic system can provide the possibility of therapeutic treatments for paraplegics. To prevent medical complications resulting from the paralysis of limbs, it is important to maintain musculature through physiotherapy. The orthosis has motors and sensors which enable it to stimulate the muscles in sequences and levels of effort in a manner which echoes natural muscular contractions as much as possible (see Figure 1.32). The *MotionMakerT* was marketed by the Swiss foundation for Cybertheses (FSC).

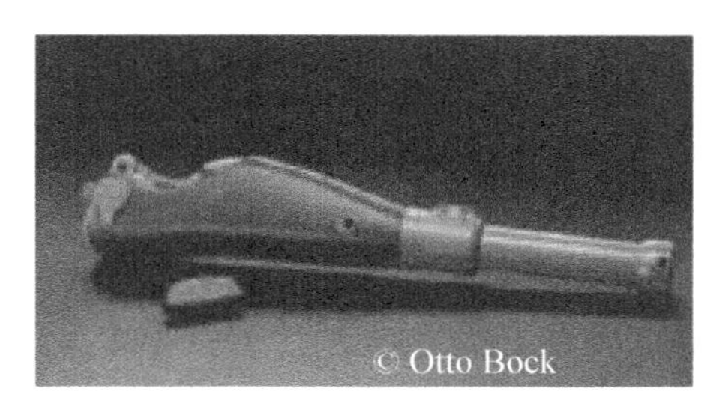

Figure 1.31. *C-Leg orthopedic prosthesis made by Otto Bock, Germany*

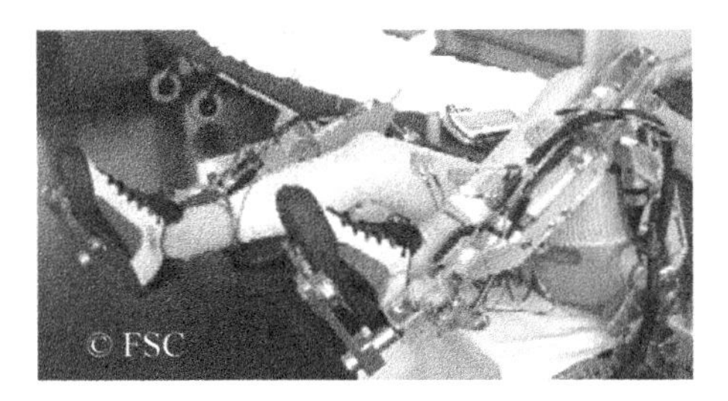

Figure 1.32. *MotionMakerT Functional rehabilitation system made by FSC, Switzerland*

1.5.6. *Surveillance robots*

The EURON report identified the need for surveillance robots, for outside security surveillance as well as for high-risk areas such as stations and museums. Robots which carry out the inspection of industrial structures, buildings or works of art also belong to this category.

Many surveillance robots have wheels or circular tracks because they move around in structured environments, on flat ground or on level surfaces. In this way, certain museums such as the one in St Petersburg, Russia, have adopted a nocturnal mobile robot surveillance system to detect intruders or other occurrences (e.g. fire or smoke) which sets off an automatic alarm. An interesting development of this aspect of inspection and surveillance is the technology of the Swedish company *Rotundus*, which also uses wheels but with the addition of an internal pendulum to create a forward movement. According to the EURON report [EUR 05], the market for this type of robot will rapidly increase in the next few years with future installations amounting to an estimated 400 million dollars. The role to be played by legged robots or bipedal robots in the surveillance sector, however, is yet to be determined.

1.6. Conclusion

After having studied robot manipulators for many years, we are now witnessing a real interest in robots which use their environment to help them to move around, such as mobile, wheeled or legged robots for example. Today there is a multitude of teams creating bipedal robots worldwide. Anthropomorphic bipedal robots are now studied by a variety of scientific disciplines such as biomechanics, mechanics, medicine and automatics.

It is not yet possible for a bipedal robot to imitate human performances, even on the joint friction level. In a healthy human subject, knee articulation friction is practically zero. Nevertheless, the variables that have been observed in the animal world and especially in humans, such as body size, energy consumption during walking and running gaits or the basic ability to move, must be taken into consideration when creating a bipedal robot. To progress, bipedal robots need to acquire dexterity, suppleness and human-like initiative to carry out surveillance, control or maintenance tasks. As far as imitating the performances of man, we will have to wait decades before robots can even get close.

1.7. Bibliography

[ALE 77] ALEXANDER R.MC.N., "Mechanics and scaling of terrestrial locomotion", in T.J. PEDLEY (ed.), *Scale Effects in Animal Locomotion*, Academic Press, London, p. 93–110, 1977.

[ALE 79] ALEXANDER R.MC.N., GOLDSPINK G., JAYES A.S., MALOIY G.M.O. and WATHUTA E.M., "Allometry of the limb bones of mammals from shrew (Sorex) to elephant (Loxodonta)", *Journal of Zoology from the Zoological Society of London,* vol. 189, 305–314, 1979.

[ALE 83] ALEXANDER R.MC.N., JAYES A.S., "A dynamic similarity hypothesis for the gaits of quadruped mammals", *Journal of Zoology London*, vol. 201, 135–152, 1983.

[ALE 84] ALEXANDER R.MC.N., "The gaits of bipedal and quadrupedal animals", *International Journal of Robotics Research*, vol. 3(2), 49–59, 1984.

[ALE 00] ALEXANDER R.MC.N., "Storage and release of elastic energy in the locomotor system and the stretch-shotening cycle", Chapter 2 in B.M NIGG, B.R.M MACINTOSH, J. MESTER (eds.), *Biomechanics and Biology of Movement*, Human Kinetics, 2000.

[ALE 04] ALEXANDER R.MC.N., "Bipedal animals, and their differences from humans", *Journal of Anatomy*, vol. 204, 321–330, 2004.

[ALE 05] ALEXANDER R.MC.N., "Models and the scaling of energy costs for locomotion", *Journal of Experimental Biology*, vol. 208, 1645–1652, 2005.

[AMA 02] AMANO H., "Present status and problems of fire fighting robots", *Proceedings of the 41st SICE Annual Conference, SICE 2002*, vol. 2, 880–885, 2002.

[AND 77] ANDRIACCHI T.P., OGLE J.A., GALANTE J.O., "Walking speed as a basis for normal and abnormal gait measurements", *Journal of Biomechanics*, vol. 10(4), 261–268, 1977.

[BEA 03] BEAUPIED H., Etude mécanique et énergétique de la marche, de la course et de la transition marche-course: influence de la spécialité athlétique, PhD Thesis, University of Rennes II, 2003.

[BEJ 06] BEJAN A., MARDEN J., "Locomotion: une même loi pour tous", *Pour la science*, vol. 346, 68–73, 2006.

[BIA 98] BIANCHI L., ANGELINI D., LACQUANITI F., "Individual characteristic of human walking mechanics", *European Journal of Sport Science*, vol. 436, 343–356, 1998.

[BOR 80] BORELLI G., *De Motu Animalium*, A. Bernabo, Rome, 1680.

[BRI 98] BRIONES L., BUSTAMANTE P., SERNA M.A., "Robicen: A wall-climbing pneumatic robot for inspection in nuclear power plants", *Robotics and Computer-Integrated Manufacturing*, vol. 11(4), 287–292, 1998.

[BRO 89] BROOKS R.A., "A robot that walks – emergent behaviors from a carefully evolved network", *Neural Computation*, vol. 1, 355–363, 1989.

[BRU 98] BRUNEAU O., Approche biomimétique pour la conception et la simulation du comportement et la simulation du comportement dynamique de robots bipeds, PhD. Thesis, Pierre and Marie Curie University, 1998.

[CAI 94] CAILLEUX M.N., Amputation de cuisse chez l'adulte actif: Plaidoyer pour le CAT-CAM, PhD Thesis, Besançon School of Medicine and Pharmacology, 1994.

[CAV 86] CAVAGNA G.A., FRANZETTI P., "The determinants of the step frequency in walking in humans", *Journal of Physiology*, vol. 373, 235–242, 1986.

[CHA 93] CHAILLET N., Etude et réalisation d'une commande position-force d'un robot bipède, PhD Thesis, University Louis Pasteur of Strasbourg, 1993.

[COL 05] COLLINS S.H., RUINA A., "Bipedal walking robot with efficient and human like gait", Proceedings of the 2005 *IEEE International Conference on Robotics and Automation*, ICRA, April 2005.

[DEL 04] DELOISON Y., *Préhistoire du Piéton. Essai sur les nouvelles origines de l'homme*, Editions Plon, Paris, 2004.

[DER 70] DE RICQLES A., "Les animaux à la conquête du ciel", *La Recherche*, vol. 317, 118–123, 1970.

[DOY 66] DOYON A., LIAIGRE L., *Jacques Vaucanson: mécanicien de génie*, Presses Universitaires de France, Paris, 1966.

[EUR 05] EURON, "Contribution to World Robotics 2005", *Report of European Robotics Network of Excellence*, p. 300–368, 2005.

[FRA 98] FRANÇOIS C., SAMSON C., "A new approach to the control of the planar one-legged hopper", *International Journal of Robotics Research*, vol. 17(11), 1150–1166, 1998.

[GAG 90] GAGE J.R., "An overview of normal walking", *AAOS Instructional Courses Lectures, American Nature*, vol. 121, 571–585, 1990.

[GEL 90] GELHAUS F.E., ROMAN H.T., "Robot applications in nuclear power plants", *Progress in Nuclear Energy*, vol. 23(1), p. 1–33, 1990.

[GRI 94] GRISHIN A.A., FORMALSKY A.M., LENSKY A.V., ZHITOMIRSKY S.V., "Dynamical walking of a vehicle with two telescopic legs controlled by two drives", *International Journal of Robotics Research*, vol. 13(2), 137–147, 1994.

[GUR 81] GURFINKEL V.S., GURFIRTKEL E.V., SCHNEIDER A.Y., DEVJANIN E.A., LENSKY A.V., SHITILMAN L.G., "Walking robot with supervisory control", *Mechanism and Machine Theory*, vol. 16, 31–36, 1981.

[HAN 64] HANAVAN E.P., "A Mathematical model of the human body", Aerospace Medical Research Laboratories, Wright-Patterson Air Force Base, Dayton, Ohio, USA, 1964.

[HIL 50] HILL A.V., "The dimensions of animals and their muscular dynamics", *Science Progress*, vol. 38, 209–230, 1950.

[HIL 67] HILDEBRAND M., "Symmetrical gaits of primates", *American Journal of Physical Anthropology*, vol. 26, 119–130, 1967.

[KAJ 96] KAJITA S., TANI K., "Experimental study of biped dynamic walking", *IEEE Control Systems Magazine*, vol. 16(1), 13–19, 1996.

[KHE 07] KHEDDAR A., LAUMOND J.P., YOKOI K., YOSHIDA E., "Increasing Robots Autonomy", JRL Scientific Report, AIST/IS-CNRS/ST2I, 2007.

[LAA 92] LAASSEL E.M., Analyse et modélisation multidimensionnelles de la marche humaine, PhD Thesis, University of Valenciennes, 1992.

[LAR 80] LARSON L.E., ODENRICK P., SANDLUND B., WEITZ P., OBERG P.A., "The phases of the stride and their interaction in human gait", *Scandinavian Journal of Rehabilitation Medicine*, vol. 12(3), 107–122, 1980.

[MAI 05] MAIN J., "Exoskeletons for Human Performance Augmentation", DARPA Project, 3701 North Fairfax Drive, Arlington, 2005.

[MAN 80] MANN R.A., HAGY J., "Biomechanics of walking, running, and sprinting", *American Journal of Sports Medicine*, vol. 8(5), 345–350, 1980.

[MAR 05] MARDEN J.H., "Scaling of maximum net force output by motors used for locomotion", *Journal of Experimental Biology*, vol. 208, 1653–1664, 2005.

[MCG 90] McGeer T., "Passive dynamic walking", *International Journal of Robotic Research*, vol. 9(2), 62–82, 1990.

[MCM 75] MCMAHON T.A., "Using body size to understand the structural design of animals: quadruped locomotion", *Journal of Applied Physiology*, vol. 39, 619–627, 1975.

[MED 06] MEDERREG L., Etude cinématique et reproduction robotique de la marche chez l'oiseau, PhD Thesis, University of Versailles, 2006.

[MIU 84] MIURA H., SHIMOYAMA I., "Dynamic walk of a biped", *International Journal of Robotics Research*, vol. 3(2), 60–74, 1984.

[MIY 02] MIYAZAWA K., "Fire robots developed by the Tokyo Fire Department", *Advanced Robotics, Brill Academic Publishers*, vol. 16(6), 553–556, 2002.

[MOH 04] MOHAMED B., Torse biofidèle pour le robot ROBIAN: Analyse et conception basées sur l'équivalence dynamique des systèmes, PhD Thesis, University of Versailles, 2004.

[MUR 84] MURPHY K.N., RAIBERT M.H., "Trotting and bounding in a planar two-legged model", *Theory and Practice of Robots and Manipulators, Proceedings of the 5th CISM-IFToMM Symposium, RoManSy'84*, 1984.

[NIL 89] NILLSON J., THORSTENSSON A., "Ground reaction forces at different speeds of human walking and running", *Acta Physiologica Scandinavica*, vol. 136, 217–227, 1989.

[NOR 06] NORDEZ A., Caractérisation et modélisation du comportement mécanique du complexe musculo-articulaire en conditions passives, Influence de protocoles d'étirements cyclique et statique, PhD Thesis, University of Nantes, 2006.

[NOV 98] NOVACHECK T.F., "The biomechanics of running", *Gait and Posture*, vol. 7, 77–95, 1998.

[OGU 06] OGURA Y., AIKAWA H., SHIMOMURA K., KONDO H., MORISHIMA A., LIM H., TAKANISHI A., "Development of a humanoid robot WABIAN-2", *Proceedings of IEEE International Conference on Robotics and Automation, ICRA '06*, Barcelona, Spain, 76–81, 2006.

[OZG 84] OZGUNER F., TSAI S.J., MCGHEE R.B., "An approach to the use of terrain-preview information in rough-terrain locomotion", *International Journal of Robotic Research*, vol. 3(2), 134–146, 1984.

[RAI 83] RAIBERT M.H., SUTHERLAND I.E., "Machines that walk", *Scientific American*, vol. 248(2), 44–53, 1983.

[RAI 86] RAIBERT M.H., *Legged Robots that Balance*, MIT Press, Cambridge, 1986.

[SAN 90] SANO A., FURUSHO J., "Realization of natural dynamic walking using the angular momentum information", *Proceedings of 1990 IEEE International Conference on Robotics and Automation*, vol. 3, p. 1476–1481, 1990.

[SAN 91] SANO A., FURUSHO J., "Control of torque distribution for the BLR-G2 biped robot", *5th International Conference on Advanced Robotics "Robots in Unstructured Environments", 91 ICAR*, vol. 1, p. 729–734, 1991.

[SCH 04] SCHILDER R.J., MARDEN J.H., "A hierarchical analysis of the scaling of force and power production by dragonfly flight motors", *Journal of Experimental Biology*, vol. 207, 767–776, 2004.

[SEL 04] SELLERS W.I., DENNIS L.A., WANG W.-J., CROMPTON R.H., "Evaluating alternative gait strategies using evolutionary robotics", *Journal of Anatomy*, vol. 204, 343–351, 2004.

[SEY 02] SEYFARTH A., GEYER H., GÜNTHER M., BLICKHAN R., "A movement criterion for running", *Journal of Biomechanics*, vol. 35, 649–655, 2002.

[TAK 84] TAKANISHI A., NAITO G., ISHIDA M., KATO I., "Realization of plane walking by the biped walking robot WL-10R", *Theory and Practice of Robots and Manipulators, Proceedings of the 5th CISM-IFToMM Symposium, RoManSy'84*, 1984.

[THY 01] THYS H., "Place de l'énergie mécanique dans le déterminisme du coût énergétique de la locomotion", *Staps*, vol. 54, 131–143, 2001.

[VAN 96] VAN EMMERIK R.E.A., WAGENAAR R.C., "Effects of walking velocity on relative phase dynamics in the trunk in human walking", *Journal of Biomechanics*, vol. 29(9), 1175–1184, 1996.

[VAU 84] VAUGHAN K.R., "Biomechanics of running gait", *Critical Reviews in Biomedical Engineering*, vol. 12(1), 1–48, 1984.

[ZOS 05] ZOSS A., KAZEROONI H., CHU A., "On the mechanical design of the Berkeley Lower Extremity Exoskeleton (BLEEX)", *2005 IEEE/RSJ International Conference on Intelligent Robots and Systems, IROS 2005*, p. 3465–3472, 2005.

Chapter 2

Kinematic and Dynamic Models for Walking

2.1. Introduction

The human locomotion system is the archetypal model for the kinematic arrangement of many bipedal robots. The objective aimed for is to move towards the ease and fluidity of human walking. However, although natural and apparently not very constrained, human gait solicits many complementary degrees of freedom (DoF) in the whole body. These prolong and improve the movement of lower limbs. The transversal and frontal rotations of the pelvis must be mentioned in particular, as well as the counter rotation of the trunk and the counter balancing of the arms [SUT 94]. Only humanoid-type robots with arms together with an appropriate pelvis–trunk articulation can perform such additional movements.

The descriptions and models presented in this chapter are based on the kinematics of the legs only. It must however be specified that the kinematic structure of the lower limbs in walking robots only replicates the basic DoF of the human legs when considering the kinematic chain of knee-shin-ankle-foot. The locomotion system of common bipedal robots is therefore limited to 12 internal DoF, which is the strict required minimum to bring about walking gaits specific to human locomotion. The models presented in this chapter have been built along these lines. In addition, the models under consideration focus on straight line walking on a flat surface. However, this approach does not exclude pivoting step modeling as well as rising or descending steps.

2.2. The kinematics of walking

2.2.1. *DoF of the locomotion system*

The kinematic layout of the lower limbs is the basic determiner for walking. The leg is made up of two upper segments which determine the length of the step and a distal segment (the foot), which is the interface between the biped and the ground. The distribution of internal DoF in this simple kinematic chain defines the basic kinematic aptitudes of a walking biped. These DoF are naturally regrouped at hip, knee and ankle type articulations.

The human hip articulation is a perfect ball-and-socket joint. It is reproduced in numerous bipedal robots with three concurrent revolute joints which are successively orthogonal, resulting in an uncoupling of the movements of flexion-extension, adduction-abduction and internal-external rotation (e.g. BIP [SAR 98] and "Johnnie" [LÖF 03] bipeds, P2 [HIR 98], ASIMO [SAK O2], HRP1 [KAN 04], [YOK 04] and HRP2 [HIR 03] humanoid robots). The order of layout for these three elementary links can vary from one biped to another. This order is no longer apparent in the hip articulation of a biped such as ROBIAN [OUE 03] which was designed in the form of a spherical link with parallel kinematics.

The human knee essentially performs the simple kinematic function of flexion-extension. Nevertheless, leg orthopedists and prosthetists know that this articulation cannot be reduced to a single axis joint defined about an anatomically fixed axis with respect to adjacent body segments. A more satisfactory approximation of the kinematic function of the human knee when considering leg prosthesis is often made in the form of a crossed four-bar linkage [RAD 94]. By putting together a four-bar knee, it is hoped that it will practically conform to the kinematics of a human functional knee. This is not constraining for a mechanical biped. The use of a single transversal axis joint is therefore the easiest solution.

The rotations in a human ankle and foot are complex [ROS 94]. It is to be noted that two rotational axes are functionally required in the ankle. This is to ensure a movement of flexion-extension as well as a rotation of adduction-abduction in conjunction with the same movement at the hip, to enable the lateral transfer of the biped's center of pressure (CoP) at every step without any unbalanced gesticulations. During walking, the human foot uses its metatarsian articulation, which enables the forefoot to prolong and complete the propulsion effect of the rear foot. This articulation is rarely reproduced in today's mechanical bipeds. When the foot is articulated, it is made up of passive compliances (e.g. ROBIAN biped [OUE 03], Wabian-2R [OGU 06]).

These six basic DoF, which are all actuated, are therefore necessary to enable the legs to generate three-dimensional human-like walking. The resulting locomotion system with 12 internal DoF will be the kinematic reference model for the study of bipedal-robot walking hereafter.

2.2.2. *Walking patterns*

Walking is the alternating repetition, from one leg to another, of the same elementary movements. Its resulting cyclic characteristic can be described in two different ways. According to orthopedists, the walking cycle coincides with the cyclic movement of one leg [SUT 94]. Conventionally, the cycle starts at heel-strike to the following heel-strike of the same foot. For human locomotion, this description is best applicable to the qualitative analysis of walking and, more particularly, to that of the identifiable dysfunctions of each leg. Dynamic analysis and control of locomotion must be approached from a different viewpoint. In this case, it is the global coordination of intersegmental movements of both legs that needs to be taken into consideration.

Walking must then be treated as a series of steps which are elementary reference movements that can be executed in a variety of ways, but with strictly defined basic kinematic characteristics. A step is made up of two main kinematically distinct phases:

– a *single support phase* (SSP) or *swing phase* during which the locomotion system evolves as an open kinematic chain;

– a *double support phase* (DSP), during which the locomotion system moves as a closed kinematic loop.

The evolution of the biped as an open or closed kinematic loop results in a mechanical system moving with a time-varying kinematic topology. This characteristic will have significant consequences on the dynamic modeling of the walking cycle, on the dynamic synthesis of movement and on the control of the biped. In the double support phase, the locomotion system is over-actuated and the dynamics of the movement are underdetermined. This situation requires the development of specific approaches in order to master the redundancies which result from it.

In Figure 2.1, the single support phase begins at toe-off and ends at heel-touch of the foot of the swing leg. This is a transfer phase. The double support starts at heel-touch of the front foot to toe-off of the rear foot. This is essentially a propulsive phase. Changes in the modes of foot-ground contact during the same phase also determine as many sub-phases, which are characterized by the loss or the gain in a DoF in the locomotion system.

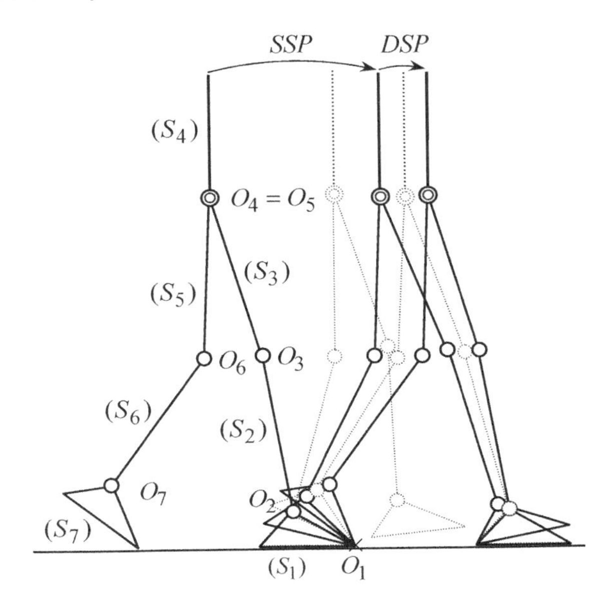

Figure 2.1. *The SSP-DSP sequence of a step of a seven-segment planar biped. The five geometric configurations represented here are configurations of transition between phases (continuous black lines) and sub-phases (dotted lines)*

The phases of contact with the ground can follow on from each other according to various scenarios which will help to define different walking patterns. The four modes of foot-ground contact presented hereafter successively define decreasingly constraining gait patterns. During the first three modes, the stance foot, through the swing phase, remains flat on the floor. Only the sub-phases of the double support therefore need to be described. The fourth mode determines a walking pattern which is close to human locomotion.

2.2.2.1. *Pattern 1*

The rear foot remains flat during the front foot's heel rotation (this is the first sub-phase of the double support phase), then rotates with the front foot maintained flat (second sub-phase) (Figure 2.2). We should note that this pattern imposes the presence of a flat foot during the whole walking cycle. It is therefore a very constraining gait which provides maximum ground support during each step. It could be used for walking at a slow or moderate velocity for a mechanical biped, with a search for balance enhancing in a maximum contact zone at every moment. We note that the transition configuration of one sub-phase to the next is therefore performed with simultaneous flat contact of both feet.

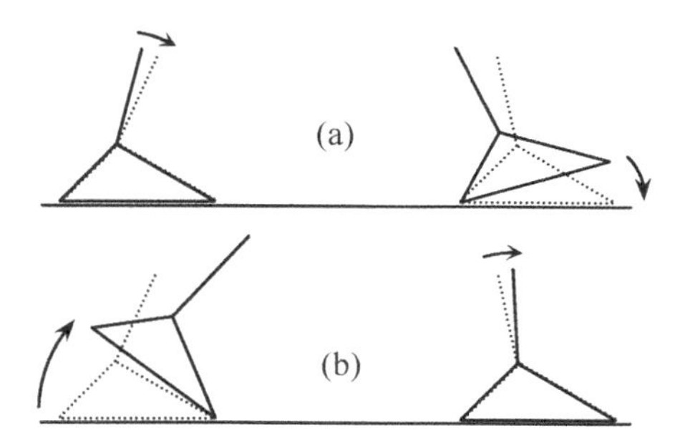

······Transition configurations of feet between both sub-phases

Figure 2.2. *Double support phase broken down into two sub-phases* (a) *and* (b) *with the rear foot then the front foot maintained flat in succession during each of the sub-phases*

2.2.2.2. *Pattern 2*

The rear foot rotates about its front edge as soon as the front foot touches down, and the end of the heel-rocker phase of the latter marks the end of the double support (Figure 2.3). The distribution of the DoF is maintained, and there are no sub-phases. The biped kinematic configuration remains unchanging during the totality of the double support phase. The result will be a simpler dynamic model and the formulation of a less constrained dynamic-synthesis problem. It should be noted that the extension of the heel-rocker movement on the overall double support does not conform to human walking, during which this movement only represents a brief episode of the double-support phase. This characteristic has been taken into account in pattern 3.

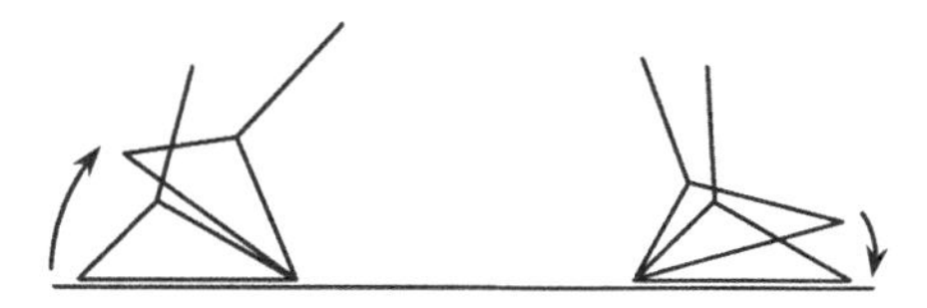

Figure 2.3. *The double support phase which evolves with simultaneous rotations of the rear foot on its front tip and of the front foot on its heel*

2.2.2.3. *Pattern 3*

The rear foot starts to rotate during a short heel rotation of the front foot (first sub-phase) and continues to rotate with the front foot flat on the ground (second sub-phase) (Figure 2.4). This movement is an imitation of human walking during the double support phase.

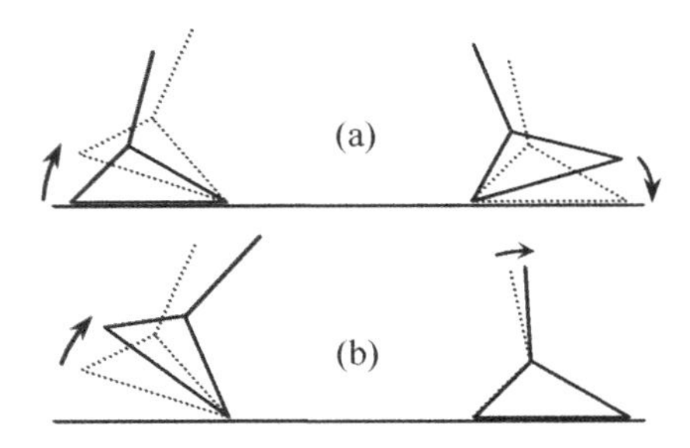

·····Transition configurations of feet between both sub-phases

Figure 2.4. *(a) Frontal hedge rotation of the rear foot during a short heel rotation sub-phase of the front foot followed by (b) flat-footed support sub-phase*

2.2.2.4. *Pattern 4*

During the swing phase of human walking, two sub-phases appear when the foot of the stance leg, when it is flat on the ground (first sub-phase), lifts its heel near the end of the phase by starting a rotation about the metatarsal axis (second sub-phase) (Figure 2.5). It must be stated that in human walking this second sub-phase is followed by a rotation of the forefoot on its frontal edge. This is a brief movement which represents a third sub-phase during which an additional DoF appears. The absence of an active foot/forefoot articulation in the models developed in the following section do not allow for this movement.

This contact pattern can be completed for a step by splitting up the double support into a heel rotation sub-phase and into a sub-phase with a front flat foot which are analogous to those of pattern 3. The result is the step division into four sub-phases such as those that have been drawn in the sagittal plane in Figure 2.1.

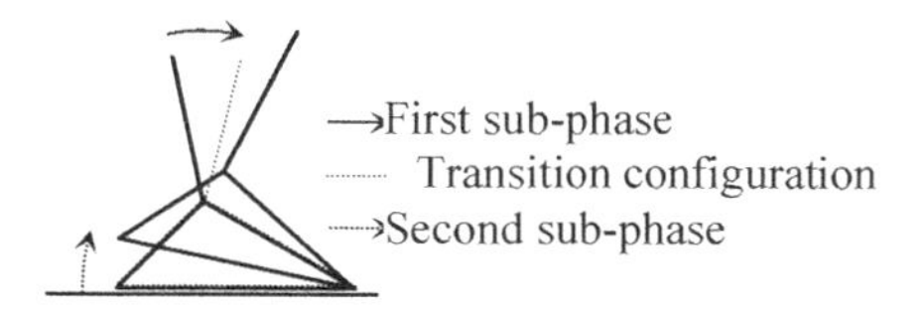

Figure 2.5. *Single-support phase division into two sub-phases: foot flat, then pivoting on its frontal edge*

Figure 2.6 shows the time slicing in a step cycle. The two main phases of single and double support occur successively over time intervals defined as $[t^i, t_2^*]$ and $[t_2^*, t^f]$ in which t^i and t^f each represent an initial and final instant of the SSP-DSP cycle. The instants t_1^* and t_3^* are each transition times between the sub-phases of single and double support if they exist. Afterwards, the division into four sub-intervals, as defined in Figure 2.6, will be favored. The following notations will be associated with this partition:

$$\begin{gathered} I_1 = [t^i(\equiv t_0^*), t_1^*];\; I_2 = [t_1^*, t_2^*];\; I_3 = [t_2^*, t_3^*];\; I_4 = [t_3^*, t^f(\equiv t_4^*)]; \\ I_{SSP} = [t^i, t_2^*];\; I_{DSP} = [t_2^*, t^f];\; I_{STEP} = [t^i, t^f] \end{gathered} \qquad [2.1]$$

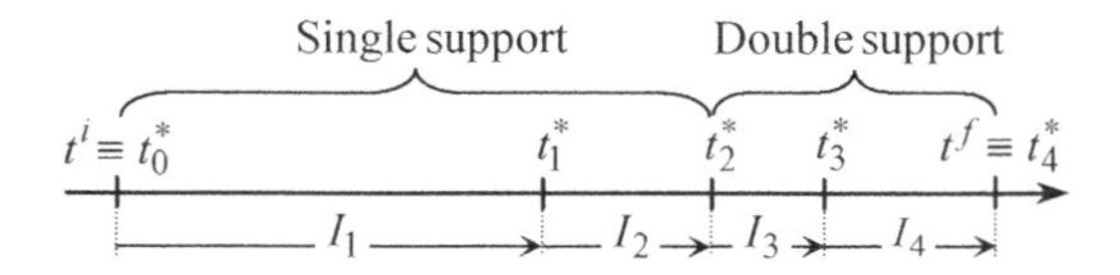

Figure 2.6. *Splitting up the SSP-DSP sequence into four time sub-intervals for a step made up of the four sub-phases of contact pattern 4*

2.2.3. *Generalized coordinates for a sagittal step*

A set of configuration coordinates can be defined according to two different approaches depending on whether the biped is considered as a kinematically free system or as a rooted kinematic chain. A constrained dynamic model will be associated with the first approach over the whole step. The second approach will be translated by a smaller number of configuration parameters and by a constrained dynamic model in the double support phase only.

2.2.3.1. *Generalized coordinates with an implicit rooted contact*

During a SSP-DSP sequence, the tip (or forefoot) of the foot which is situated in the central position remains motionless (Figure 2.7). This contact linkage, which is permanent through the step-sequence considered (Figure 2.6), brings us to define joint coordinates of a rooted (to the floor) system by following the kinematic tree-structure from the proximal segment which is the central foot (body segment S_1 in Figure 2.7), towards the extremities of the terminal chains: swing leg, pelvis-trunk-head system and eventually arms.

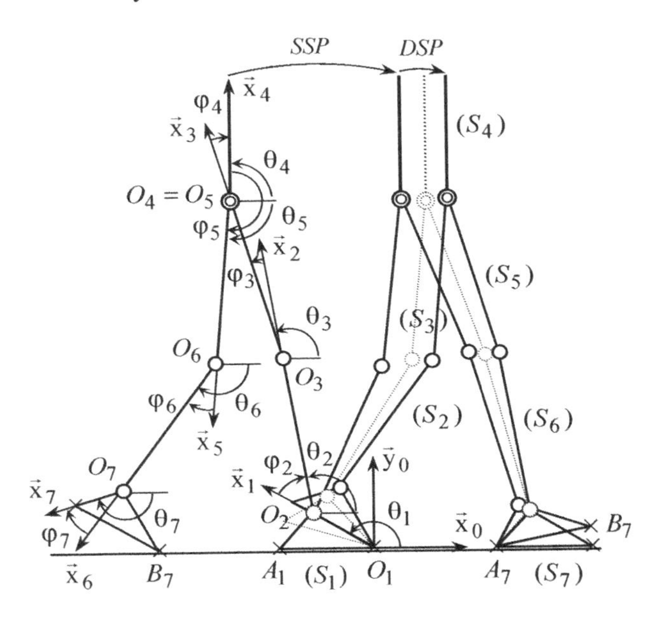

Figure 2.7. *The biped is rooted at the tip O_1 of the stance foot (S_1). The sagittal movement can be described by a set of seven generalized coordinates which may be the absolute or relative rotations (θ_is and φ_is, respectively) of the seven body segments*

In a sagittal model, two types of configuration variables can be simply defined when considering the relative rotations of body segments, or absolute rotations. By referring to the notation of Figure 2.7, we can successively define:

$$\vec{x}_i = \overrightarrow{O_iO}_{i+1} / \left\| \overrightarrow{O_iO}_{i+1} \right\|, \; 1 \leq i \leq 6 \; ; \; \vec{x}_7 = \overrightarrow{O_7A_7} / \left\| \overrightarrow{O_7A_7} \right\|$$

$$\theta_i = (\vec{x}_0, \vec{x}_i)_{\vec{z}_i} \, , \; \varphi_i = (\vec{x}_{i-1}, \vec{x}_i)_{\vec{z}_i} \, , \; 1 \leq i \leq 7 \;\; (\vec{y}_i = \vec{z}_i \wedge \vec{x}_i)$$

The result is a configuration vector made up of seven generalized coordinates such that:

$$q = (q_1, \ldots, q_{n_q})^T,\ q_i = \theta_i \text{ or } \varphi_i,\ \ i \leq n_q\ (n_q = 7) \qquad [2.2]$$

that are not necessarily independent.

If the foot is maintained flat in the single support phase, then the condition $q_1 = cste$ can be treated as a holonomous constraint associated with configuration [2.2] or, more simply, q_1 can be eliminated from equation [2.2] in order to reduce the configuration coordinates to six independent parameters.

The double-support phase necessitates the introduction of constraint equations which specify the closure conditions for the kinematic chain at the front foot. These express that the heel point A_7 is in contact with the ground, that is, on flat ground:

$$t \in I_{DSP}\ ,\ \begin{cases} \phi_1(q(t)) := \overrightarrow{B_1 A_7}(q(t)) \cdot \vec{x}_0 - L_{step} + l_{foot} = 0 \\ \phi_2(q(t)) := \overrightarrow{B_1 A_7}(q(t)) \cdot \vec{y}_0 = 0 \end{cases} \qquad [2.3]$$

where L_{step} and l_{foot} represent the step-length and the length of the foot respectively.

In addition, a third condition must be introduced over the second sub-phase in order to convey that the foot is maintained flat on the ground, i.e.:

$$t_3^* \in]t_2^*, t^f[,\ t \in I_4,\ \phi_3(q(t)) := \overrightarrow{A_7 B_7}(q(t)) \cdot \vec{y}_0 = 0 \qquad [2.4]$$

The number of independent variables of configuration [2.2] is in this way reduced to five by relations [2.3], then to four when condition [2.4] is added.

The choice of such generalized coordinates, which can be found in [BES 04, BES 05, SEG 05], has the advantages of restraining the configuration to the joint rotations and to the rotation of the front foot about its tip axis.

2.2.3.2. *Generalized coordinates without implicit constraint*

The biped is now considered as a free system, without any direct kinematic links with its environment (Figure 2.8). This is the approach used in [CHE 01, MUR 03, SAI 03]. As hip point O_3 is at the center of the tree-like structure, the Cartesian coordinates of this point are naturally accounted for, together with the absolute rotation of the trunk. The set of generalized coordinates may then be completed as is indicated in Figure 2.8 as follows:

$$\overrightarrow{OO_3} = x\vec{x}_0 + y\vec{y}_0$$

$$\varphi_3 = (\vec{x}_0, \vec{x}_3)_{\vec{z}_0}, \; \varphi_7 = (\vec{x}_3, \vec{x}_7)_{\vec{z}_0}$$

$$\varphi_j = (\vec{x}_{j-1}, \vec{x}_j)_{\vec{z}_0}, \; i = 4, 5, 6, 8, 9$$

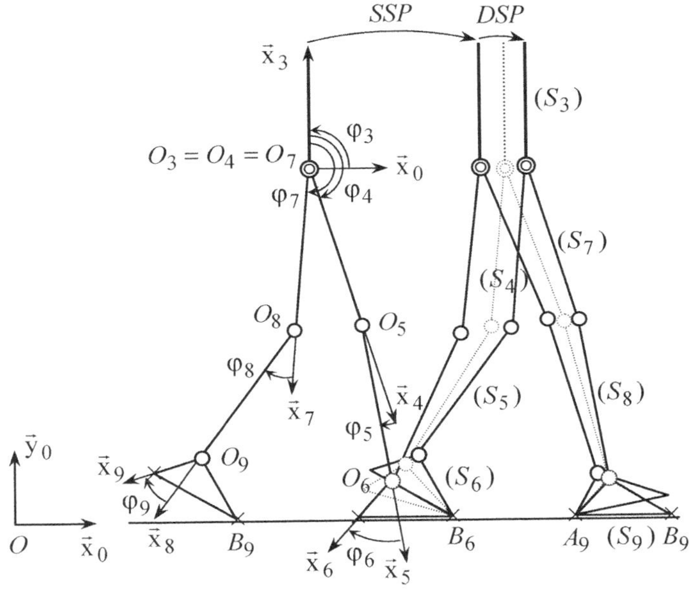

Figure 2.8. *The biped is considered as a kinematically free system with nine DoF. A set of generalized coordinates may then include, for example, the Cartesian coordinates x and y of the hip point O_3, the absolute rotation angle of the trunk and the relative joint-rotations*

We therefore obtain the configuration with nine q_i coordinates such that:

$$q_1 = x\,;\; q_2 = y\,;\; q_k = \varphi_k\,,\; 3 \le k \le 9 \quad [2.5]$$

The constraint equations resulting from contact pattern 4 can be expressed in the following way (walking on flat horizontal ground):

$$t \in I_1,\ \phi_1(q(t)) := \overrightarrow{A_6B_6}(q(t)) \cdot \vec{y}_0 = 0 \qquad [2.6]$$

$$t \in I_{STEP},\ \begin{cases} \phi_2(q(t)) := \overrightarrow{OB_6}(q(t)) \cdot \vec{x}_0 - x_{init} = 0 \\ \phi_3(q(t)) := \overrightarrow{OB_6}(q(t)) \cdot \vec{y}_0 - y_{init} = 0 \end{cases} \qquad [2.7]$$

$$t \in I_{DSP},\ \begin{cases} \phi_4(q(t)) := \overrightarrow{B_6A_9}(q(t)) \cdot \vec{x}_0 - L_{step} + l_{foot} = 0 \\ \phi_5(q(t)) := \overrightarrow{B_6A_9}(q(t)) \cdot \vec{y}_0 = 0 \end{cases} \qquad [2.8]$$

$$t_3^* \in]t_2, t^f[,\ t \in I_4,\ \phi_6(q(t)) := \overrightarrow{A_9B_9}(q(t)) \cdot \vec{y}_0 = 0 \qquad [2.9]$$

In equation [2.7], x_{init} and y_{init} are the given coordinates of a reference point of contact.

The three relations of equations [2.6] and [2.7], the two relations of equation [2.7], the four relations of equations [2.7] and [2.8] and the five relations of equation [2.7], [2.8] and [2.9] must be fulfilled in the four successive sub-intervals of time, as defined in Figure 2.6. The increase in the dimension of the configuration vector $q(t)$ is naturally accompanied, in each phase of movement, by as many supplementary constraint equations which need to be fulfilled.

2.2.4. *Generalized coordinates for three-dimensional walking*

The locomotion system has, in this case, 12 internal DoF, to be compared to the six DoF of the sagittal model. The three-dimensional model cannot benefit from the simplifications provided by the introduction of absolute rotation angles for the planar model. A great complexity of relationships must therefore be expected, whether these consist of geometric, kinematic and sthenic constraints, or more particularly, dynamics equations to be developed for a dynamic synthesis of the walking cycle.

2.2.4.1. *Rooted-system configuration*

Figure 2.9 is a transposition of Figure 2.7 in the case of straight-line, three-dimensional walking on a flat horizontal surface. The locomotion system is made up (in the same way as in the planar biped-model), of seven main body segments. Its 12 internal DoF are distributed as is indicated in the figure, along the lines of the architecture described in section 2.2.1. Due to the fact that over the sequence SSP-DSP the front edge of the stance foot S_1 remains fixed on the ground, a 13-

dimensional configuration vector q can be defined to describe the biped movements. Its components q_i $(1 \leq i \leq 13)$ are naturally distributed along a simple kinematic chain made up of the biped-model's seven body segments which are then indexed as S_1 to S_{13}, as indicated in Figure 2.9.

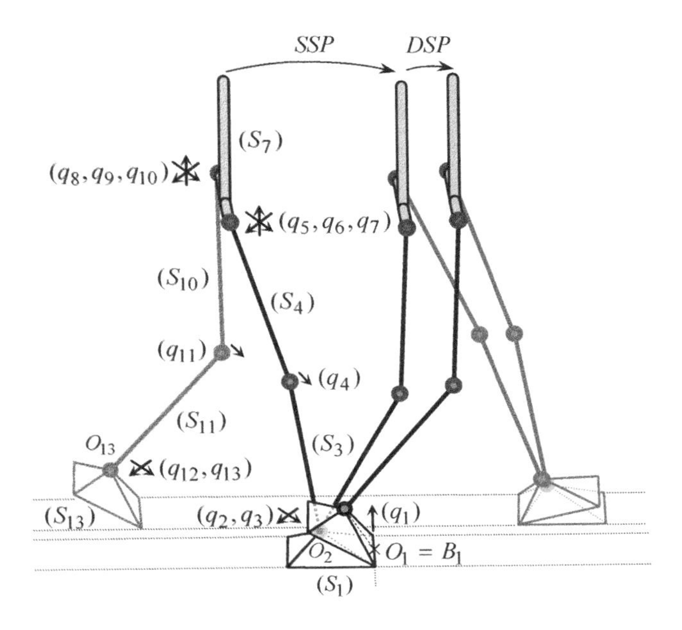

Figure 2.9. *Suggested sequence of generalized coordinates for an SSP-DSP three-dimensional walking step, exploiting the fact that the stance foot (S_1) remains rooted along its frontal edge, over the whole step*

An explicit determination of the q_is requires the definition of local frames attached to the successive links of the articulated chain. The construction of Denavit-Hartenberg [DEN 55] using Khalil-Kleinfinger's convention [KHA 86] is a perfectly adapted means for the kinematic parameterization of three-dimensional multibody systems. This can be seen in the construction (Figure 2.10) used for the biped in Figure 2.9. The q_i generalized coordinates are therefore defined as:

$$q_i = (\vec{x}_{i-1}, \vec{x}_i)_{\vec{z}_i}, \; i = 1, \ldots, 13$$

The kinematic constraint equations are to be taken into account during the double support phase, and have been formulated in section 2.2.4.3.

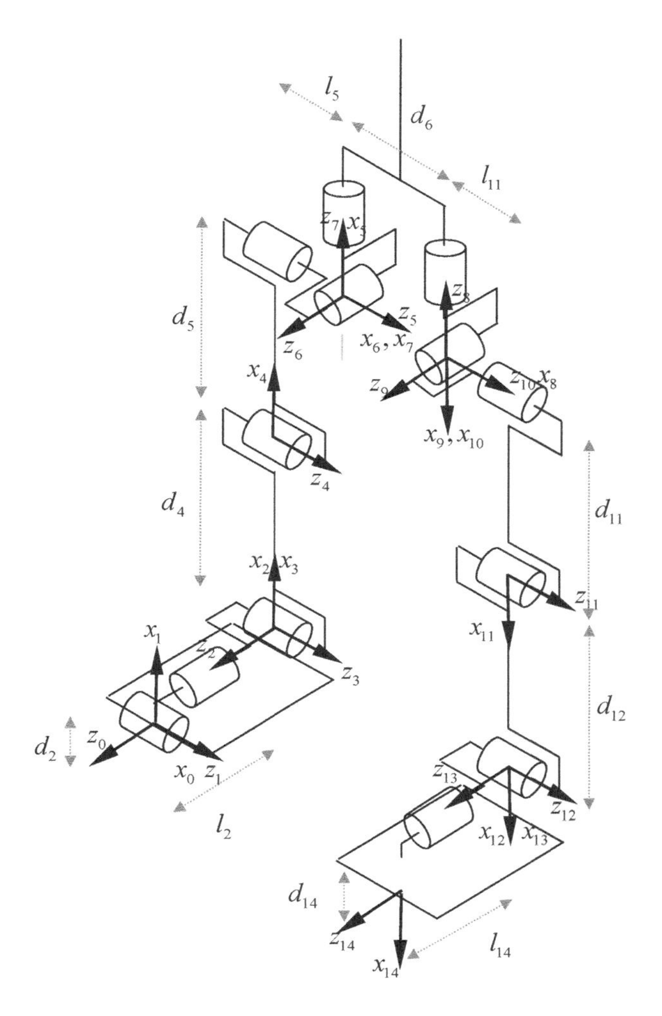

Figure 2.10. *Local frames linked to the biped's body segments, obeying the Khalil-Kleinfinger convention*

2.2.4.2. *Free-system configuration*

Figure 2.11 presents a possible distribution of generalized coordinates describing the kinematics of the biped considered as a free system. The central body S_6 (the pelvis-trunk) receives a free-solid kinematic parameterization consisting of the three Cartesian coordinates of one of its points (e.g. the median point O_6 of the connecting segment between the two joint centers of the hips; Figure 2.11) and its Euler angles with respect to the fixed reference frame. We obtain, in this way, the first six q_i coordinates of the sought-after configuration vector.

The kinematic parameterization set-up can then be followed up by using the Khalil-Kleinfinger construction along the tree-structure which makes up the locomotion limbs. The result is an 18-dimensional configuration-vector with its q_i components distributed as shown in Figure 2.11.

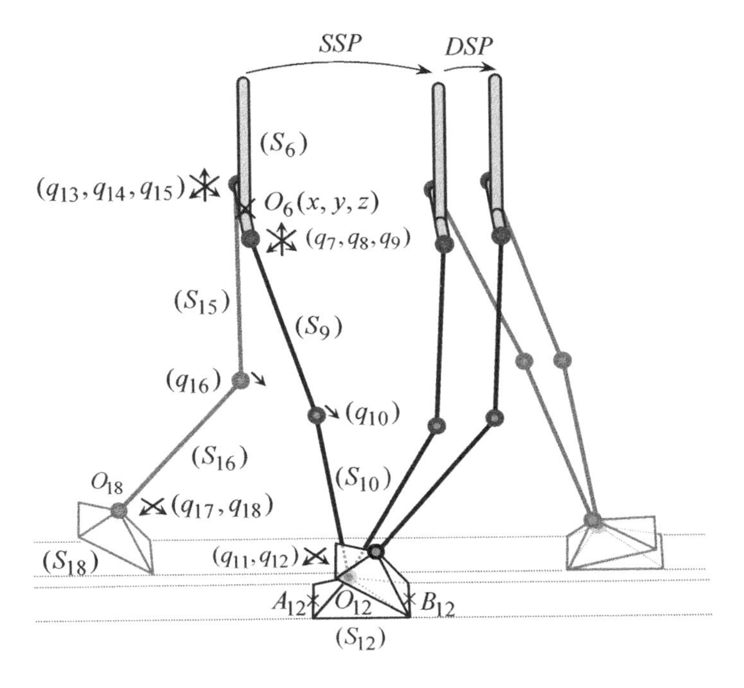

Figure 2.11. *Kinematic parameterization of a biped in free system configuration with 18 DoF*

This set of generalized coordinates could be extended to a trunk-head-arms system, which would lead to the introduction of numerous supplementary coordinates. For example, the biped BIP [SAR 98] has a spherical pelvis-trunk linkage, which brings the number of configuration coordinates required to describe its movements in this second mode of kinematic modeling to 21.

2.2.4.3. *Constraint equations*

The assigned foot-positioning on the ground results in geometric relationships defining a constrained set of generalized coordinates. In particular, during the double support, the geometric closure conditions of the locomotion system must express a foot contact mode according to, for example, one of the patterns described in section 2.2.2.

To begin with, some geometric characteristics of the foot must be specified. To simplify, and without altering the basis of the problem, we can consider feet with flat soles and straight-edged sides as is suggested in Figures 2.9 and 2.11. Let us also suppose that the frontal and rear edges are parallel, and that they remain orthogonal to the biped's sagittal plane when the biped is walking in a straight line.

Figure 2.12 represents the feet in contact with the ground in the transition configurations between the phases and sub-phases of an SSP-DSP-SSP sequence. The elements which are attached to the stance-propulsive foot (the rear foot in Figure 2.12) will be designated using the subscript p (pth step) which becomes $p + 1$ when considering those attached to the front foot. We have introduced the following data and notations, specified for the S_p foot, to be transcribed for S_{p+1} by changing the subscript p into $p + 1$:

– A local frame $(O_p; \vec{x}_p, \vec{y}_p, \vec{z}_p)$ is attached to S_p (Figure 2.13):

- O_p is the joint center of the ankle;

- the axis $(O_p; \vec{x}_p)$ is parallel to the front and rear edges of the foot;

- $(O_p; \vec{z}_p)$ is the adduction-abduction axis of the ankle.

This axis is parallel to the sole, and its orthogonal projection on the sole is the segment A_pB_p which joins the posterior and anterior edges of the foot (Figure 2.12).

– The contact footprint of S_p is attached to a targeted footprint frame, $(Q_p; \vec{n}_p, \vec{u}_p, \vec{v}_p)$ (Figure 2.12) such that:

- $\vec{n}_p$ is the ascending vector normal to the contact plane $(Q_p; \vec{u}_p, \vec{v}_p)$;

- $(Q_p; \vec{u}_p)$ defines the A_pB_p median line of the footprint; and

- the axis $(Q_p; \vec{v}_p)$ carries the posterior edge of the footprint.

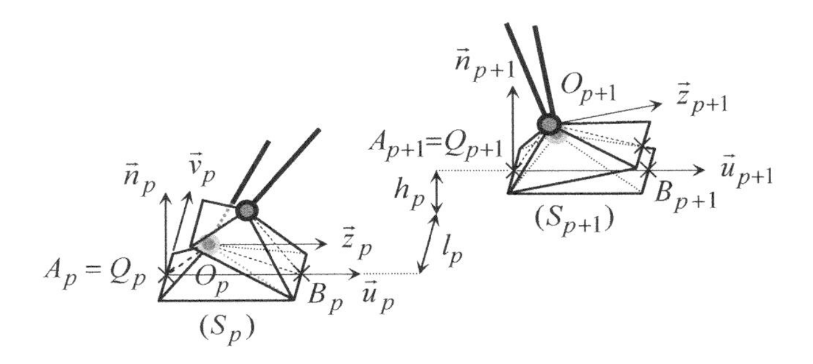

Figure 2.12. *Relative position of the rear foot S_p and the front foot S_{p+1} in the double support phase*

The two contact planes under consideration, $(Q_p; \vec{u}_p, \vec{v}_p)$ and $(Q_{p+1}; \vec{u}_{p+1}, \vec{v}_{p+1})$, are not necessarily horizontal and parallel. If they are horizontal and parallel, their relative height can be shifted positively or negatively and denoted h_p (Figure 2.12). This can represent a doorstep height or a shelf to reach stairs, for example.

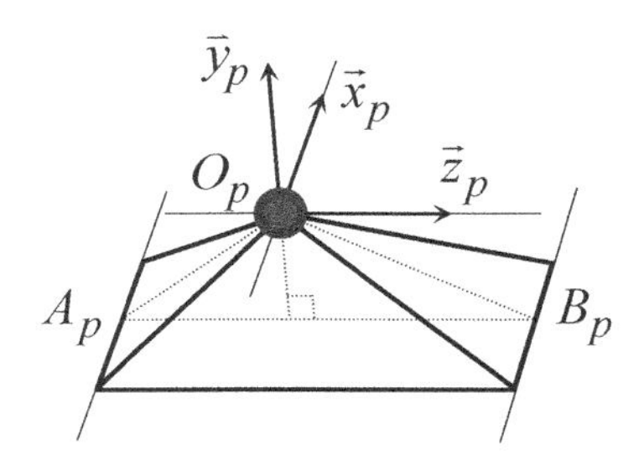

Figure 2.13. *Local frame $(O; \vec{x}_p, \vec{y}_p, \vec{z}_p)$ attached to the foot (S_p)*

As a reference example, let us consider a biped locomotion system with free configuration submitted to the contact mode of pattern 4 of section 2.2.2.

The constraints which specify the geometric conditions of contact during each phase and sub-phase of the step must be formulated according to the successive time intervals as defined by the step-time slicing, as depicted in Figure 2.6 (notation [2.1]).

2.2.4.3.1. First single-support sub-phase: $t \in I_1$

Only the stance foot S_p is under consideration here. During the first sub-phase, six conditions describe how the foot frame is linked to the footprint frame, i.e.:

$$\begin{cases} \phi_1^1(q(t)) := \overrightarrow{Q_p B_p}(q(t)) \cdot \vec{n}_p = 0 \\ \phi_2^1(q(t)) := \overrightarrow{Q_p B_p}(q(t)) \cdot \vec{u}_p - l_{foot} = 0 \\ \phi_3^1(q(t)) := \overrightarrow{Q_p B_p}(q(t)) \cdot \vec{v}_p = 0 \end{cases} \qquad [2.10]$$

$$\begin{cases} \phi_4^1(q(t)) := -\vec{x}_p(q(t)) \cdot \vec{u}_p = 0 \\ \phi_5^1(q(t)) := \vec{x}_p(q(t)) \cdot \vec{n}_p = 0 \\ \phi_6^1(q(t)) := \vec{y}_p(q(t)) \cdot \vec{u}_p = 0 \end{cases} \qquad [2.11]$$

Conditions [2.10] express the coincidence of point B_p of S_p with its assigned position in the contact plane (Figure 2.12). The second group of conditions, [2.11], define the foot's appropriate angular orientations. Let us note that the introduction of a minus sign in the expression ϕ_4^1 of equation [2.11] aims to correct the sign in the identification of the multiplier associated with this constraint, with the corresponding moment of contact forces (section 2.3.1.2).

Considering a free system with 18 DoF, the six constraining equations [2.10] and [2.11] must be accounted for. In the configuration with 13 generalized coordinates (section 2.2.4.1), conditions [2.10] and the first two conditions of [2.11] are implicitly fulfilled. In this case, only the last equation of [2.11] is non-trivial and must be taken into account explicitly.

These conditions are represented in the general form as:

$$t \in I_1, \Phi^1(q((t)) = 0 \ (\in \Re^{n_1}, n_1 = n_q - 12) \qquad [2.12]$$

in which:

$$\Phi^1(q(t)) = (\phi_1^1(q(t),..., \phi_{n_1}^1(q(t))^T$$

2.2.4.3.2. Second single-support sub-phase: $t \in I_2$

The contact of the stance foot S_p takes place at its frontal edge, along a specified contact segment (Figure 2.12). This geometric constraint is expressed by the first five conditions of equations [2.10] and [2.11] prolonged over the interval I_2. The constraint relationships which need to be fulfilled are:

$$t \in I_2, \ \phi_i^2(q(t)) \equiv \phi_i^1(q(t)) = 0, \quad 1 \le i \le 5 \qquad [2.13]$$

By reuniting, as was done previously, functions ϕ_i^2 can be expressed as a vector function such that:

$$\Phi^2(q(t)) = (\phi_1^2(q(t),..., \phi_{n_2}^2(q(t))^T, \ (n_2 = 5)$$

Conditions [2.13] are summarized in the interval I_2 as the vector equation:

$$t \in I_2, \Phi^2(q((t)) = 0 \ (\in \Re^{n_2}, n_2 = n_q - 13) \qquad [2.14]$$

During this sub-phase, the 13 generalized coordinates which describe the movement of the biped implicitly considered as a rooted system (section 2.2.4.1) are independent. They are therefore not constraining equations in this case.

2.2.4.3.3. First double-support sub-phase: $t \in I_3$

The stance foot S_p continues its rotation movement around its frontal edge during the totality of the double support phase. Consequently, conditions [2.14] are carried over to the whole time interval I_{DA} which enables us to define the following first five constraints:

$$t \in I_3, \ \phi_i^3(q(t)) \equiv \phi_i^2(q(t)) = 0, \quad 1 \le i \le 5 \qquad [2.15]$$

For the front foot S_{p+1} it is useful to pre-define a relative positioning of the contact point Q_{p+1} with respect to Q_p by defining:

$$\overrightarrow{Q_p Q_{p+1}} = L_p \vec{u}_p + l_p \vec{v}_p + h_p \vec{n}_p \qquad [2.16]$$

where L_p, l_p and h_p are a step-length, a step-width and a step-height for the pth step (Figure 2.12), respectively.

The permanence of contact of S_{p+1} along the heel's edge during the first sub-phase is expressed by the five following relationships, analogous to the conditions [2.13] taken from [2.10] and [2.11]:

$$\begin{cases} \phi_6^3(q(t)) := \overrightarrow{Q_{p+1}A_{p+1}}(q(t)) \cdot \vec{n}_{p+1} = 0 \\ \phi_7^3(q(t)) := \overrightarrow{Q_{p+1}A_{p+1}}(q(t)) \cdot \vec{u}_{p+1} = 0 \\ \phi_8^3(q(t)) := \overrightarrow{Q_{p+1}A_{p+1}}(q(t)) \cdot \vec{v}_{p+1} = 0 \end{cases} \qquad [2.17]$$

$$\begin{cases} \phi_9^3(q(t)) := -\vec{x}_{p+1}(q(t)) \cdot \vec{u}_{p+1} = 0 \\ \phi_{10}^3(q(t)) := \vec{x}_{p+1}(q(t)) \cdot \vec{n}_{p+1} = 0 \end{cases} \qquad [2.18]$$

By reuniting the functions ϕ_i^3 in the vector function such that

$$\Phi^3(q) = (\phi_1^3(q),\ldots,\phi_{n_3}^3(q))^T$$

the 10 conditions [2.15], [2.17] and [2.18] are summarized, on the interval I_3 under the global form:

$$t \in I_3,\ \Phi^3(q(t)) = 0\ (\in \Re^{n_3}, n_3 = n_q - 8) \qquad [2.19]$$

If the set of generalized coordinates is that of section 2.2.4.1, only closure conditions [2.17] and [2.18] formulated at the front foot are to be taken into account.

2.2.4.3.4. Second double-support sub-phase: $t \in I_4$

The rear foot S_p finishes its rotation along the frontal edge and the front foot is flat on the ground (Figure 2.12).

Constraint equations [2.15], [2.17] and [2.18] (summarized in equation [2.19]) must be fulfilled during the whole double support and must, therefore, be transcribed identically over the interval I_4. The following new condition is added:

$$\phi_{11}^4(q(t)) := \vec{y}_{p+1}(q(t)) \cdot \vec{u}_{p+1} = 0 \qquad [2.20]$$

which, with equation [2.18], conveys that the sole of the front foot S_{p+1} is maintained parallel to the contact plane.

Finally, defining the vector function

$$\Phi^4(q) = ((\Phi^3(q))^T, \phi_{11}^4(q))^T$$

conditions [2.19] and [2.20] are summarized, on the interval I_4, by the vector relationship:

$$t \in I_4,\ \Phi^4(q(t)) = 0\ (\in \Re^{n_4}, n_4 = n_q - 7) \qquad [2.21]$$

2.2.5. *Transition conditions*

The transition conditions between phases and sub-phases mainly depend on the mechanical characteristics of contact at heel-touch, and whether the contact occurs with impact or not. They also depend on the type of step succession to be taken into consideration, e.g. a purely cyclic step, accelerated or decelerated step, start or stop step or turning step. It is the kinematic aspect of this problem which is of interest here, in the form of the initial and final conditions to be fulfilled by the successive steps and their sub-phases.

2.2.5.1. *Initialization constraints of a walking step*

The initialization conditions of the step under consideration are linked to the final kinematic constraints of the preceding step, which can be different from the step underway. Therefore, in order to specify the escape conditions of the rear foot noted as S_{p-1} in Figure 2.14, at the initial instant t^i of the single support phase, the position of its frontal edge must be specified in relation to the stance foot S_p. This includes the given local topography of the ground and lateral dissymmetry due to the change in foot. The footprint of S_{p-1} is defined for the (p–1)th step by the frame $(Q_{p-1}; \vec{u}_{p+1}, \vec{v}_{p-1}, \vec{n}_{p-1})$ with a relative position of Q_{p-1} and Q_p as defined by the relationship:

$$\overrightarrow{Q_{p-1}Q_p} = L_{p-1}\vec{u}_{p-1} + l_{p-1}\vec{v}_{p-1} + h_{p-1}\vec{n}_{p-1} \qquad [2.22]$$

where L_{p-1}, l_{p-1} and h_{p-1} represent a length, a width and height of the (p–1)th step, which are all geometric parameters which can be different from their counterparts subscripted p in equation [2.16] for the pth step (note that the algebraic value of l_{p-1} is of opposite sign to l_p in all cases).

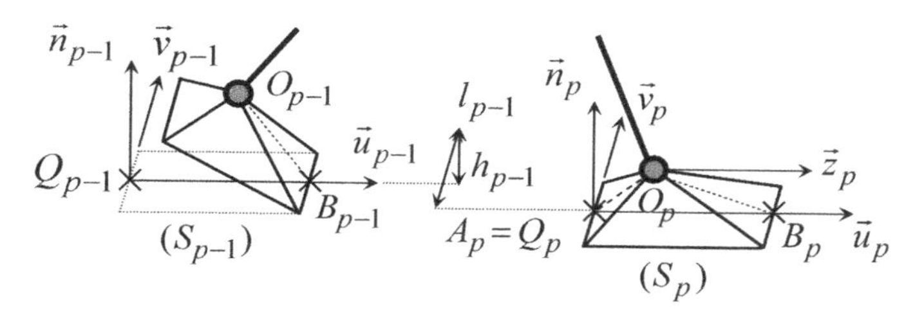

Figure 2.14. *Relative position of both feet at the end of double-support during the (p–1)th step*

When toeing off of the foot in position S_{p-1} at instant t^i, the contact takes place along the frontal edge. This situation can be expressed by the relationships:

$$\overrightarrow{Q_{p-1}B}_{p-1}(q(t^i)) - l_{foot}\,\vec{u}_{p-1} = \vec{0} \tag{2.23}$$

$$\begin{cases} \vec{x}_{p-1}(q(t^i)) \cdot \vec{u}_{p-1} = 0 \\ \vec{x}_{p-1}(q(t^i)) \cdot \vec{n}_{p-1} = 0 \end{cases} \tag{2.24}$$

It is to be noted that these conditions are of identical form to that of the first five constraints of [2.10] and [2.11] formulated for the foot S_p when it is in contact on its frontal edge.

The continuity of the joint velocities at toeing-off requires taking into account a kinematic screw of S_{p-1} at instant t_i, characterized by the following relationships:

$$\vec{V}(B_{p-1}(q(t^i))) = \vec{0} \tag{2.25}$$

$$\begin{cases} \vec{\Omega}(S_{p-1}(q(t^i)) \cdot \vec{n}_{p-1} = 0 \\ \vec{\Omega}(S_{p-1}(q(t^i)) \cdot \vec{u}_{p-1} = 0 \end{cases} \tag{2.26}$$

Condition [2.25] signifies that the foot toes off without slipping, its tip having a normal component of velocity equal to zero. Condition [2.26] expresses the fact that the initial rotation of the foot S_{p-1} can only take place along the frontal edge of which direction is defined by the axis $(B_{p-1};\vec{v}_{p-1})$. We note that equation [2.25] could be deduced from equation [2.23] by formal differentiation with respect to t^i. In the same way, the relationships of equation [2.26] could be derived from relations [2.24] at t_i.

In addition, these conditions are analogous to those which appear at the end of the double-support at instant t^f, and that can be deduced from the constraint equations formulated on interval I_4.

Equations [2.23] to [2.26] represent 10 punctual (or instantaneous) constraints, which will be referred to in the following general form:

$$g_0(t_0^*) = 0 \ (\in \Re^{10}) \qquad [2.27]$$

where $t_0^* \equiv t^i$ (see equation [2.1]).

2.2.5.2. *Transition conditions at shifting times*

The conditions which apply to the transition configurations $q(t_1^*)$, $q(t_2^*)$ and $q(t_3^*)$ are explicitly contained in the constraint equations each formulated in [2.12], [2.14] and [2.19] on the intervals I_1, I_2 and I_3. The corresponding conditions must then be added for the velocities.

At instant t_1^*, the stance foot S_p starts a rotation on its frontal edge, beginning from a flat-footed position. Consequently, its initial rotation velocity in the second single-support sub-phase is equal to zero. In other words, as the foot is fixed during the interval I_1, it is still fixed at instant t_1^* of heel lift-off. The result is that the kinematic screw of S_p is zero at t_1^*, i.e.

$$\begin{cases} \vec{V}(B_p(q(t_1^*)) = \vec{0} \\ \vec{\Omega}(S_p(q(t_1^*)) = \vec{0} \end{cases}$$

This condition expresses the continuity of the velocities at t_1^* or, more exactly, at t_1^{*+} and can be derived from equation [2.12] or in an equivalent way from equations [2.10] and [2.11], in the form:

$$\frac{\partial \phi_i^1}{\partial q}(q(t_1^*))\, \dot{q}(t_1^*) = 0\,, \quad 1 \le i \le 6 \qquad [2.28]$$

Instant t_2^* is that of heel-touch which may be carried out with or without impact. If the contact of the heel with the ground occurs without impact, there is a continuity of velocities. This condition, which is to be specified at the end of the single-support, is written as:

$$\begin{cases} \vec{V}(A_{p+1}(q(t_2^*))) = \vec{0} \\ \vec{\Omega}(S_{p+1}(q(t_2^*))) \cdot \vec{n}_{p+1} = \vec{0} \\ \vec{\Omega}(S_{p+1}(q(t_2^*))) \cdot \vec{u}_{p+1} = \vec{0} \end{cases}$$

It can be formally derived from the five last constraint equations of equation [2.19] (or, in an equivalent manner, from equation [2.17] and [2.18]) formulated on I_3, i.e.

$$\frac{\partial \phi_i^3}{\partial q}(q(t_2^*))\,\dot{q}(t_2^*) = 0\,,\ 6 \le i \le 10 \qquad [2.29]$$

If there is impact, velocity jumps can appear according to the adopted dynamic impact model. This problem is considered in section 2.3.3.

Instant t_3^* marks the end of the front foot heel rotation. Again, the flat contact of the sole with the ground can take place with or without impact. In the absence of impact, a translation needs to be made of the fact that the foot's rotation (which occurs at the heel's edge) is canceled at instant t_3^* or, more precisely, at t_3^{*-} when the foot is laid flat. As at t_2^* previously, this condition can be derived from constraint equation [2.20] in the form:

$$\frac{\partial \phi_{11}^4}{\partial q}(q(t_3^*))\,\dot{q}(t_3^*) = 0 \qquad [2.30]$$

The case of contact with impact is examined in section 2.3.3. Following equation [2.27], constraints [2.28] to [2.30] are formally summarized:

$$\begin{cases} g_1(t_1^*) = 0 \ (\in \Re^6) \\ g_2(t_2^*) = 0 \ (\in \Re^5) \\ g_3(t_3^*) = 0 \ (\in \Re) \end{cases} \qquad [2.31]$$

2.2.5.3. *Closure constraints of the step cycle*

When the step is purely cyclic, i.e. when it is reproduced symmetrically by shifting the limb configurations for the next step, the final configuration vector $q(t^f)$ and its time-derivative vector $\dot{q}(t^f)$ are deduced from their initial counterparts $q(t^i)$ and $\dot{q}(t^i)$ by permutation of their components. This operation is formally summarized by the equation (with notation t_4^* for t^f, see equation [2.1]):

$$g_4(t_4^*) = 0 \quad (\in \Re^{2n_q}) \qquad [2.32]$$

If the step is not replicated identically, the link to be established between final and initial configurations will depend on the conditions of step ending which need to be taken into consideration e.g. accelerating or decelerating step, downward or upward step. The implicit dependence of g_4 in equation [2.32] with respect to $q(t^i)$ and $\dot{q}(t^i)$ can then disappear partially or completely.

2.3. The dynamics of walking

One of the most striking characteristics of walking is the instability of its equilibrium. If this aspect of movement is perfectly mastered in the living world, it is still extremely delicate to manage when mechanical bipeds are concerned. The dynamics of walking must naturally combine gravity, foot-ground contact efforts and the coordination of joint movements in order to maintain the balance of the biped while ensuring the appropriate propulsive effect. To generate a walking step, we can extract a feasible solution from a dynamic model of the biped, i.e. a solution satisfying all the constraints which characterize the movement to be computed. The unilateral nature of the ground-foot contact and the non-sliding requirement are the most restrictive conditions to be applied to the sought-after solution.

With this in mind, the first basic choice to make is which dynamic model to use. The Lagrange energetic model offers an appropriate algebraic structure for the construction of a solution by using a technique of parametric optimization. However, it must also be noted that the calculation complexity of the Lagrange equations is $O(n_q^3)$ [KHA 02] depending on the n_q number of DoF of the mechanical system. Lengthy formulations are therefore to be expected, so that they can define a three-dimensional system where the number of DoF is greater than 12.

The Newton-Euler equations are easier to program. They enable us to calculate in a simple way the inverse dynamic model, and can be used to formulate the direct dynamic model by using the structure of the dynamic equations which result from Lagrange's formalism.

2.3.1. *Lagrangian dynamic model*

The reference movement to be modeled is a walking step, as defined in section 2.2. The various phases of movement are described using the same set of non-independent configuration coordinates:

$$t \in [t^i, t^f],\ q(t) = (q_1(t),..., q_{n_q}(t))^T\ (\in \Re^{n_q}) \qquad [2.33]$$

where n_q is equal to 13 or 18 for three-dimensional walking, according to the kinematic modeling employed (section 2.2.4).

The q_is fulfill the constraint equations which were derived in detail in section 2.2.3 for a sagittal step and in section 2.2.4 for a three-dimensional walking step.

In the following, we will focus on the constraints represented by the general expressions [2.12], [2.14], [2.19] and [2.21], formulated on the intervals I_1, I_2, I_3 and I_4 (defined in [2.1]) and globally respresented:

$$t \in I_\alpha,\ \Phi^\alpha(q(t)) = 0\ (\in \Re^{n_\alpha}, n_\alpha < n_q),\ \alpha = 1, 2, 3, 4 \qquad [2.34]$$

Relation [2.34] represents, on each time interval I_α, a system of holonomic independent constraints.

2.3.1.1. *Lagrange equations*

Let us note that the Lagrangian of the mechanical system is commonly written as [GAR 94]:

$$L(q, \dot{q}) = T(q, \dot{q}) - V(q) \qquad [2.35]$$

where T represents the kinetic energy:

$$T(q, \dot{q}) = \tfrac{1}{2}\dot{q}^T M(q)\,\dot{q} \qquad [2.36]$$

and V is the potential energy of the conservative force fields, later reduced to gravity.

The Lagrange equations are formally written as the n_q vector equation:

$$\frac{d}{dt}\left(\frac{\partial L}{\partial \dot{q}}\right) - \frac{\partial L}{\partial q} = Q + J^T \lambda \qquad [2.37]$$

where the vector of the generalized forces Q is here restrained to the dissipative forces together with the joint actuating torques. By taking the latter remark and the representations [2.35] and [2.36] into account, equation [2.37] can be developed as:

$$M(q)\ddot{q} + C(q,\dot{q}) + G(q) = D(q,\dot{q}) + A^{\tau}\tau + J(q)^T \lambda \qquad [2.38]$$

where M is the symmetric positive-definite inertia matrix which defines the kinetic energy [2.36]; C is an n_q-dimensional vector which groups together the centrifugal and Coriolis inertia terms; G represents the terms of gravity, $G = -\partial V / \partial q$; D is a vector which regroups the terms of dissipative forces at joints; A^{τ} is an $(n_q \times n_{\tau})$ matrix which depends on the chosen kinematic modeling; τ is the n_{τ}-dimensional vector of joint actuating torques τ_i with subscript i varying from 7 to 18 in free system modeling and from 2 to 13 in rooted system modeling $(n_{\tau} = 12)$; J is the Jacobian $(n_{\tau} \times n_q)$ matrix of the constraint equations grouped together in the vector $\Phi(q)$ so that $J(q) = \partial\Phi/\partial q$; and λ is the vector of Lagrange multipliers.

Indeed, equation [2.38] must be specifically formulated according to the time interval I_{α} under consideration, where the corresponding constraint equation formulated in equation [2.34] must be taken into account. As a result, the biped's dynamics are formulated for each movement's sub-phase by an algebraic-differential system such that:

$$t \in I_{\alpha}, \begin{cases} M(q(t))\ddot{q}(t) + N(q(t),\dot{q}(t)) = A^{\tau}\tau(t) + (J^{\alpha}(q(t)))^T \lambda^{\alpha}(t), \\ \Phi^{\alpha}(q(t)) = 0\ ; \quad \alpha = 1,2,3,4, \end{cases} \qquad [2.39]$$

where, on the basis of formulations [2.38] and [2.34],

$$N(q,\dot{q}) := C(q,\dot{q}) + G(q) - D(q,\dot{q})$$

$$\left.\begin{array}{l} J^{\alpha} := \partial\Phi^{\alpha}/\partial q \\ \lambda^{\alpha} = (\lambda_1^{\alpha},...,\lambda_{n_\alpha}^{\alpha})^T \end{array}\right\}, \ \alpha \in \{1,2,3,4\}$$

It should be noted that in the three-dimensional kinematic model (section 2.2.4), matrix A^{τ} admits one of the following simple structures depending on whether the biped is considered as a rooted system (section 2.2.4.1) or free system (section 2.2.4.2), respectively:

$$n_q = \begin{cases} n_\tau + 1, \ (A^{\tau})_{n_q \times n_\tau} = \begin{pmatrix} O_{1\times n_\tau} \\ I_{n_\tau \times n_\tau} \end{pmatrix} \\ n_\tau + 6, \ (A^{\tau})_{n_q \times n_\tau} = \begin{pmatrix} O_{6\times n_\tau} \\ I_{n_\tau \times n_\tau} \end{pmatrix} \end{cases} \quad [2.40]$$

This representation does not depend on the phase considered during the walking cycle under consideration.

2.3.1.2. *Identifying the multipliers*

In [2.39], the components of vector λ^{α} represent a set of forces and moments to be applied to the feet to maintain them in the contact positions assigned by the constraint equations.

During the first sub-phase of single support, the stance foot S_p (section 2.2.4.3) is submitted to a contact wrench which can be represented at point B_p (see Figures 2.12 and 2.13) with $\alpha = 1$, as:

$$t \in I_1, \begin{cases} \vec{F}_p^1 := \vec{R}(S_0 \to S_p) \\ \vec{M}_p^1(B_p) := \vec{M}(B_p; S_0 \to S_p) \end{cases} \quad [2.41]$$

where the resultant $\vec{F}_p^1$ and the moment $\vec{M}_p^1$ are exerted by the ground S_0 on the foot S_p.

By denoting the multiplier of λ^1 by λ_p^1 with reference to S_p, the virtual power of the contact forces enables the expression:

$$W^* = \left((J^1)^T \lambda_p^1\right)^T \dot{q}^* \qquad [2.42]$$

and the formulation:

$$W^* = \vec{F}_p^1 \cdot \vec{V}^*(B_p) + \vec{M}_p^1(B_p) \cdot \vec{\Omega}^*(S_p) \qquad [2.43]$$

where the notation $\dot{q}^*$, $\vec{V}^*(B_p)$ and $\vec{\Omega}^*$ represent the vector of the generalized virtual velocities, the virtual velocity of B_p and the rate of the virtual rotation vector of S_p, respectively. The expression of W^* is obviously separate from the chosen reduction center, B_p.

The differentiation of equation [2.12] through [2.10] and [2.11] establishes the formal identity:

$$\frac{\partial \Phi^1}{\partial q} \dot{q}^* \equiv \begin{pmatrix} \vec{V}^*(B_p) \\ \vec{\Omega}^*(S_p) \end{pmatrix}$$

Representation [2.42] becomes:

$$W^* = (\lambda_p^1)^T \begin{pmatrix} \vec{V}^*(B_p) \\ \vec{\Omega}^*(S_p) \end{pmatrix} \qquad [2.44]$$

The comparison of formulations [2.43] and [2.44] therefore leads to the identification of the multiplier λ_p^1 in terms of components, in the form:

$$\lambda_{p1}^1 = \vec{F}_p^1 \cdot \vec{n}_p \ ; \ \lambda_{p2}^1 = \vec{F}_p^1 \cdot \vec{u}_p \ ; \ \lambda_{p3}^1 = \vec{F}_p^1 \cdot \vec{v}_p \qquad [2.45]$$

$$\begin{cases} \lambda_{p4}^1 = \vec{M}_p^1(B_p) \cdot \vec{n}_p \\ \lambda_{p5}^1 = \vec{M}_p^1(B_p) \cdot \vec{u}_p \\ \lambda_{p6}^1 = \vec{M}_p^1(B_p) \cdot \vec{v}_p \end{cases} \qquad [2.46]$$

The choice of formulations [2.10] and [2.11] therefore leads to a simple interpretation of the six multipliers λ_{pi}^1 $(1 \le i \le 6)$; they are the components of the contact wrench exerted by the ground on the foot S_p and projected onto the local base $(\vec{n}_p, \vec{u}_p, \vec{v}_p)$ linked to the footprint.

During time interval I_2 of the second sub-phase of single support, the components of the contact forces exerted by the ground on the stance foot S_p are formulated by adopting notations analogous to those of equation [2.41]:

$$t \in I_2, \begin{cases} \vec{F}_p^2 := \vec{R}(S_0 \to S_p) \\ \vec{M}_p^2(B_p) := \vec{M}(B_p; S_0 \to S_p) \end{cases} \tag{2.47}$$

The foot S_p is in non-actuated rotation along its frontal edge. The moment of contact forces is therefore equal to zero along this edge, namely:

$$\vec{M}_p^2(B_p) \cdot \vec{v}_p = 0$$

As stated previously, we come to the following identification of the multiplier λ^2, denoted once again as λ_p^2 with reference to S_p in terms of contact forces as they are represented in equation [2.47]:

$$\left.\begin{aligned} \lambda_{p1}^2 &= \vec{F}_p^2 \cdot \vec{n}_p \\ \lambda_{p2}^2 &= \vec{F}_p^2 \cdot \vec{u}_p \\ \lambda_{p3}^2 &= \vec{F}_p^2 \cdot \vec{v}_p \\ \lambda_{p4}^2 &= \vec{M}_p^2(B_p) \cdot \vec{n}_p \\ \lambda_{p5}^2 &= \vec{M}_p^2(B_p) \cdot \vec{u}_p \end{aligned}\right\}, \text{ Foot } (S_p) \tag{2.48}$$

During the first sub-phase of double support, which takes place during the time interval I_3, the rear foot S_p and front foot S_{p+1} (Figure 2.11) are submitted to contact forces of which resultant and moment can be represented, for the rear foot, at point B_p:

$$t \in I_3, \begin{cases} \vec{F}_p^3 := \vec{R}(S_0 \to S_p) \\ \vec{M}_p^3(B_p) := \vec{M}(B_p; S_0 \to S_p) \end{cases} \tag{2.49}$$

and for the front foot, at point A_{p+1}:

$$t \in I_3, \begin{cases} \vec{F}_{p+1}^3 := \vec{R}(S_0 \to S_{p+1}) \\ \vec{M}_{p+1}^3(A_{p+1}) := \vec{M}(A_{p+1}; S_0 \to S_{p+1}) \end{cases} \qquad [2.50]$$

The pivotal moments about the contact edges are zero:

$$\vec{M}_p^3(B_p) \cdot \vec{v}_p = 0 \; ; \; \vec{M}_{p+1}^3(A_{p+1}) \cdot \vec{v}_{p+1} = 0$$

The result is the following identification of multiplier λ^3 in terms of contact efforts, as defined in equations [2.49] and [2.50]:

$$\left.\begin{array}{l} \lambda_{p1}^3 = \vec{F}_p^3 \cdot \vec{n}_p \\ \lambda_{p2}^3 = \vec{F}_p^3 \cdot \vec{u}_p \\ \lambda_{p3}^3 = \vec{F}_p^3 \cdot \vec{v}_p \\ \lambda_{p4}^3 = \vec{M}_p^3(B_p) \cdot \vec{n}_p \\ \lambda_{p5}^3 = \vec{M}_p^3(B_p) \cdot \vec{u}_p \end{array}\right\}, \text{ Foot } (S_p) \qquad [2.51]$$

$$\left.\begin{array}{l} \lambda_{p+1,1}^3 = \vec{F}_{p+1}^3 \cdot \vec{n}_{p+1} \\ \lambda_{p+1,2}^3 = \vec{F}_{p+1}^3 \cdot \vec{u}_{p+1} \\ \lambda_{p+1,3}^3 = \vec{F}_{p+1}^3 \cdot \vec{v}_{p+1} \\ \lambda_{p+1,4}^3 = \vec{M}_{p+1}^3(A_{p+1}) \cdot \vec{n}_{p+1} \\ \lambda_{p+1,5}^3 = \vec{M}_{p+1}^3(A_{p+1}) \cdot \vec{u}_{p+1} \end{array}\right\}, \text{ Foot } (S_{p+1}) \qquad [2.52]$$

where the components of the multiplier λ^3 are those of the two vectors λ_p^3 and λ_{p+1}^3 such that:

$$\lambda^3 = ((\lambda_p^3)^T, (\lambda_{p+1}^3)^T)^T \qquad [2.53]$$

Representations [2.51] and [2.52] must be carried forward to interval I_4 of the second sub-phase of double support by replacing the superscript 3 with superscript 4, which results in the relationship (as in equation [2.53]):

$$\lambda^4 = ((\lambda_p^4)^T, (\lambda_{p+1}^4)^T)^T \qquad [2.54]$$

The supplementary multiplier $\lambda_{p+1,6}^4$ which is associated with the constraint [2.20] is introduced in the form:

$$\lambda_{p+1,6}^4 = \vec{M}_{p+1}^4 \cdot \vec{v}_{p+1} \qquad [2.55]$$

When a rooted system configuration with 13 generalized coordinates is used (section 2.2.4.1), the constraint equations are reduced to the conditions of closure under the front foot (S_{p+1}) in the double support. Only the representations [2.52] during interval I_{DA} and equation [2.55] during interval I_4 therefore remain. The components of contact forces under the foot will have to be determined, in this case, by complementary formulations obtained by using the Newton-Euler equations for the global system in the following way:

$$\begin{cases} \vec{F}_p^\alpha = M(\vec{\gamma}(G) + \vec{g}) + P(\alpha)\vec{F}_{p+1}^\alpha \\ \vec{M}_p^\alpha(B_p) = \vec{\delta}(B_p) + M\overrightarrow{B_pG} \wedge \vec{g} + P(\alpha)[\vec{M}_{p+1}^\alpha(A_{p+1}) + \overrightarrow{B_pA}_{p+1} \wedge \vec{F}_{p+1}^\alpha] \end{cases} \qquad [2.56]$$

where the polynomial function: $P(\alpha) = (\alpha - 1)(\alpha - 2)(9 - 2\alpha)/6$ fulfils the conditions: $P(1) = P(2) = 0$ and $P(3) = P(4) = 1$ to make distinctions between the cases which correspond to the values of $\alpha = 1, 2, 3$ and 4. In addition, M, $\vec{\gamma}(G)$ and $\vec{\delta}(B_p)$ denote the mass, acceleration vector of the mass center G and the biped's rate of moment of momentum calculated at point B_p, respectively. Vector $\vec{g}$ represents the gravitational acceleration.

For $\alpha = 3, 4$ it should be noted that the components of the force $\vec{F}_{p+1}^\alpha$ and moment $\vec{M}_{p+1}^\alpha(A_{p+1})$ of equation [2.56] are multipliers $\lambda_{p+1,i}^\alpha$ which correspond to equations [2.52] and [2.55].

2.3.1.3. *Inverse dynamics*

Treating movement equations using the inverse dynamics approach is necessary for two major applications. In the order of the following chapters, the first is dynamic synthesis of walking when it is performed through some parametric optimization techniques (Chapter 4). The second application is relative to the design of control laws (Chapter 5). To carry out the dynamic synthesis of movement, it is necessary to solve the algebraic system [2.39] either with respect to the actuating torques only (section 2.3.1.3.1) or with respect to both the actuating torques and multipliers (sections 2.3.1.3.2 and 2.3.1.3.3).

2.3.1.3.1. Inverse dynamics in τ with parameterization of the multipliers

This is the hypothesis where the multipliers are parameterized, i.e. represented by given time-dependent functions which are also defined by shaping parameters to be optimized (see section 4.5.3). This approach will amount to dealing with multipliers as forces and moments considered as given time-functions. The Lagrange equation [2.39] is therefore considered in the form of the linear equation in τ:

$$A^{\tau}\tau = b^{\alpha} \quad [2.57]$$

in which

$$b^{\alpha} = M(q)\ddot{q} + N(q,\dot{q}) - (J^{\alpha}(q))^{T}\lambda^{\alpha *} \quad [2.58]$$

where the superscript * of $\lambda^{\alpha *}$ is introduced to signify that the multipliers are considered as parameterized in this formulation. The matrix A^{τ} is detailed in equation [2.40]. In the kinematic configuration comprising 18 generalized coordinates (section 2.2.4.2) we obtain the following solution for $\alpha \in \{1,2,3,4\}$, with the dependence type indicated as:

$$b_i^{\alpha}(q,\dot{q},\ddot{q},\lambda^{\alpha *}) = 0,\ 1 \le i \le 6 \quad [2.59]$$

$$\tau_i^{\alpha} = b_i^{\alpha}(q,\dot{q},\ddot{q},\lambda^{\alpha *}),\ 7 \le i \le 18 \quad [2.60]$$

Equalities [2.59] appear as conditions of compatibility to fulfill so that solution [2.60] can be validated. This first approach calls for the following two remarks.

Remark 1

The parameterization of all the λ_i^{α} s will add numerous supplementary unknown variables to the parametric optimization problem to be formulated (section 4.5.3). For this reason, it is useful to work along the basis of the reduced kinematic configuration obtained by considering the biped as a rooted system. Indeed, this reduces the number of Lagrange multipliers to be parameterized, which are limited to $\lambda_{p+1}^{\alpha *}$ for $\alpha = 3, 4$. This choice leads to the following solution with the type of dependence indicated for the actuating torques τ_i^{α}:

$$\alpha = 1, 2 \begin{cases} b_1^{\alpha}(q, \dot{q}, \ddot{q}) = 0 \\ \tau_i^{\alpha}(q, \dot{q}, \ddot{q}) = b_i^{\alpha}\ ;\ 2 \leq i \leq 13 \end{cases} \quad [2.61]$$

$$\alpha = 3, 4 \begin{cases} b_1^{\alpha}(q, \dot{q}, \ddot{q}, \lambda_{p+1}^{\alpha *}) = 0 \\ \tau_i^{\alpha}(q, \dot{q}, \ddot{q}, \lambda_{p+1}^{\alpha *}) = b_i^{\alpha}\ ;\ 2 \leq i \leq 13 \end{cases} \quad [2.62]$$

It should be noted that in these formulations only one equality-type constraint appears in equation [2.61] then in equation [2.62], instead of the six in equation [2.59]. Reducing the number of this type of constraint helps to reduce the order of complexity of the associated optimization problem (Chapter 4). Let us add that the contact forces under the stance foot, defined in equation [2.56] by the vectors $\vec{F}_p^{\alpha}$ and $\vec{M}_p^{\alpha}(B_p)$ of which components are regrouped in the vectors λ_p^{α}s as for τ_i^{α}s, allow for the dependence:

$$\lambda_p^{\alpha} = \begin{cases} \Lambda_p^{\alpha}(q, \dot{q}, \ddot{q});\ \alpha = 1, 2 \\ \Lambda_p^{\alpha}(q, \dot{q}, \ddot{q}, \lambda_{p+1}^{\alpha *});\ \alpha = 3, 4 \end{cases} \quad [2.63]$$

Remark 2

As soon as the contact forces are considered as given efforts, the Newton-Euler model is a natural choice because of its simplicity and because of the fact that it enables the equality constraints on b_i^{α} and b_1^{α} as they appear in equations [2.59], [2.61] and [2.62] to be removed. We will, for the same reasons as before, select the kinematic model with 13 generalized coordinates. The torques τ_i^{α}s and the vectors λ_p^{α}s are therefore obtained with the same type of dependence as in equations [2.61], [2.62] and [2.63]. The complementary formulations [2.56] are of no use in this case, as the results which they give are contained in the iterative formulation of the Newton-Euler dynamic model (section 2.3.2).

2.3.1.3.2. Inverse dynamics in τ and λ by pseudo-inversion

The Lagrange multipliers are considered now as unspecified time functions to be determined as the τ_i^{α}s, as functions depending on the movement kinematics (or more precisely, depending on the biped kinetics). We are therefore looking to extract a solution in (τ, λ) from equation [2.39]. By defining

$$B(q, \dot{q}, \ddot{q}) := M(q)\ddot{q} + N(q, \dot{q})$$

$$A^{\alpha}(q) = (A^{\tau}, (J^{\alpha}(q))^T) \qquad [2.64]$$

$$u^{\alpha} = (\tau^T, (\lambda^{\alpha})^T)^T \qquad [2.65]$$

the Lagrange equation takes the form:

$$t \in I_{\alpha},\ A^{\alpha}(q(t))\, u^{\alpha}(t) = B(q(t), \dot{q}(t), \ddot{q}(t)) \qquad [2.66]$$

This must then be resolved algebraically in $u^{\alpha}(t)$ for any fixed time t. As the matrix A^{α} of the system to be solved is not generally square, it is important to first analyze its dimensions and rank. By taking into account the composition of the matrix A^{τ} given in equation [2.40], the matrix A^{α} of equation [2.66], with dimensions $n_q \times (n_{\tau} + n_{\alpha})$, verifies the following matrix-block structure:

$$A^{\alpha} = \begin{pmatrix} 0_{(n_q - n_{\tau}) \times n_{\tau}} & \\ & (J^{\alpha})^T_{n_q \times n_{\alpha}} \\ I_{n_{\tau} \times n_{\tau}} & \end{pmatrix} \qquad [2.67]$$

The dimensions of this matrix are specified for each sub-phase in Table 2.1. Note that the biped with 12 active joints $(n_{\tau} = 12)$, parameterized in a free system with $n_q = 18$, presents a number n_{α} of constraint equations successively equal to six, five, ten and eleven (as indicated in Table 2.1 during a walking cycle with four sub-phases). The result is the detail of the corresponding dimensions of A^{α}. In all cases, this matrix is of full rank. Similar indications relative to the rooted system configuration have also been added to the table.

During the first sub-phase of single support, the algebraic system to be solved has as many equations as it has unknowns. The matrix A^{α} is invertible as it is of maximal rank. There is therefore a unique solution in u^{α}. During the second sub-phase of single support, the number of equations is superior to that of the unknown variables: the matrix A^{α} is a high rectangular matrix, and the system is over-determined. The biped is therefore under-actuated.

In the double support phase, system [2.66] has more unknowns than equations: the matrix A^{α} is wide rectangular and the system is under-determined. In this case, the biped is over-actuated. This leads to the problem of constructing or choosing a solution. This essential point of inverse dynamics will be examined in the following sections.

Free System	n_q	$n_{\tau} + n_{\alpha}$	$dim(A^{\alpha})$	$rk(A^{\alpha})$
First single support sub-phase	18	12 + 6	18 × 18	18
Second single support sub-phase	18	12 + 5	18 × 17	17
First double support sub-phase	18	12 + 10	18 × 22	18
Second double support sub-phase	18	12 + 11	18 × 23	18
Rooted System				
First single support sub-phase	13	12 + 1	13 × 13	13
Second single support sub-phase	13	12 + 0	13 × 12	12
First double support sub-phase	13	12 + 5	13 × 17	13
Second double support sub-phase	13	12 + 6	13 × 18	13

Table 2.1. *Dimensions and rank of matrix* A^{α} *during the successive phases of a walking step*

The problem of resolving linear systems of any dimension is largely discussed in numerous linear algebra works [CHA 88, GAL 01, STR 80]. Whatever the dimension and the rank of the matrix of the system under consideration, the method called *singular value decomposition* leads to the construction of a *pseudoinverse matrix* which is unique and written as $(A^{\alpha})^{+}$ for the particular problem that we are interested in, and results in a solution in the form of:

$$u^{\alpha} = (A^{\alpha})^{+} B \qquad [2.68]$$

in which the matrix $(A^{\alpha})^{+}$ is determined by the factorization:

$$(A^\alpha)^+ = VD^+U^T$$

where V (respectively, U) is the orthogonal matrix in which the columns are the eigenvectors of $A^\alpha(A^\alpha)^T$ (respectively $(A^\alpha)^T A^\alpha$); $D^+ = \text{diag}(d_1,..., d_r,0,...,0)$; $d_i = 1/\sqrt{\mu_i}$; and μ_i is the ith decreasing non-zero eigenvalue of $A^\alpha(A^\alpha)^T$, $1 \le i \le r$ (r, rank of $A^\alpha(A^\alpha)^T$).

In fact, as the matrix A^α is of maximal rank in all the cases described below, it can therefore be shown that its pseudoinverse takes the form [STR 80]:

$$\begin{cases} A^\alpha, \text{square matrix}, & (A^\alpha)^+ = (A^\alpha)^{-1} \\ A^\alpha, \text{high rectangular}, & (A^\alpha)_g^+ = ((A^\alpha)^T A^\alpha)^{-1}(A^\alpha)^T \\ A^\alpha, \text{wide rectangular}, & (A^\alpha)_d^+ = (A^\alpha)^T (A^\alpha (A^\alpha)^T)^{-1} \end{cases} \qquad [2.69]$$

where, by designating r as the rank of A^α (in Table 2.1, $r = n_q$ or $n_q - 1$), $(A^\alpha)_g^+$ is the left pseudoinverse of A^α: $(A^\alpha)_g^+ A^\alpha = I_{r \times r}$ and $(A^\alpha)_d^+$ is the right pseudoinverse: $A^\alpha (A^\alpha)_d^+ = I_{r \times r}$.

Each of these matrices gives a solution in equation [2.68], the signification of which is to be specified according to the relative dimensions of A^α and therefore according to the sub-phase of the walking step under consideration.

Let us first examine the case of the single-support sub-phases. During the first sub-phase ($\alpha = 1$), the matrix A^α is invertible. The result is the unique solution:

$$u^\alpha = (A^\alpha)^{-1} B$$

In the second sub-phase ($\alpha = 2$), A^α is a high rectangular matrix. By considering the equality:

$$u^\alpha = (A^\alpha)_g^+ B \qquad [2.70]$$

it is shown that u^α is the least-squares solution of equation [2.66], or that it is the minimizing element of the function (considered at any fixed time t):

$$u \rightarrow \left\| A^{\alpha} u - B \right\|^2$$

This solution of lesser deviation does not necessarily fulfill equation [2.66].

It is possible, in this case, to proceed in a more direct way. In the kinematic configuration with 18 parameters, the structure of matrix A^{α} as it appears in the general representation [2.67] shows that a first sub-system with six equations and five unknowns λ_1^{α} to λ_5^{α} :

$$\sum_{j=1}^{5} J_{ji}^{\alpha}(q)\lambda_j^{\alpha} = B_i \,,\; 1 \le i \le 6$$

can be solved in the form of:

$$\lambda_j^{\alpha} = \Lambda_j^{\alpha}(q, \dot{q}, \ddot{q}) \,,\; 1 \le j \le 5 \qquad [2.71]$$

$$\Psi_6^{\alpha}(q, \dot{q}, \ddot{q}) := \sum_{j=1}^{5} J_{j6}^{\alpha}(q)\Lambda_j^{\alpha}(q, \dot{q}, \ddot{q}) - B_6(q, \dot{q}, \ddot{q}) = 0 \qquad [2.72]$$

Substituting Λ_j^{α} of equation [2.71] for λ_j^{α} in the 12 remaining equations of [2.66] immediately leads to the determination of the τ_i s which can then be written formally as:

$$\tau_i = \mathrm{T}_i(q, \dot{q}, \ddot{q}) \,,\; 7 \le i \le 18 \qquad [2.73]$$

After the explicit determination of the λ_j^{α}s and τ_is , it must be noted that equation [2.72] presents itself as a compatibility relation to be fulfilled by the kinematics of the movement to be generated.

In the case where the biped is considered as rooted, there are no multipliers during this second sub-phase. The first of the 13 equations which are therefore included in equation [2.66] does not depend explicitly on the τ_i s (subscripted from 2 to 13, section 2.3.1.1). It is written simply as:

$$\Psi_1^{\alpha}(q, \dot{q}, \ddot{q}) := B_1(q, \dot{q}, \ddot{q}) = 0 \qquad [2.74]$$

This can be seen as a compatibility equation analogous to its counterpart in equation [2.72] for the case where the biped is described as a free system.

During the two sub-phases of double support ($\alpha = 3$ or 4), we should note that the matrix A^{α} is wide rectangular and of full rank. In this case, it is known that the solution given by the right pseudoinverse defined in equation [2.69] (third line), namely:

$$u^{\alpha} = (A^{\alpha})_d^{+} B \qquad [2.75]$$

is a minimal quadratic norm solution.

We can also further the representation of u^{α} by considering the expression:

$$u_Z^{\alpha} = (A^{\alpha})_d^{+} B + (I - (A^{\alpha})_d^{+} A^{\alpha}) Z \qquad [2.76]$$

where Z is an arbitrary vector. It can be shown that this general solution of equation [2.66] has a minimal quadratic deviation from Z (e.g. [KHA 02]), i.e.:

$$\left\| u_Z^{\alpha} - Z \right\|^2 = \underset{(A^{\alpha} u^{\alpha} = B)}{Min} \left\| u^{\alpha} - Z \right\|^2$$

The complementary term in Z of equation [2.76] can be used to minimize another criterion. For example, the minimization of a criterion $\Psi(q)$ using the gradient method can be carried out by defining $Z = -\alpha \, \Delta\Psi(q)$ where $\Delta\Psi(q)$ is the gradient of Ψ with respect to q.

It is important to note that all the preceding solutions in u^{α}, with $u^{\alpha} = (\tau^T, (\lambda^{\alpha})^T)^T$, are obtained with the following type of dependence in $q, \dot{q}$ and $\ddot{q}$:

$$(\tau^T(q,\dot{q},\ddot{q}), (\lambda^{\alpha}(q,\dot{q},\ddot{q}))^T)^T = (A^{\alpha}(q))^{+} B(q,\dot{q},\ddot{q}) \qquad [2.77]$$

where the matrix $(A^{\alpha}(q))^{+}$ is either the inverse matrix, the left pseudoinverse or the right pseudoinverse of A^{α}.

2.3.1.3.3. Inverse dynamics in τ and λ by differentiating the constraint equations

The interest of this approach is to use the second order derivatives of constraint equations, which allows for the extraction of feasible accelerations which can be used for solving system [2.29] in λ and then in τ successively. As before, this determination of τ and λ, in relation to the kinematics of movement, enables us to avoid the parameterization of multipliers. This technique results in heavy calculations, however.

The operation of eliminating multipliers in Lagrange equations is discussed in the literature. The reader can refer to works such as [GAR 94, HUS 90, WIT 77]. This approach is based on the second-order time-differentiation of the constraint equations [2.34], namely:

$$i \leq n_\alpha,\ t \in I_\alpha,\ \begin{cases} (\text{Grad}\,\Phi_i^\alpha(q))\,\dot{q} = 0 \\ \dot{q}^T\,(\text{Hess}\,\Phi_i^\alpha(q))\,\dot{q} + (\text{Grad}\,\Phi_i^\alpha(q))\,\ddot{q} = 0 \end{cases} \qquad [2.78]$$

where $\text{Hess}\,\Phi_i^\alpha$ and $\text{Grad}\,\Phi_i^\alpha$ represent the Hessian and the gradient of Φ_i^α, respectively.

As:

$$J^\alpha = \begin{pmatrix} \text{Grad}\,\Phi_1^\alpha \\ \cdots \\ \text{Grad}\,\Phi_{n_\alpha}^\alpha \end{pmatrix}$$

and defining in succession:

$$H_i^\alpha = \text{Hess}\,\Phi_i^\alpha\,;\ h_i^\alpha = \dot{q}^T H_i^\alpha \dot{q}\,;\ h^\alpha = (h_1^\alpha,\ldots,h_{n_\alpha}^\alpha)^T$$

the velocity and acceleration constraints [2.78] can be condensed to:

$$t \in I_\alpha,\ \begin{cases} J^\alpha(q)\,\dot{q} = 0 \\ h^\alpha(q,\dot{q}) + J^\alpha(q)\,\ddot{q} = 0 \end{cases} \qquad [2.79]$$

As the mass matrix M is invertible, we can obtain $\ddot{q}$ from equation [2.39] in the form:

$$\ddot{q} = M^{-1}[-N + A^{\tau}\tau + (J^{\alpha})^{T}\lambda^{\alpha}]$$

and introduce this expression into the second equation of [2.79] which becomes:

$$h^{\alpha} + J^{\alpha}M^{-1}(-N + A^{\tau}\tau) + J^{\alpha}M^{-1}(J^{\alpha})^{T}\lambda^{\alpha} = 0$$

In the third term of the left-hand member, the matrix:

$$K^{\alpha} := J^{\alpha}M^{-1}(J^{\alpha})^{T}$$

is invertible because J^{α} is of full rank. The result is the following expression for the multiplier λ^{α}:

$$\lambda^{\alpha} = -(K^{\alpha})^{-1}[h^{\alpha} + J^{\alpha}M^{-1}(A^{\tau}\tau - N)] \qquad [2.80]$$

that we introduce in equation [2.39] to obtain:

$$W^{\alpha}A^{\tau}\tau^{\alpha} = M\ddot{q} + (J^{\alpha})^{T}(K^{\alpha})^{-1}h^{\alpha} + W^{\alpha}N \qquad [2.81]$$

where

$$W^{\alpha} := I - (J^{\alpha})^{T}(K^{\alpha})^{-1}J^{\alpha}M^{-1}$$

Equation [2.81] remains to be solved with respect to τ^{α}. It is possible to proceed directly using the singular value decomposition approach:

$$\tau^{\alpha} = (W^{\alpha}A^{\tau})^{+}[M\ddot{q} + (J^{\alpha})^{T}(K^{\alpha})^{-1}h^{\alpha} + W^{\alpha}N] \qquad [2.82]$$

This solution is different from those previously obtained, but it is theoretically more satisfactory as it is explicitly formulated on the basis of accelerations which are compatible with the constraint equations. We can also use the regularity of matrix W^{α} to write equation [2.81] in the form:

$$A^{\tau}\tau^{\alpha} = b^{\alpha} \qquad [2.83]$$

where the right-hand member is defined as

$$b^{\alpha} := (W^{\alpha})^{-1}[M\ddot{q} + (J^{\alpha})^{T}(K^{\alpha})^{-1}h^{\alpha}] + N$$

The structure of the matrix A^{τ} as it appears in equation [2.40] therefore shows that:

$$\Psi_i^{\alpha}(q,\dot{q},\ddot{q}) := b_i^{\alpha} = 0\,,\ 1 \le i \le 6 \qquad [2.84]$$

$$\tau_i^{\alpha} = b_i^{\alpha}\,,\ 7 \le i \le 18 \qquad [2.85]$$

which is a formulation which is formally identical to that of equations [2.59] and [2.60]. Once again, the equalities [2.84] can be considered as compatibility constraints.

To conclude, let us note that in the presence of kinematic loops, the problem of inverse dynamics is undetermined. There are therefore an infinite amount of solutions. In this respect, it must also be noted that the operations of inverse dynamics which have been presented here lead to as many distinct representations of possible solutions to the set problem. These representations will be particularly useful to carry out the numerical synthesis of movements in terms of parametric optimization techniques (Chapter 4).

2.3.2. *Newton-Euler's dynamic model*

The Newton-Euler formalism is based on the general theory of mechanics. It enables us to obtain the n_q equations which link accelerations, torques and contact forces, without having to explicitly calculate the matrices and vectors M, C, G, D, A^{τ}, J (see section 2.3.1.1 for the notation definitions). The method by Luh *et al.* [LUH 80] which was modified by Khalil and Kleinfinger for the arborescent chains [KHA 87] is based on double recurrence. This resulted in highly efficient calculation methods for manipulative robots and for all articulated systems in general.

2.3.2.1. *Principle*

For each link indexed as i, the equations which describe the equilibrium of a rigid link are written:

$$\begin{aligned} &\sum \vec{F}(\rightarrow C_j) = m_j\, \dot{\vec{V}}(G_j) \\ &\sum \vec{M}(G_j;\rightarrow C_j) = I_j\, \dot{\vec{\Omega}}(C_j) + \vec{\Omega}(C_j) \times (I_j \vec{\Omega}(C_j)) \end{aligned} \qquad [2.86]$$

where m_j is the mass of link C_j, I_j is the link's inertia wrench around its center of gravity G_j, $\sum \vec{F}(\rightarrow C_j)$ is the sum of the applied forces on link C_j and $\sum \vec{M}(G_j;\rightarrow C_j)$ is the sum of the moments around G_j applied to the link C_j.

Luh *et al.* [LUH 80] wrote these equations in the form of double recurrence. The robot is described in the form of a tree structure of $n + 1$; link 0 is the reference link for the description. The other links are numbered in an increasing order. Link j is connected to its antecedent $a(j)$ by a joint with one DoF q_j according to Khalil and Kleinfinger's method [KHA 02].

The first recurrence enables us to calculate the velocities, accelerations and dynamic wrenches for each link (from link 0 to link n).

The second recurrence based on the equilibrium of each link, from link n to link 0, allows us to deduce the forces and torques exerted by the joints and the ground.

2.3.2.1.1. Forward recursive calculation from link 0 to the end of the feet

For link j with the antecedent link i, the velocity is:

$$\begin{aligned} \vec{\Omega}(C_j) &= \vec{\Omega}(C_i) + \bar{s}_j \dot{q}_j \, \vec{a_j} \\ \vec{V}(G_j) &= \vec{V}(G_i) + \vec{\Omega}(C_i) \times \vec{L_j} + s_j \dot{q}_j \, \vec{a_j} \end{aligned} \qquad [2.87]$$

with $s_j = 0$ for a revolute joint of $s_j = 1$ for a prismatic joint ($\bar{s}_j = 1 - s_j$), $\vec{a_j}$ is a unitary vector defined on the joint axis, $\vec{L_j}$ is the vector $\overrightarrow{O_i O_j}$ defined between the origins of the frames R_i and R_j associated with links i and j respectively.

The acceleration of link j is:

$$\begin{aligned} \dot{\vec{\Omega}}(C_j) &= \dot{\vec{\Omega}}(C_i) + \bar{s}_j (\ddot{q}_j \, \vec{a_j} + \vec{\Omega}(C_i) \times \dot{q}_j \vec{a_j}) \\ \dot{\vec{V}}(O_j) &= \dot{\vec{V}}(O_i) + \dot{\vec{\Omega}}(C_i) \times \vec{L_j} + \vec{\Omega}(C_i) \times (\vec{\Omega}(C_i) \times \vec{L_j}) + s_j \, (\ddot{q}_j \vec{a_j} + 2\vec{\Omega}(C_i) \times \dot{q}_j \vec{a_j}) \end{aligned}$$

$$[2.88]$$

The linear acceleration of the mass center of this link is:

$$\dot{\vec{V}}(G_j) = \dot{\vec{V}}(O_j) + \dot{\vec{\Omega}}(C_i) \times \overrightarrow{S_j} + \vec{\Omega}(C_i) \times \vec{\Omega}(C_i) \times \overrightarrow{S_j} \qquad [2.89]$$

with $\overrightarrow{S_j}$ being the vector which defines the position of the mass center G_j of link j with respect to O_j ($\overrightarrow{S_j} = \overrightarrow{O_j G_j}$).

The dynamic efforts which are exerted on the mass center of link j are:

$$\begin{aligned} \overrightarrow{F_j} &= m_j \dot{\vec{V}}(G_j) \\ \overrightarrow{N_j} &= I_j\, \dot{\vec{\Omega}}(C_i) + \vec{\Omega}(C_i) \times (I_j \vec{\Omega}(C_i)) \end{aligned} \qquad [2.90]$$

Remark 1

The antecedent of link 0 is not defined. For the iteration of link 0, only equations [2.88] and [2.89] are useful.

Remark 2

The initial conditions of calculation for a walking robot are:

$$\dot{\vec{V}}(O_0) = \dot{\vec{V}}(O_0) - g \qquad [2.91]$$

where g is the gravitational acceleration in order to take into account the forces of gravity and $\dot{\vec{V}}(O_0)$ in the right-hand term is the real acceleration of the center frame 0.

2.3.2.1.2. Backward recursive calculation from the feet extremities to link 0

The equilibrium of link j (we do not take into account the gravitational forces grouped with the initial accelerations) gives:

$$\begin{aligned} \overrightarrow{F_j} &= \overrightarrow{f_j} - \sum_{k/a(k)=j} \overrightarrow{f_k} \\ \overrightarrow{N_j} &= \overrightarrow{n_j} - \overrightarrow{S_j} \times (\overrightarrow{f_j} - \sum_{k/a(k)=j} \overrightarrow{f_k}) - \sum_{k/a(k)=j} (\overrightarrow{n_k} + \overrightarrow{L_k} \times \overrightarrow{f_k}) \end{aligned} \qquad [2.92]$$

where $a(k)$ represents the antecedent of k, and f_j and n_j represents the force and moment exerted on link C_j by its antecedent.

Note that according to the definition of the tree structure, the ground can be considered as either an antecedent or as a successive link. This is relative to the links which are in contact with the ground.

If link j is in contact with the ground and the ground is its antecedent, then $\overrightarrow{f_j}, \overrightarrow{n_j}$ is the ground reaction wrench on link $\overrightarrow{f_j} = \overrightarrow{F^\alpha}, \overrightarrow{n_j} = \overrightarrow{M^\alpha}$.

If a link j is in contact with the ground and the ground is the successive link, then $\overrightarrow{f_k}, \overrightarrow{n_k}$ with $(k) = j$ therefore represents the opposite of the contact force: $\overrightarrow{f_k} = \overrightarrow{-F^\alpha}, \overrightarrow{n_k} = \overrightarrow{-M^\alpha}$.

If link 0 is not in contact with the ground, then as link 0 has no antecedent, we have: $\overrightarrow{f_0} = \vec{0}, \overrightarrow{n_0} = \vec{0}$.

We obtain:

$$\begin{aligned} \overrightarrow{f_j} &= \overrightarrow{F_j} + \sum_{k / a(k)=j} \overrightarrow{f_k} \\ \overrightarrow{n_j} &= \overrightarrow{N_j} + \sum_{k / a(k)=j} (\overrightarrow{n_k} + \overrightarrow{L_k} \times \overrightarrow{f_k}) + \overrightarrow{S_j} \times \overrightarrow{F_j} \end{aligned} \qquad [2.93]$$

If we assume that there is neither friction or inertia of the articulations, the torque (or the force) τ_j is defined as the projection of $\overrightarrow{n_j}$ (or $\overrightarrow{f_j}$) along the joint axis j:

$$\tau_j = \overrightarrow{a_j}.(s_j \overrightarrow{f_j} + \overline{s_j} \overrightarrow{n_j}) \qquad [2.94]$$

The parameterization adapted to the dynamic model with the Newton-Euler formalism consists of using the joint variables to describe the movements of the different links with respect to their antecedent, and the velocities and acceleration of link 0 are expressed in a fixed reference frame.

Parameterization without an implicit joint can be used, as it has been suggested in sections 2.2.3.2 and 2.2.4.2. In the 2D plane case (see Figure 2.8), we simply have: $\vec{V}(0_3) = \dot{q}_1 \overrightarrow{x_0} + \dot{q}_2 \overrightarrow{y_0}, \vec{\Omega}(C_0 = S_3) = \dot{q}_3 \overrightarrow{z_0}$. In the 3D case (see Figure 2.11), the linear velocity of link $C_0 = S_6$ is calculated directly from the time derivative coordinates of the frame origin. The joint velocity $\vec{\Omega}(C_0 = S_6)$ is not equal to the time derivative of the Euler (or Cardan's) angles which enables this reference link to be orientated. However, there is a transformation which enables us to go from the vector $\vec{\Omega}(C_0 = S_6)$ to the derivatives of the orientation variables [KHA 02]. This transformation has a particular orientation, but it is easily avoidable in the case of robot walkers, as the orientation variations of the pelvis are limited in walking.

Parameterization with implicit joint contact, as has been suggested in sections 2.2.3.1 and 2.2.4.1, is well adapted to the transcription of a dynamic model using the Newton-Euler method. In this case, the reference link C_0 is the stance foot. If the foot is fixed then $\vec{V}(0_0) = \vec{0}, \vec{\Omega}(C_0) = \vec{0}$, and the forces exerted by the ground on this foot are obtained during the backward recurrence $\vec{f_0} = \overrightarrow{F^\alpha}, \vec{n_0} = \overrightarrow{M^\alpha}$. If a foot rotates along the frontal edge, then $\vec{V}(0_1) = \vec{0}, \vec{\Omega}(C_0) = \dot{q}_1 \vec{v}$.

2.3.2.2. *Contact efforts*

When modeling the contact of the foot with the ground, it is possible to use a model based on the deformation of the ground. For example, the normal component of contact force can obey a linear [FRE 91] or nonlinear spring/damper [HUN 75] model:

$$F_z = -\lambda |z|^n v_z - k|z|^{n-1} z \text{ for } z < 0$$

where λ and k represent the physical nature of the ground (the damping and the stiffness respectively), n defines the contact geometry (n = 2/3 for a sphere in collision with a plane) and z and v_z represent the penetration depth of the foot and the normal velocity of the penetration, respectively.

For the tangential component, it is possible to use a friction model. Generally, these models are relatively complex and take into account the linear slide velocity for the two contact solids, as well as their relative displacements. Information about friction models can be found in [ARM 94, CAN 95]. One difficulty of these models is the necessity of defining a large number of physical parameters linked to the nature of contact surfaces.

This type of model, depending on the positions and the relative velocities of contact solids (penetration distances in the ground and slide velocities), entirely determines the contact efforts. Hypotheses on the distribution of efforts on the stance foot contact surface enable us to define the complete wrench. These models enable us to directly calculate the wrench of contact efforts $\overrightarrow{F^\alpha}, \overrightarrow{M^\alpha}$ depending on the situation and the velocities of the different links. The exterior effort wrenches can therefore be determined in the forward recurrence. An advantage of this model is that it unifies all the phases of movement, be it the airborne phase, single support, multiple support, impact or roll off. It also reduces the number of necessary operations needed for a direct dynamic model solution. However, the hypothesis of contact deformation between the foot and the ground is completely incompatible with the hypothesis of kinematic constraint. The location of link 0 is not found by solving the constraint equations, but by integrating the dynamic model. This approach can be used for the simulation of the robot, but it is not used for the development of a control law. For generating a trajectory, the forces can be parameterized (and are therefore known) or they can be determined by kinematic constraint equations. They are rarely determined by a foot/ground contact model.

A second model simply supposes that all links in contact with each other are rigid. Therefore, if a link is in contact with the ground, and it cannot move forward, we consider that there is no penetration, no slide and no roll off. This property is translated to each walking phase by an equation of the type [2.34].

As these constraints are fulfilled at each instant, for a given walking phase, the derivative of these constraints must also be zero. This enables us to deduce velocity and acceleration constraints [2.79]. For $q, \dot{q}$ and $\ddot{q}$ which fulfill the constraints [2.34] and [2.79], the dynamic model will enable us to calculate the contact forces and the torque to be applied and/or the accelerations obtained. Depending on whether we consider the dynamic model to be direct or inversed, and depending on the movement phases under study, the solution of the dynamic model differs. These differences will be presented in the following sections.

One of the interesting aspects of describing the dynamic model using Newton-Euler formalism is the systematic calculation of the efforts between the links and the contact efforts with the ground. An evaluation of the joint efforts between links is useful for establishing the dimensions of the robot. The calculation of the contact efforts produced by the ground is very important in order to check that the contact hypotheses are respected, as will be developed in section 2.4.

2.3.2.3. *Direct dynamic model*

The objective of the direct dynamic model is to calculate the wrench efforts of foot/ground contact and the robot's acceleration, with known position, velocities and joint torques.

Walker and Orin's method [WAL 82] is based on the Newton-Euler formulation developed by Luh *et al.* [LUH 80] and was originally used for the calculation of the inverse dynamic model. This method is based on the Lagrangian knowledge of the dynamic model form, where the different matrix elements are reconstructed column by column.

2.3.2.3.1. Parameterization of a free system

The dynamic model is written in the form of equation [2.38] where it was shown in section 2.3.1.2 that λ is the non-null vector of the wrench efforts of foot/ground contact. With parameterization adapted to the use of the Newton-Euler method, the following equation holds:

$$M_E(q)\begin{pmatrix}\dot{V}(0_0)\\ \dot{\Omega}(C_0)\\ \ddot{q}_{7,n_q}\end{pmatrix}+N_E(q,\dot{q})=\begin{pmatrix}0_{6\times n_T}\\ I_{n_T\times n_T}\end{pmatrix}\tau+J_E^{\alpha}(q)^T\lambda \qquad [2.95]$$

With the double recurrence of the Newton-Euler method, for the known configurations and velocities, for a given acceleration $\dot{V}(0_0),\dot{\Omega}(C_0),\ddot{q}_{7,n_q}$ and for given contact forces λ, we can calculate the joint torques and the $\overrightarrow{f_0},\overrightarrow{n_0}$ which would ensure the equilibrium if $\overrightarrow{f_0},\overrightarrow{n_0}$ existed. As a consequence, this double iteration enables us to obtain:

$$\begin{pmatrix}\overrightarrow{f_0}\\ \overrightarrow{n_0}\\ \tau\end{pmatrix}=M_E(q)\begin{pmatrix}\dot{V}(0_0)\\ \dot{\Omega}(C_0)\\ \ddot{q}_{7,n_q}\end{pmatrix}+N_E(q,\dot{q})-J_E^{\alpha}(q)^T\lambda \qquad [2.96]$$

Consequently, if in equation [2.96] we define:

$$\begin{pmatrix}\dot{V}(0_0) & \dot{\Omega}(C_0) & \ddot{q}_{7,n_q}\end{pmatrix}'=e_i\,',\dot{q}=0,g=0,\lambda=0$$

where e_i is a unitary vector of dimension n_q, where the element i is equal to 1 and the others equal to 0, the vector $\begin{pmatrix}\overrightarrow{f_0} & \overrightarrow{n_0} & \tau\end{pmatrix}'$ is equal to the transposition of the ith column of $M_E(q)$. By giving the acceleration vector all the possible combinations of vector e_i, we can then determine matrix $M_E(q)$ column by column. The vector $N_E(q,\dot{q})$ is determined by canceling the joint accelerations, as well as the contact forces $\begin{pmatrix}\dot{V}(0_0) & \dot{\Omega}(C_0) & \ddot{q}_{7,n_q}\end{pmatrix}' = 0', \dot{q} \neq -9.81, g = 0, \lambda = 0$. The Jacobian matrix can be calculated by using a traditional calculation method of the Jacobian matrix [KHA 02] or by applying the method with $\begin{pmatrix}\dot{V}(0_0) & \dot{\Omega}(C_0) & \ddot{q}_{7,n_q}\end{pmatrix}' = 0'$ $\dot{q} = 0, g = 0, \lambda = e_i$.

The accelerations must also fulfill constraint equations [2.79]. By taking into account the parameterization in use, the constraint equation which concerns the acceleration of the robot is written:

$$J_E^{\alpha}(q)\begin{pmatrix}\dot{V}(0_0)\\ \dot{\Omega}(C_0)\\ \ddot{q}_{7,n_q}\end{pmatrix} + h^{\alpha}(q,\dot{q}) = 0 \qquad [2.97]$$

By combining equations [2.95] and [2.97] we obtain a system of $n_q + n_\lambda$ equations to unknown $n_q + n_\lambda$ which enables us to calculate accelerations and non-zero elements of the wrench of the ground contact efforts:

$$\begin{pmatrix}\dot{V}(0_0)\\ \dot{\Omega}(C_0)\\ \ddot{q}_{7,n_q}\\ \lambda\end{pmatrix} = \begin{pmatrix} M_E(q) & -J_E^{\alpha}(q)^T \\ J_E^{\alpha}(q) & 0 \end{pmatrix}^{-1}\left(\begin{pmatrix}0_{6\times n_T}\\ I_{n_T\times n_T}\\ 0_{n_\alpha \times n_T}\end{pmatrix}\tau - \begin{pmatrix}N_E(q,\dot{q})\\ h^{\alpha}(q,\dot{q})\end{pmatrix}\right) \qquad [2.98]$$

2.3.2.3.2. Parameterization with an implicit rooted contact

The parameterization as defined in section 2.2.4.1 enables us to simplify the model by taking into account the fact that for a flat stance foot we have $\overrightarrow{\dot{V}(0_0)} = \vec{0}, \overrightarrow{\dot{\Omega}(C_0)} = \vec{0}$ and for a foot which allows for a rotation along the frontal edge we have $\overrightarrow{\dot{V}(0_0)} = \vec{0}, \overrightarrow{\dot{\Omega}(C_0)} = \ddot{q}_1 \vec{v}$.

In the first case, the model is written:

$$M'_E(q)\ddot{q} + N'_E(q,\dot{q}) = \tau + J_E^{\alpha}(q)^T \lambda \qquad [2.99]$$

where λ describes the contact forces with are exerted on the foot which is not linked to C_0 if there are such contact forces.

The model is obtained in the same way as before by building the matrices column by column, by using an appropriate choice of acceleration, force and velocity in the double iteration of the dynamic model by using the fact that

$$\tau = M'_E(q)\ddot{q} + N'_E(q,\dot{q}) - J_E^{\alpha}(q)^T \lambda \qquad [2.100]$$

In the second case, the model is written:

$$M'_E(q)\ddot{q} + N'_E(q,\dot{q}) = \begin{pmatrix} 0_{1\times n\tau} \\ I_{n_\tau \times n_\tau} \end{pmatrix} \tau + J_E^{\alpha}(q)^T \lambda \qquad [2.101]$$

where λ describes the contact forces with are exerted on the foot which is not linked to C_0 if there are such contact forces. The model is obtained in the same way as before: by building the matrices column by column, by using an appropriate choice of velocity, acceleration and force in the double iteration of the dynamic model and by taking into account the fact that

$$\begin{pmatrix} \vec{v}.\overrightarrow{n_0} \\ \tau \end{pmatrix} = M'_E(q)\ddot{q} + N'_E(q,\dot{q}) - J_E^{\alpha}(q)^T \lambda \qquad [2.102]$$

Indeed, the torque to be applied along the passive articulation carried by vector $\vec{v}$ to obtain the acceleration $\overrightarrow{\dot{\Omega}(C_0)} = \ddot{q}_1 \vec{v}$ is $\vec{v}.\overrightarrow{n_0}$.

2.3.2.4. *Inverse dynamic model*

The objective of the inverse dynamic model is to calculate the wrench foot/ground contact efforts and the joint torques to obtain the desired acceleration of the robot, when the position and joint velocities are known. The desired acceleration of the robot must be compatible with the contact constraints.

The Newton-Euler algorithm is established from the given $\overrightarrow{\dot{V}(0_0)}, \overrightarrow{\dot{\Omega}(C_0)}$, although these accelerations can easily be transcribed whether we use parameterization with implicit joint or not. It is useful that link C_0 corresponds to a stance foot in the single-support phase to simplify the calculations.

To simplify the notation we will assume, in what follows, that parameterization with implicit joint has been used. For the study of robot walking in Table 2.1, or for the different walking sub-phases here, an inverse dynamic model can have a single solution, an infinite number of solutions or no solution. In the case of no solution, the desired acceleration will not be attainable if it does not satisfy extra constraints.

We are therefore going to consider, in succession, the different sub-phases for each of these possible cases.

2.3.2.4.1. Inversed dynamic model in the first single support sub-phase

The robot is grounded with a flat foot. By taking into account the contact constraints, we can obtain $\overrightarrow{\dot{V}(0_0)} = \overline{0}, \overrightarrow{\dot{\Omega}(C_0)} = \vec{0}$. The torques to be applied are obtained in the second iteration τ. The wrench efforts $\overrightarrow{f_0}, \overrightarrow{n_0}$ correspond to the effort that the ground must exert on the stance foot to enable the equilibrium of the robot.

2.3.2.4.2. Inverse dynamic model in the second single support sub-phase

The stance foot turns along the frontal edge. By taking into account the contact constraints, we can obtain $\overrightarrow{\dot{V}(0_0)} = \overline{0}, \overrightarrow{\dot{\Omega}(C_0)} = \ddot{q}_1 \vec{v}$.

The acceleration $\ddot{q}_1$ cannot be chosen because the ground does not exert the torque along the frontal edge: $\vec{v}.\overrightarrow{n_0} = 0$.

Two iterations of the Newton Euler algorithm enable us to determine the torques τ and the acceleration $\ddot{q}_1$ which corresponds to a desired acceleration vector $\ddot{q}_{7,n_q}$.

For the first iteration, the acceleration is $\ddot{q} = (0 \quad \ddot{q}_{7,n_q}{}^T)^T$ and we obtain:

$$\begin{pmatrix} \overrightarrow{f_0}^1 \\ \overrightarrow{n_0}^1 \\ \tau^1 \end{pmatrix} = NE(q, \dot{q}, (0 \quad \ddot{q}_{7,n_q}{}^T)^T) \qquad [2.103]$$

For the second iteration, the acceleration is $\ddot{q} = (1 \quad \ddot{q}_{7,n_q}{}^T)^T$ and we obtain:

$$\begin{pmatrix} \overrightarrow{f_0}^2 \\ \overrightarrow{n_0}^2 \\ \tau^2 \end{pmatrix} = NE(q, \dot{q}, (1 \quad \ddot{q}_{7,n_q}{}^T)^T) \qquad [2.104]$$

As the dynamic model is linear with respect to the acceleration components, we know that the dynamic model is written:

$$\begin{pmatrix} \overrightarrow{f_0} \\ \overrightarrow{n_0} \\ \tau \end{pmatrix} = \begin{pmatrix} \overrightarrow{f_0}^1 \\ \overrightarrow{n_0}^1 \\ \tau^1 \end{pmatrix} + \ddot{q}_1 \begin{pmatrix} \overrightarrow{f_0}^2 \\ \overrightarrow{n_0}^2 \\ \tau^2 \end{pmatrix} \qquad [2.105]$$

As the acceleration $\ddot{q}_1$ is such that $\vec{v}.\overrightarrow{n_0} = 0$, we obtain:

$$\ddot{q}_1 = -\frac{\vec{v}.\overrightarrow{n_0}^1}{\vec{v}.\overrightarrow{n_0}^2}$$

$$\begin{pmatrix} \overrightarrow{f_0} \\ \overrightarrow{n_0} \\ \tau \end{pmatrix} = \begin{pmatrix} \overrightarrow{f_0}^1 \\ \overrightarrow{n_0}^1 \\ \tau^1 \end{pmatrix} - \frac{\vec{v}.\overrightarrow{n_0}^1}{\vec{v}.\overrightarrow{n_0}^2} \begin{pmatrix} \overrightarrow{f_0}^2 \\ \overrightarrow{n_0}^2 \\ \tau^2 \end{pmatrix} \qquad [2.106]$$

2.3.2.4.3. Inverse dynamic model in the double support phase

In the double support phases, any desired acceleration can be generated. There is an infinite number of solutions which will correspond to different distributions of feet/ground contact efforts.

In order to describe the solutions as a set, we can use many iterations of the Newton-Euler algorithm to write the dynamic model in the form of:

$$\begin{pmatrix} \overrightarrow{f_0} \\ \overrightarrow{n_0} \\ \tau \end{pmatrix} = \begin{pmatrix} \overrightarrow{f_0}^1 \\ \overrightarrow{n_0}^1 \\ \tau^1 \end{pmatrix} - J_E^{\alpha}(q)^T \lambda \text{ where } \begin{pmatrix} \overrightarrow{f_0}^1 \\ \overrightarrow{n_0}^1 \\ \tau^1 \end{pmatrix} = NE(q, \dot{q}, \ddot{q}, \lambda = 0) \qquad [2.107]$$

$J_E^{\alpha}(q)^T$ is determined either by many iterations of the Newton-Euler algorithm, or by traditional calculation methods of the Jacobian method, as in the case of the direct dynamic model.

If the rear foot turns along the frontal edge, we have $\vec{v}.\overrightarrow{n_0} = 0$, which is translated by a constraint on λ. Depending on the sub-phase, λ is a vector with 5 or 6 components. We can choose λ to fulfill the objective to minimize the torques or distribute the efforts between the front foot and the rear foot. Equation [2.107] clearly shows the effect of λ on the joint torques and reaction efforts of the ground on link C_0.

2.3.3. *Impact model*

2.3.3.1. *Modeling an impact between two rigid links*

In written work, we traditionally find two modes for impacts between solids [BRA 89, KEL 86, ORH 94]: an *impulsional* (or distributional) model, used when contact solids are supposedly rigid (no deformation at the contact points) and a *continuous* model, used when we wish to take into account the deformation of links [ORH 94]. As in our study, the contacts between the foot and the ground are treated by considering them to be rigid links; the impulsional impact model is therefore favored, and will be discussed here.

Impact between two links is seen as a phenomenon of very short duration and implies an abrupt change in the velocities of the links. This leads to a significant mechanical interaction at the moment of collision. The impacts are generally considered as very high amplitude forces during an infinitely small duration Δt. We therefore represent the impulsional force or impulsion as being the integral of the contact force during this interval of time Δt. Moreover, we suppose that Δt is sufficiently small for the variation of position during impact to be non-existent [BAU 90]. Indeed, Orhant [ORH 94] shows that the impact between two solids can be seen as a variation of velocity without a variation of position, as the impulsion is then modeled by a function of Dirac amplitude equal to the value of impulsion.

The impact solution consists of determining the velocity of the link after impact, as well as the impulsional forces, by supposing that the velocity is known before impact. This solution is based on *a priori* knowledge of the coefficients of the restitution of impacts e, and of friction f [BRA 89]. The hypotheses that we define are relative to the physical nature of the impact. We must consider the way in which the foot rebounds from the ground, as well as the way it slides at impact.

The rebound for a link in relation to another is traditionally characterized by the coefficient of the restitution of impacts e [BAU 90]. This coefficient enables us (with Newton's law of impact) to define the normal component of the relative velocity of the two links after impact, depending on the normal component of the relative velocity of the two links before impact. In the case of impact of a mobile link with a fixed link, this relation is transcribed as:

$$\mathrm{V}_{\mathrm{in}}^{+} = -e\, \mathrm{V}_{\mathrm{in}}^{-} \qquad [2.108]$$

where V_{in} is the normal velocity component for foot i at its point of contact. In the case of walking, the mobile link is the foot, and the fixed link is the ground.

The accepted values of e are known and depend on the nature of the materials in contact: $0 \leq e \leq 1$. The value ($e = 0$) corresponds to a plastic impact, and ($e = 1$) to an elastic impact. In the case of walking robots, $e = 0$ is generally used and justified by experimental studies [FUR 95]. Indeed, there is a great damping in the materials used for the extremities of the feet, the ground, as well as in the joints.

The slide of the two links in contact which each other is traditionally characterized by the ratio μ of the tangential component on the normal component, and by the friction coefficient between the materials, noted as f (Coulomb's law of friction [BAM 81]). The analogy between the traditional models of friction and the slide of solid links at impact can be used to distinguish two distinct cases [BAU 90, WHI 04]:

– there is no sliding: the tangential velocity relative to the two links after impact is zero and the absolute value of μ must be inferior to f;

– there is sliding: the tangential velocity relative to the two links after impact is not zero and we therefore have $|\mu| = f$.

It has been shown [BRA 89, KAN 85] that this treatment can lead to a gain of energy during impact.

2.3.3.2. *Impulsional model for a biped*

Starting with dynamic model equation [2.38], we can obtain an impulsional dynamic model that describes the impact phase. The impulsional form of this equation is written by integrating of the dynamic model around the impact, with a duration of the impact phase that converges towards zero. The integration of vectors N and τ, which have finite values, gives zero vectors. The acceleration of the robot is taken as infinite and by integrating it we obtain the velocity variation $\dot{q}^+ - \dot{q}^-$. The superscript (+) signifies post-impact and (−) pre-impact. The contact efforts are infinite and integrating them gives the amplitude of the impulsional efforts I_{λ_a}. The matrices M and J^α are taken as constant during the collision time as they depend only on the configuration of the robot. Consequently, the impulsional model is written:

$$M(q)\left(\dot{q}^+ - \dot{q}^-\right) = J^\alpha(q)^T I_{\lambda_a} \qquad [2.109]$$

In this dynamic model, as was detailed previously, J^α describes both the way in which the foot/ground contact forces intervene in the dynamic model and the kinematic constraints. During the impact phase, it will be important to distinguish the roles of this matrix or, more exactly, to distinguish the instant at which the contact forces and kinematic constraints are considered. The parameterization which is chosen to describe the robot during the impact phase must enable us to describe the kinematic constraints before and after impact. For example, we cannot use parameterization which corresponds to implicit liaisons if these liaisons can be breached during the impact phase.

The kinematic constraints before impact are known; the difficulty is determining the kinematic constraints after impact. In the same way when there are contact forces, there are associated dynamic constraints (no roll-off and no slide; see section 2.4); the conditions related to the impulsional efforts will also be defined.

Different cases must be envisaged. The hypothesis and constraints associated with these cases must be studied. The higher the number of contact points and/or impact points, the higher the number of possible cases. The existence and the unicity of the solutions are still an open problem [GEN 98]. Cases of inconsistency (absence of solution) or indetermination (many solutions) can arise.

There are different models which correspond to different behavioral hypotheses. Moreau's model [MOR 88] is a direct extension of Newton's law in the case of multiple groundings, as long as we do not take friction into account. It is written in the form of a linear complementary problem (LCP) and can be solved numerically.

In the same way as we considered the possibility of sliding during the support phases, we will consider the possibility of sliding during impact.

In the modeling used here, we consider the impact phase as a unique phase and the forces applied during the impact phase correspond to the kinematic constraints obtained after impact. For example, if a stance foot takes off during impact, there are no applied impulsional forces at this point. If a slide is observed, the tangential impulsional force acts against the slide. There are more complex models based on decomposing impact into many phases [KEL 86, PFE 96].

To simplify the presentation, we have considered the case of a planar robot (Figure 2.7) in single support and rotating around point B6. The free moving foot makes impact and hits the ground at point A9. Prior to impact, constraint equations [2.7] are fulfilled. After impact, there are six possible cases:

1. contact occurs at B6 and A9;
2. contact occurs at B6, point A9 slides along the ground;
3. contact at B6 is interrupted (the rear foot takes off), there is contact at A9;
4. contact at B6 is interrupted, (the rear foot takes off), point A9 slides;
5. point B6 slides, there is contact at A9; and
6. points B6 and A9 slide.

Contact at A9 cannot be interrupted as we assume that the normal velocity of an impact point is zero after the impact.

Two general cases can be envisaged: cases 1 and 3. In case 1, there is a double support phase (as described at the beginning of this chapter). In case 3, the robot starts the single support phase on the one foot, and impact enables the rear foot to spontaneously take off.

Let us consider case 1. The dynamic model [2.109] is accompanied by equation:

$$J^{\alpha}(q)\dot{q}^{+} = 0 \qquad [2.110]$$

where $J^{\alpha}(q)$ is the gradient for constraint equations [2.7] and [2.8] which correspond to the conditions of non-take-off and non-slide for the different points of contact or of impact. The first two lines of $J^{\alpha}(q)\dot{q}^{+}$ describe the velocity of point B6. The next two lines describe the velocity of A9.

As the inertia matrix is invertible, the velocity after impact fulfills:

$$\dot{q}^{+} = \dot{q}^{-} + M(q)^{-1} J^{\alpha}(q)^{T} I_{\lambda_a} \qquad [2.111]$$

The velocities of points B6 and A9 are deduced from it:

$$J^{\alpha}(q)\dot{q}^{+} = J^{\alpha}(q)\dot{q}^{-} + J^{\alpha}(q)M(q)^{-1} J^{\alpha}(q)^{T} I_{\lambda_a} \qquad [2.112]$$

When matrix J^{α} is of full rank, this equation enables us to calculate I_{λ_a} for a given behavioral hypothesis [2.110]:

$$I_{\lambda_a} = \left(J^{\alpha}(q)M(q)^{-1} J^{\alpha}(q)^{T}\right)^{-1} J^{\alpha}(q)\dot{q}^{-} \qquad [2.113]$$

We can then check whether the associated conditions of contact persistence are fulfilled for I_{λ_α} (see section 2.4.1.6). If they are fulfilled, the impulsional forces are known and the velocity after impact is deduced from equation [2.111]:

$$\dot{q}^{+} = \dot{q}^{-} + M(q)^{-1} J^{\alpha}(q)^{T} \left(J^{\alpha}(q)M(q)^{-1} J^{\alpha}(q)^{T}\right)^{-1} J^{\alpha}(q)\dot{q}^{-} \qquad [2.114]$$

Hurmuzlu and Chang propose a sequential exploration of the solutions [HUR 92], followed by a validation of the associated constraints. The method starts with a non-slide hypothesis. If the relation μ_0 between the tangential force and the normal force is superior to the friction coefficient f, different slide possibilities are then explored for one of the two feet. The constraint which is associated with tangential zero velocity is eliminated, and is replaced by a hypothesis about the relationship between the impulsional force components: $I_1 = f\text{sign}(\mu_0)I_2$ where I_1 and I_2 correspond to the normal and tangential components of the impulsional forces at the slide point. In this way we have a model which has as many equations as unknown variables to be resolved.

This approach can be extended to the case where there are more contact points, or to the case of contacts distributed over a surface or along a line. The number of possible cases increases and foot rotations are possible.

To use an impact model in the context of generating an optimal trajectory, the user chooses kinematic constraints after impact and the impulsional force conditions are deduced from this. We will not explore all of the possible cases. For certain robot morphologies, a double support phase after impact (i.e. after a discontinuity in velocity) can be difficult to obtain for normal walking [BOU 04, MIO 04], in this case the impact must be avoided by a zero velocity of the swing foot that touches the ground.

2.4. Dynamic constraints

The term *dynamic constraints* refers to a set of conditions to be fulfilled by the biped's interaction forces with its physical environment. These external forces are inherently linked to the dynamics of movement. We have only taken into account the ground reactions in what follows. Consequently, there are two types of constraints which must be respected: that which ensure there is no sliding on the ground, and that which express that the contacts are unilateral and properly located. The notion of unilaterality of mechanical contacts is linked to the notion of CoP which will enable us to deal efficiently with the problem of contact conditions including the contact area assignment for each foot.

We must first define the notion of CoP and its associated characteristics. The result is a formulation of restrictions for the ground reaction wrenches. These must guarantee effective mechanical contact in the appropriate contact areas. This section will conclude with the formulation of necessary conditions for non-sliding.

2.4.1. *CoP and equilibrium constraints*

The following presentation is applicable to the case of walking on a horizontal plane. It could also be used for generating walking steps on different supporting levels as for walking up and down stairs. We assume that the feet have rigid soles and are in contact with firm ground.

2.4.1.1. *The existence of a CoP*

The preceding hypothesis does not exclude the possible presence of a moderately compressible contact layer acting as a damper. The contact forces can be distributed over a surface or else along a line when contact occurs along an edge. If a contact zone is closely centered on a determined point we can, for the sake of simplicity, consider it as punctual. At every point P of the contact area, it is possible to define a surface density (or a density per unit length or point-shaped density) of contact forces exerted by the ground on the stance feet. This density can be resolved into a normal component and a tangential component (see Figure 2.15):

$$\forall P \in S,\ \vec{\sigma}(P) = \nu(P)\vec{n} + \vec{\tau}(P) \qquad [2.115]$$

where S represent the contact areas, not necessarily adjoining (Figure 2.15), $\nu(P)$, normal component of $\vec{\sigma}(P)$, is the contact pressure at point P which satisfies $\nu(P) > 0$ and $\vec{\tau}(P)$ is the density of tangential forces such that $\vec{\tau}(P) \cdot \vec{n} = 0$.

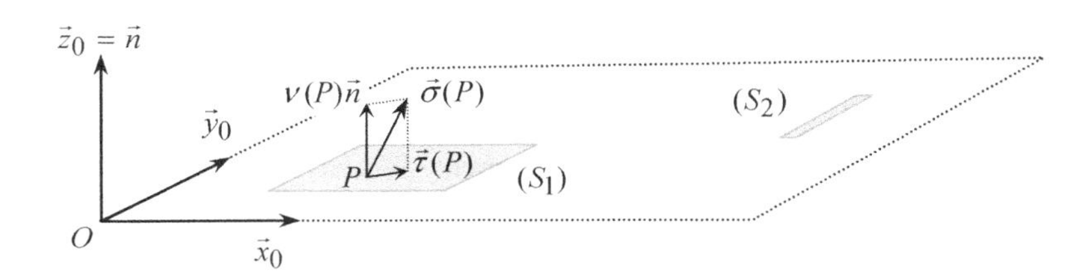

Figure 2.15. *Contact footprints on a horizontal level surface referred to the frame* $(O; \vec{x}_0, \vec{y}_0, \vec{z}_0)$ *; the axis* $(O; \vec{z}_0)$ *is vertical and the contact area is* $S = S_1 \cup S_2$

These two force fields each define a pressure wrench and a friction wrench. Their resultant and moment about point O are each transcribed as:

$$\begin{cases} \vec{R}_p = N_p\, \vec{z}_0, \quad N_p = \int\limits_{P \in S} \nu(P) dP \\ \vec{M}_p(O) = \left(\int\limits_{P \in S} \nu(P)\overrightarrow{OP}\, dP \right) \wedge \vec{z}_0 \end{cases} \quad [2.116]$$

$$\begin{cases} \vec{R}_f = \int\limits_{P \in S} \vec{\tau}(P)\, dP \\ \vec{M}_f(O) = \int\limits_{P \in S} \overrightarrow{OP} \wedge \vec{\tau}(P)\, dP \end{cases} \quad [2.117]$$

where the subscripts p and f are each relative to the (normal) pressure forces and to the (tangential) friction forces.

Representations [2.115], [2.116] and [2.117] imply that

$$\vec{R}_p \cdot \vec{z}_0 > 0; \ \vec{M}_p(O) \cdot \vec{z}_0 = 0 \quad [2.118]$$

and that

$$\vec{R}_f \cdot \vec{z}_0 = 0; \ \vec{M}_f(O) \wedge \vec{z}_0 = \vec{0} \quad [2.119]$$

It must also be stated that the moment of the pressure forces has an axis of action situated on the contact plane, whereas the friction forces result in a pivotal moment about a normal direction.

In addition, equation [2.116] immediately reveals that the pressure wrench satisfies

$$\vec{R}_p \cdot \vec{M}_p(O) = 0$$

This simple result indicates that this wrench is represented by a sliding vector with an action axis normal to the supporting ground since its direction is given by $\vec{R}_p$. The field of moments is zero about every point of this axis and, in particular, at the intersection point with the supporting plane. This point denoted C_p, about which the moment of pressure forces is zero, is called the center of pressure (CoP) of the contact forces.

We note that at point C_p the moment of contact forces is reduced to the pivoting moment of friction forces about the normal to the ground. In other words, the biped's sagittal and frontal tipping moments are zero. We can therefore consider C_p, according to its location on the contact, as an indicator of the biped's balance on its footholds.

This emphasizes the interest in determining the CoP. Since this controls its position, we can control the biped's dynamic balance.

2.4.1.2. *Determining the CoP*

The CoP was defined in the preceding section as the point of the contact plane about which the moment of pressure forces is zero.

Point C_p is therefore determined by:

$$\vec{M}_p(C_p) = \vec{0}$$

As $\vec{M}_p(C_p) = \vec{M}_p(O) + \vec{R}_p \wedge \overrightarrow{OC}_p$, the vector $\overrightarrow{OC}_p$ appears as the solution of the vector equation:

$$\vec{R}_p \wedge \overrightarrow{OC}_p = -\vec{M}_p(O)$$

The solution must fulfill $\overrightarrow{OC}_p \cdot \vec{R}_p = 0$ and therefore is written:

$$\overrightarrow{OC}_p = (\vec{R}_p \wedge \vec{M}_p(O)) \Big/ \left\|\vec{R}_p\right\|^2 \qquad [2.120]$$

By using the representation of $\vec{R}_p$ as it appears in equation [2.116], we obtain the simplified formulation:

$$\overrightarrow{OC}_p = \frac{1}{N_p}\ \vec{z}_0 \wedge \vec{M}_p(O) \qquad [2.121]$$

This result can also be formulated with respect to the resultant and the moment of global contact forces, as defined by:

$$\begin{aligned} \vec{R}_c &= \vec{R}_p + \vec{R}_f \\ \vec{M}_c(O) &= \vec{M}_p(O) + \vec{M}_f(O) \end{aligned} \qquad [2.122]$$

Indeed, equations [2.118] and [2.119] imply:

$$\begin{aligned} &\vec{R}_c \cdot \vec{z}_0 = \vec{R}_p \cdot \vec{z}_0 \ (= N_p) \\ &\vec{z}_0 \wedge \vec{M}_f(O) = \vec{0} \end{aligned}$$

As a result, expression [2.121] can be transcribed as:

$$\overrightarrow{OC}_p = \frac{\vec{z}_0 \wedge \vec{M}_c(O)}{\vec{z}_0 \cdot \vec{R}_c} \qquad [2.123]$$

2.4.1.3. *Experimental determination of the CoP*

Formulations [2.121] and [2.123] give us the means of calculating the position of the CoP on the basis of measurement of the pressure forces and the wrench of contact forces, respectively.

The experimental determination of the wrench of contact forces is achieved on bipeds such as Johnnie [PFE 03], ASIMO [SAK 02], HRP1 and HRP2 [KAN 04] by using a universal force sensor. The latter is put in the foot between a rigid sole and the ankle joint. The acquisition of the six components of $(\vec{R}_c, \vec{M}_c(O))$ therefore enables the calculation of $\overrightarrow{OC}_p$ in equation [2.123]. Let us note, in addition, that this very same acquisition is used for the experimental dynamic analysis for human locomotion by using force-plates which play the role of universal force sensor, with a supporting surface situated at ground level and on which the foot touches down.

Formulation [2.121] shows that the measurement of contact forces can be limited to that of the pressure forces. In this case, it is sufficient to measure normal forces at three non-aligned points on the sole (the collected data from two points is sufficient in the case of a contact along an edge). A measurement set-up of this type using three one-dimensional force sensors as indicated in Figure 2.16 [SAR 98] was created for the BIP biped's foot.

Figure 2.16. *BIP biped's foot: implantation of three one-dimensional sensors in order to determine the CoP*

The field of pressure forces (the weight of the sole not being considered) is therefore equivalent to the three measured normal forces. The determination of the equivalent pressure wrench results from this, as does that of $\overrightarrow{OC}_p$ subsequently in equation [2.121].

2.4.1.4. *CoP existence area*

The existence area of C_p is independently defined by the two following results.

Proposition 1

The CoP C_p belongs to the convex hull of the set S of contact areas: $C_p \in \text{Conv}(S)$.

Proof

Defining successively:

$$P \in S\,,\ \overrightarrow{OP} = x_P\,\vec{x}_0 + y_P\,\vec{y}_0$$

$$\overrightarrow{OC}_p = X_C\,\vec{x}_0 + Y_C\,\vec{y}_0$$

$$X = \int_{P \in S} \nu(P) x_P\, dP$$

$$Y = \int_{P \in S} \nu(P) y_P\, dP$$

the expression of the moment of pressure forces in equation [2.116] and that of the CoP in equation [2.121] become:

$$\vec{M}_p(O) = Y\vec{x}_0 - X\vec{y}_0$$

$$\overrightarrow{OC}_p = \frac{1}{N_p}(X\vec{x}_0 + Y\vec{y}_0)$$

resulting in the representations:

$$X_C = \frac{X}{N_P} \equiv \frac{\int_{P \in S} \nu(P) x_P\, dP}{\int_{P \in S} \nu(P)\, dP}\,,\ Y_C = \frac{Y}{N_P} \equiv \frac{\int_{P \in S} \nu(P) y_P\, dP}{\int_{P \in S} \nu(P)\, dP} \qquad [2.124]$$

In addition, the set S is bounded i.e.

$$\exists (x_{\min}, x_{\max}), (y_{\min}, y_{\max}) / \forall P \in S, \begin{cases} x_{\min} \le x_P \le x_{\max} \\ y_{\min} \le y_P \le y_{\max} \end{cases}$$

This results in obvious lower and upper limits of the upper integrals in both equations of [2.124]. These reveal, after simplification, that:

$$x_{\min} \le X_C \le x_{\max}\,,\ y_{\min} \le Y_C \le y_{\max}$$

These inequalities show that C_p is situated between four half-planes, tangent to the set S. The four tangents are parallel (in twos) to the axes of coordinates in the reference frame. This result remains true regardless of the chosen frame in the contact plane. Consequently, C_p is situated between all straight-lines tangent to S. The set which is limited by these lines is precisely the convex hull of S, i.e. $C_p \in \text{Conv}(S)$.

Proposition 2

Let two imprints of simultaneous contacts be S_1 and S_2. Let C_1 and C_2 be their respective centers of pressure and N_1, N_2 the associated normal components of contact forces. The global CoP C_p is then correlated to C_1 and C_2 by the barycentric relationship:

$$(N_1 + N_2)\overrightarrow{OC_p} = N_1\,\overrightarrow{OC_1} + N_2\,\overrightarrow{OC_2} \qquad [2.125]$$

Proof

In evident notations, the moment about O and the resultant of the global pressure forces are expressed as:

$$\vec{M}_p(O) = \vec{M}_1(O) + \vec{M}_2(O)$$
$$N_p = N_1 + N_2$$

Introducing these expressions into equation [2.121], we obtain:

$$\overrightarrow{OC_p} = \frac{\vec{z}_0 \wedge (\vec{M}_1(O) + \vec{M}_2(O))}{N_1 + N_2}$$

From formulation [2.121], we can also obtain:

$$\vec{z}_0 \wedge \vec{M}_1(O) = N_1\overrightarrow{OC_1},\ \vec{z}_0 \wedge \vec{M}_2(O) = N_2\overrightarrow{OC_2}$$

By injecting into the preceding expression, the result is equation [2.125].

We notice that barycentric relationship [2.125] establishes Proposition 1 for the contact area $S_1 \cup S_2$. It can clearly be generalized to any finite number of zones and points of contact.

2.4.1.5. *The CoP and the dynamics of movement*

The moment $\vec{M}_c(O)$ and the resultant $\vec{R}_c$ of the ground reaction forces allow the definition of C_p in the form of equation [2.123]. These two vector functions can be extracted from the Newton-Euler equations formulated for the whole biped as:

$$m\dot{\vec{V}}(G) = \vec{R}_c + m\vec{g} \quad (\vec{g} = -g\vec{z}_0) \qquad [2.126]$$

$$\dot{\vec{H}}(G) = \vec{M}_c(G) \qquad [2.127]$$

where $\dot{\vec{V}}(G)$ is the acceleration vector of the center of mass G and $\dot{\vec{H}}(G)$ is the rate of moment of momentum about G of the biped with respect to the frame linked to the ground. In equation [2.126], m is the mass of the biped and g is the gravitational acceleration.

Using the relationship:

$$\vec{M}_c(G) = \vec{M}_c(O) + \vec{R}_c \wedge \overrightarrow{OG}$$

together with equations [2.126] and [2.127], the expressions of $\vec{R}_c$, $\vec{M}_c(O)$ and $\overrightarrow{OC}_p$ as defined in equations [2.123] can be calculated following the computing sequence:

$$\begin{aligned} &- \vec{R}_c = m(\vec{\gamma}(G) - \vec{g}) \\ &- \vec{M}_c(O) = \vec{\delta}(G) + \overrightarrow{OG} \wedge \vec{R}_c \\ &- \overrightarrow{OC}_p = \frac{\vec{z}_0 \wedge \vec{M}_c(O)}{\vec{z}_0 \cdot \vec{R}_c} \end{aligned} \qquad [2.128]$$

This formulation therefore expresses the location of C_p as a function of the movement dynamics.

A similar evaluation of the local centers of pressure cannot be directly carried out for the double support. In double support, the dynamics of movement are underdetermined and the two ground reaction wrenches are indeterminate. Nevertheless, it is possible to overcome this indeterminate state by using any of the techniques of inverse dynamics presented in section 2.3.1.3; the first technique is based on the parameterization of one or both ground reaction wrenches. The result is a transitory determination of these two wrenches. Their components are represented by the corresponding Lagrange multipliers (section 2.3.1.2) as outcomes of the parameterization and/or of the inversed dynamics operation which has been used. The calculation of the locations of the centers of pressure is then immediately possible. This is the procedure adopted in Chapter 4.

In conclusion, there are two possible determinations of a CoP, be it local or global:

– one is the result of force measurements using sensors;

– the other is the result of a calculation with respect to the dynamics of movement, when the latter is known or assumed known at the instant under consideration.

It is this latter situation which appears in dynamics synthesis of movement (Chapter 4). It is also used in control problem-solving for walking robot.

When a CoP is determined using the dynamics of movement, it is generally referred to in specialist literature as the zero moment point (ZMP). This notion, which was introduced by Vukobratovic in [VUK 72], is further developed in [VUK 90, VUK 04]. In this latter work, the authors establish a distinction between the ZMP and the CoP. It must be noted, however, that for the contents of this chapter, it is the notion of CoP that has been favored.

2.4.1.6. *Contact constraints*

Let us restate that the pressure forces (which are exerted on each foot at its footprint level) are reducible to a normal resultant which is directed towards the foot and is applied to the CoP. When we use inverse dynamics to determine the wrenches, we must therefore consider the following:

– prescribe the positivity of the resulting normal force for each foot. This expresses the reality of mechanical contact (and no longer just geometric contact);

– constrain each of the resulting centers of pressure so that it is situated in its designated footprint area, which means that the contact zone is limited by this imprint (with a supposed convex contour).

We then observe the footprint cases (and therefore the soles) with a polygonal contour with n_H vertices, which is generally a rectangular contour for humanoid robots. These footprints are located in a local planar frame as indicated in Figure 2.17 (with, in this case, pentagonal imprints).

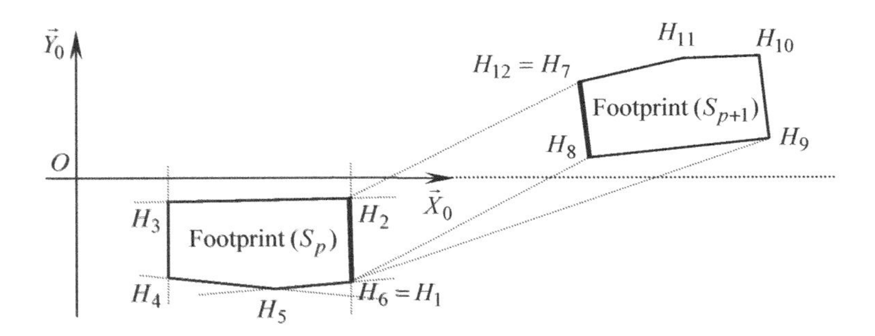

Figure 2.17. *Polygonal contour of footprints. The oblique position of the front imprint suggests the start of a turning step. The dotted lines which link the two footprints separate the travel domains of the global CoP during the sub-phases in double support*

Let us note, in addition, that the sub-phases of movement are successively designated by $\alpha = 1$, 2, 3 and 4 in equation [2.34]. In this way we introduce the following notations (relative to Figure 2.17):

– C_p^1, CoP of the stance foot (S_P) during the whole step;

– C_p^2, CoP of the front foot (S_{P+1}) during the double support;

– (X_1^α, Y_1^α), variable coordinates of C_p^1 in frame $(O; \vec{X}_0, \vec{Y}_0)$ during the four sub-phases $(1 \le \alpha \le 4)$;

– (X_2^α, Y_2^α), coordinates of C_p^2 for $\alpha = 3, 4$;

– (x_k, y_k), fixed coordinates of vertices H_k, $1 \le k \le 11$;

– l_k, length of segment $H_k H_{k+1}$, $1 \le k \le 11$.

Let us first try to characterize how C_p^1 and C_p^2 belong to their respective contact polygons. To do this, let us consider a closed planar polygonal contour with n_H vertices subscripted counter-clockwise (Figure 2.18) and a point C of the plane with coordinates X and Y. Two cases need to be distinguished depending on whether C is to be situated on the inside of the polygon or on its contour.

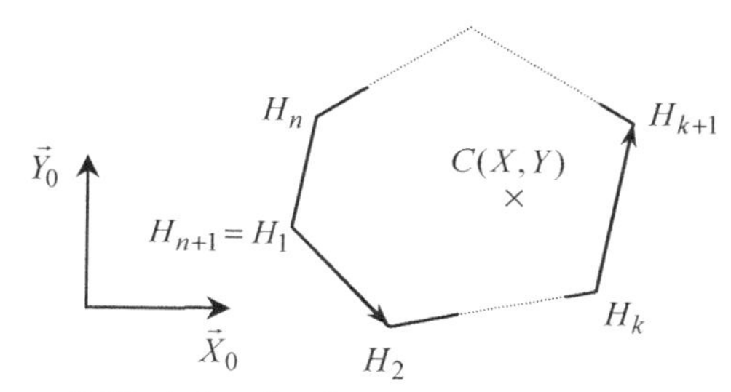

Figure 2.18. *A closed and oriented polygonal contour which surrounds a point C in the contact plane*

For the first case, we notice that the axis defined by $(H_k; \vec{Z}_0 \wedge \overrightarrow{H_k H_{k+1}})$ ($\vec{Z}_0$ is normal to the figure plane) is pointing towards the inside of the polygon. In that case, C is in the polygon if and only if the projection of $\overrightarrow{CH_k}$ on $\vec{Z}_0 \wedge \overrightarrow{H_k H_{k+1}}$ is negative for every subscript $k \leq n_H$. That is to say, if all the following triple scalar products are negative,

$$1 \leq k \leq n_H \text{ , } (\overrightarrow{CH}_k, \vec{Z}_0, \overrightarrow{H_k H_{k+1}}) < 0$$

This condition is developed in the form of the constraint functions:

$$f_k(X,Y) := -(X - x_k)(y_{k+1} - y_k) + (Y - y_k)(x_{k+1} - x_k) < 0 \qquad [2.129]$$

When C is on the support contour, we can state that it belongs to an edge, e.g. $H_k H_{k+1}$, by stating that the vectors $\overrightarrow{H_k C}$ and $\overrightarrow{H_k H_{k+1}}$ are parallel, that is:

$$\overrightarrow{H_k C} \wedge \overrightarrow{H_k H_{k+1}} = \vec{0}$$

Moreover, by limiting the projection of the first onto the second by 0 and l_k (l_k being the length of $H_k H_{k+1}$):

$$0 < \overrightarrow{H_k C} \cdot \overrightarrow{H_k H_{k+1}} < l_k \qquad [2.130]$$

The first condition is transcribed in the form of the constraint function:

$$g_k(X,Y) := (X - x_k)(y_{k+1} - y_k) - (Y - y_k)(x_{k+1} - x_k) = 0 \qquad [2.131]$$

The scalar product of the second is developed in the form of:

$$h_k(X,Y) = (X - x_k)(x_{k+1} - x_k) - (Y - y_k)(y_{k+1} - y_k) \qquad [2.132]$$

Double inequality [2.130] results in the two constraints:

$$\begin{cases} -h_k(X,Y) < 0 \\ h_k(X,Y) - l_k < 0 \end{cases} \qquad [2.133]$$

The preceding conditions must then be transposed to the contact imprints. The outcome of the foot/ground interaction forces in equations [2.45]–[2.55] will be formulated for each of the subsequent sub-phases.

2.4.1.6.1. First sub-phase ($\alpha = 1$)

The stance foot S_p is flat on the ground. The contact surface is the polygon $H_1 \ldots H_5$ (the choice of $n_H = 5$ is arbitrary here). The coordinates (X_1^1, Y_1^1) of the CoP C_p^1 depend on vector λ_p^1 which regroups the components of the wrench's ground reaction (see section 2.3.1.2). The constraints conveying that C_p^1 belongs to the contact polygon are expressed by the set of functions f_k of equation [2.129]. We formulate them generically in the form of functions such that:

$$\alpha = 1, \begin{cases} k = 1,\ldots,5;\ h_{pk}^1(\lambda_p^1) := f_k(X_1(\lambda_p^1), Y_1(\lambda_p^1)) < 0, \\ h_{p6}^1(\lambda_p^1) := -\lambda_{p1}^1 < 0 \end{cases} \qquad [2.134]$$

where the supplementary constraint function h_{p6}^1 imposes positivity on the normal contact component.

2.4.1.6.2. Subsequent sub-phases for the foot $S_p (\alpha = 2,3,4)$

The foot rotates about its frontal edge. The contact zone is reduced to the segment $H_1 H_2$. The constraints to be fulfilled are of the type [2.131] and [2.133] with $k = 1$. The result is the introduction of the following constraint functions:

$$\alpha = 2,3,4; \begin{cases} g_p^{\alpha}(\lambda_p^{\alpha}) := g_1(X_1^{\alpha}(\lambda_p^{\alpha}), Y_1^{\alpha}(\lambda_p^{\alpha})) = 0 \\ h_{p1}^{\alpha}(\lambda_p^{\alpha}) := -h_1(X_1^{\alpha}(\lambda_p^{\alpha}), Y_1^{\alpha}(\lambda_p^{\alpha})) < 0 \\ h_{p2}^{\alpha}(\lambda_p^{\alpha}) := h_1(X_1^{\alpha}(\lambda_p^{\alpha}), Y_1^{\alpha}(\lambda_p^{\alpha})) - l_k < 0 \\ h_{p3}^{\alpha}(\lambda_p^{\alpha}) := -\lambda_{p1}^{\alpha} < 0 \end{cases} \quad [2.135]$$

where, as in equation [2.134], the last condition expresses the unilaterality of the contact.

2.4.1.6.3. First double support sub-phase for the front foot S_{p+1} ($\alpha = 3$)

The contact zone is the heel edge H_7H_8. The conditions to be transcribed are a transposition of equation [2.135] with $k = 7$ and the passage from S_p to S_{p+1}:

$$\alpha = 3; \begin{cases} g_{p+1}^{3}(\lambda_{p+1}^{3}) := g_7(X_2^{3}(\lambda_{p+1}^{3}), Y_2^{3}(\lambda_{p+1}^{3})) = 0 \\ h_{p+1,1}^{3}(\lambda_{p+1}^{3}) := -h_7(X_2^{3}(\lambda_{p+1}^{3}), Y_2^{3}(\lambda_{p+1}^{3})) < 0 \\ h_{p+1,2}^{3}(\lambda_{p+1}^{3}) := h_7(X_2^{3}(\lambda_{p+1}^{3}), Y_2^{3}(\lambda_{p+1}^{3})) - l_k < 0 \\ h_{p+1,3}^{3}(\lambda_{p+1}^{3}) := -\lambda_{p+1,1}^{3} < 0 \end{cases} \quad [2.136]$$

2.4.1.6.4. Second double support sub-phase for the front foot ($\alpha = 4$)

The foot is flat. The contact zone is defined by the polygon $H_7 ... H_{11}$ (Figure 2.17). The conditions to be respected are analogous to that of equation [2.134]:

$$\alpha = 4, \begin{cases} k = 7,...,11;\ h_{p+1,k}^{4}(\lambda_{p+1}^{4}) := f_k(X_2(\lambda_{p+1}^{4}), Y_2(\lambda_{p+1}^{4})) < 0 \\ h_{p+1,12}^{1}(\lambda_{p+1}^{4}) := -\lambda_{p+1,1}^{4} < 0 \end{cases} \quad [2.137]$$

When the CoP migrates to the boundary of its assigned area, the contact takes place only on the points along an edge, or even on a vertex of the polygonal contour. The biped is then either in an unstable state of equilibrium, or about to fall. This is what happens during the swing phase, when the CoP moves from the rear edge H_3H_4 to the frontal edge H_1H_2 on which it is to be found during the second sub-phase (Figure 2.19). The contact stops being controllable in the sagittal plane, but remains so transversally on edge H_1H_2 if the CoP is not situated at one of the extremities of this edge. The biped finds itself in the position of a forward fall. The repositioning of the rear foot to the front allows for the biped's sagittal balance recovery.

If the CoP reaches a lateral border of the contact imprint, however, the correlative disequilibrium cannot be caught up without modifying the walking gait. This is done by either making a side step, or by risky gesticulations. This situation must therefore be avoided.

A simple way to avoid this is to limit the lateral migrations of the CoP by reducing its travel domain as indicated in Figure 2.19, for example. We notice that the introduction of lateral exclusion strips helps to maintain the CoP within the limits of the frontal and rear edge of the foot when it is in contact on its heel edge H_3H_4 or on its front edge H_1H_2.

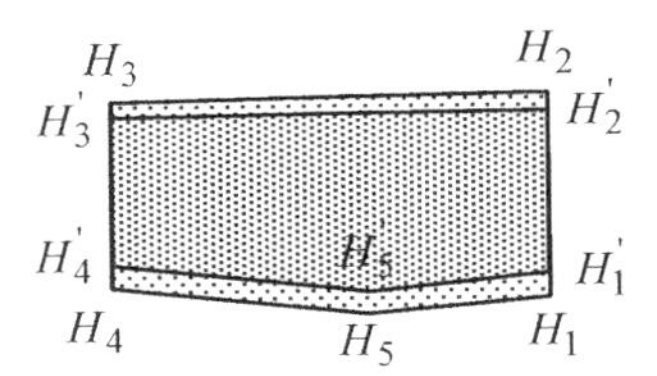

Real contact area defined by the outline $H_1...H_5$

Reduced area assignment for the CoP limited by the outline $H_1'...H_5'$

Figure 2.19. *Width reduction of the CoP's assignment area, avoiding migrations to the foot's lateral borders*

2.4.2. *Non-sliding constraints*

The stance foot S_p is subjected, during the pth step, to the contact force $\vec{F}_p^{\alpha}$ of which components are represented by the multipliers λ_{pi}^{α}, $1 \leq i \leq 3$ (see section 2.3.1.2).

Considering a Coulombian friction law, a condition of non-sliding in translation can then be written as:

$$\left\| \lambda_{p2}^{\alpha} \vec{u}_p + \lambda_{p3}^{\alpha} \vec{v}_p \right\| < f \left| \lambda_{p1}^{\alpha} \right|$$

$$(\lambda_{p1}^{\alpha} \equiv \vec{F}_p^{\alpha} \cdot \vec{n}_p,\ \lambda_{p2}^{\alpha} \equiv \vec{F}_p^{\alpha} \cdot \vec{u}_p,\ \lambda_{p3}^{\alpha} \equiv \vec{F}_p^{\alpha} \cdot \vec{v}_p)$$

Equivalently, this condition can be transcribed under the form of the following non-sliding constraint function:

$$h_p^{ngl,\alpha}(\lambda_p^{\alpha}) := (\lambda_{p2}^{\alpha})^2 + (\lambda_{p3}^{\alpha})^2 - f^2 \times (\lambda_{p1}^{\alpha})^2 < 0 \; ; \alpha \in \{1,2,3,4\} \quad [2.138]$$

In equation [2.138], the constant factor f represents a friction coefficient of the foot-ground contact. The value of this constant can vary from one step to another.

The contact of the swing-front foot S_{p+1} occurs during the double support. The components of the contact force are represented by the multipliers $\lambda_{p+1,i}^{\alpha}$ for $\alpha = 3,4$ and $i = 6,7,8$ (section 2.3.1.2). By analogy with equation [2.138], a translatory non-sliding constraint for the front foot can then be detailed in the form:

$$h_{p+1}^{ngl,\alpha}(\lambda_{p+1}^{\alpha}) := (\lambda_{p+1,7}^{\alpha})^2 + (\lambda_{p+1,8}^{\alpha})^2 - f^2 \times (\lambda_{p+1,6}^{\alpha})^2 < 0 \; ; \alpha = 3,4 \quad [2.139]$$

These conditions do not ensure non-sliding for pivoting.

2.5. Complementary feasibility constraints

A solution which is extracted from the dynamic model is said to be feasible if it complies with certain limitations which are formulated about the unknowns to be determined. For a multibody system, these unknowns first include the Lagrangian phase variables or state variables, which describe the positions and velocities of the body segments. Second, they also include the actuating torques which make up the system's physical control variables. The feasible set of these variables is limited for technological reasons (or anatomical and physiological reasons for biomechanical systems). To this are added constraints which are linked to the morphological structure of the system, as well as to the topography of the ground on which the biped progresses.

2.5.1. *Respecting the technological limitations*

Let us first mention the limited capacities of the actuators, which result in bounded joint torques such that:

$$t \in [t^i, t^f],\ \tau_{i^*+i}^{\min} \leq \tau_{i^*+i}(t) \leq \tau_{i^*+i}^{\max},\ 1 \leq i \leq n_\tau \qquad [2.140]$$

where $i^* = 1$ or 6 depending on the type of parameterization chosen (see section 2.3.1.1). The torque reversibility of the actuators generally implies the equality $\tau_k^{\min} = -\tau_k^{\max}$.

The double set of constraints [2.140] can be redefined by introducing the functions:

$$h_i^{\tau+}(\tau) := \tau_{i^*+i} - \tau_{i^*+i}^{\max}$$

$$h_i^{\tau-}(\tau) := -\tau_{i^*+i} + \tau_{i^*+i}^{\min}$$

$$h^\tau = (h_1^{\tau+}, h_1^{\tau-}, \ldots, h_{n_\tau}^{\tau+}, h_{n_\tau}^{\tau-})^T$$

$$t \in [t^i, t^f],\ h^\tau(\tau(t)) < 0\ (\in \Re^{2n_\tau}) \qquad [2.141]$$

Each articulation has its own joint movement limits, defined in an analogous way to equation [2.140] by:

$$t \in [t^i, t^f],\ q_{i^*+i}^{\min} \leq q_{i^*+i}(t) \leq q_{i^*+i}^{\max},\ 1 \leq i \leq n_q - i^*(= n_\tau) \qquad [2.142]$$

By proceeding in the same way as before, we can define in succession

$$h_i^{q+}(q) := q_{i^*+i} - q_{i^*+i}^{\max},\ h_i^{q-}(q) := -q_{i^*+i} + q_{i^*+i}^{\min}$$

$$h^q = (h_1^{q+}, h_1^{q-}, \ldots, h_{n_\tau}^{q+}, h_{n_\tau}^{q-})^T$$

$$t \in [t^i, t^f],\ h^q(q(t)) < 0\ (\in \Re^{2n_\tau}) \qquad [2.143]$$

The given bounds can have the same sign i.e. $q_k^{\min} \times q_k^{\max} > 0$, to forbid, for example, hyperextension at the knee or at the ankle during its movement of flexion-extension.

It could also be useful to consider reduced limitations in comparison to the preceding bounds, so that the movement generated can conform to a pre-determined gait type.

The actuators also have transmitted velocity limitations. As in equations [2.140] and [2.142], it is useful to introduce these limits at joint level:

$$t \in [t^i, t^f],\ \dot{q}_{i^*+i}^{\min} \leq \dot{q}_{i^*+i}(t) \leq \dot{q}_{i^*+i}^{\max},\ 1 \leq i \leq n_q - i^* \qquad [2.144]$$

In a similar way as above, equation [2.144] can be constructed in the form:

$$t \in [t^i, t^f],\ h^{\dot{q}}(\dot{q}(t)) < 0\ (\in \Re^{2n_\tau}) \qquad [2.145]$$

Regarding the torques, the indicated bounds can have opposite values:

$$\dot{q}_k^{\min} = -\dot{q}_k^{\max}$$

2.5.2. *Non-collision constraints*

There are two types of non-collision constraints (which only concern the locomotion system here). First, there is the possibility of collision between the lower limbs during the single support phase when the swing leg crosses the stance leg. This is the risk of *internal* collision. Through the same swing phase, the foot of the swing leg could bump the ground. This is the risk of *external* collision. The latter risk is also encountered when stepping over an obstacle or a threshold, as well as going up or down stairs.

2.5.2.1. *Internal non-collision*

The possibility of internal collision concerns the lower part of the swing leg, represented by body segments S_{11} and S_{13} in Figure 2.20. If there are no excessive gesticulations at hip level, and no pivoting of the foot S_{13} about the axis carried by $O_{11}O_{13}$, then keeping this axis at a distance from O_3O_4 greater than a minimal and correctly chosen value $\delta^{\min}$ can be a sufficient condition for non-collision.

The constraint to be fulfilled is then expressed as:

$$t \in I_{SA},\ h^{nci}(q(t)) := -\delta(O_3O_4, O_{11}O_{13}) + \delta^{\min} < 0 \qquad [2.146]$$

where the distance δ between the two axes can be calculated by carrying out the following operation:

$$\delta(O_3O_4, O_{11}O_{13}) = \frac{\left|\overrightarrow{O_4O_{11}} \cdot (\overrightarrow{O_3O_4} \wedge \overrightarrow{O_{11}O_{13}})\right|}{\left\|\overrightarrow{O_3O_4} \wedge \overrightarrow{O_{11}O_{13}}\right\|}$$

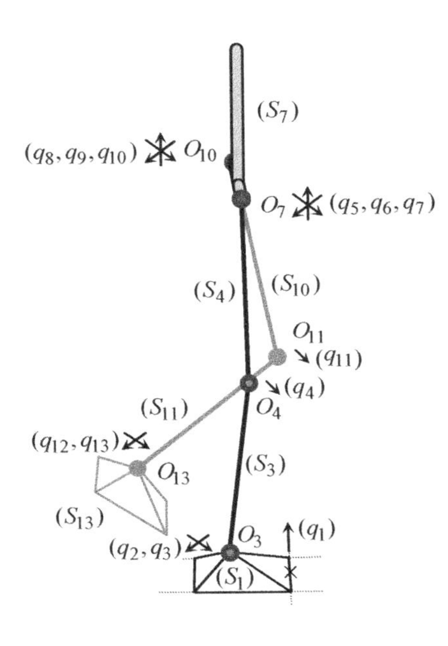

Figure 2.20. *Single support phase: illustration of the movement of the swing leg crossing the stance leg*

In this expression, we note that the dependence with respect to the configuration vector q is only effective for the q_i from q_4 to q_{11} (see Figure 2.20). In addition, by referring to the construction in Figure 2.10, we observe that $\overrightarrow{O_3O_4} = d_4\vec{x}_3$, $\overrightarrow{O_{11}O_{13}} = d_{12}\vec{x}_{11}$. This leads to the relationship:

$$\delta(O_3O_4, O_{11}O_{13}) = \frac{\left|\overrightarrow{O_4O_{11}} \cdot (\vec{x}_3 \wedge \vec{x}_{11})\right|}{\left\|\vec{x}_3 \wedge \vec{x}_{11}\right\|}$$

in which the unit vectors $\vec{x}_3,\ldots,\vec{x}_{11}$ can be determined with respect to the configuration parameters. This can be carried out using the homogenous transformation matrices associated with the Denavit-Hartenberg construction.

2.5.2.2. *External non-collision*

It is essential to avoid all contact of the swing foot with the ground during the single support, when the biped's balance is particularly precarious. The foot must distance itself sufficiently from the ground (or from an obstacle to step over), so that it can avoid any collision which could destabilize it. To this effect, taking into account conditions to be fulfilled at the median section of the foot (as seen in the sagittal plane, Figure 2.21) may be considered as satisfactory.

A simple approach to this consists of introducing avoidance curves which define exclusion zones which the foot must remain out of (Figures 2.21 and 2.22). The easiest construction uses circular arcs of appropriate radius, going through suitable predetermined ground points [SAI 03]. Bell-shaped curves allow for a more progressive movement to and from the ground or an obstacle. Simple polynomial functions have this characteristic and can be used as in [BES 04].

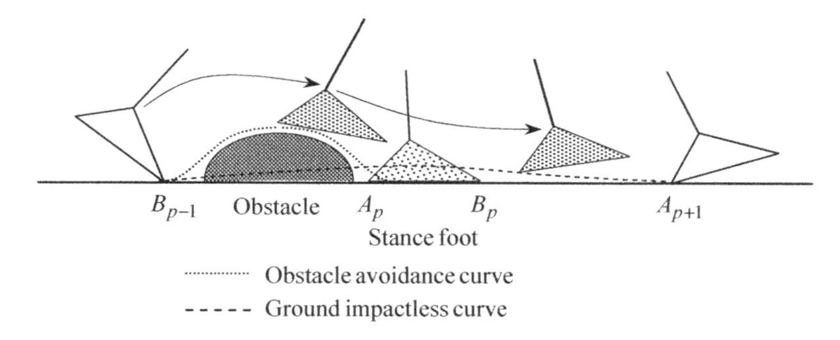

Figure 2.21. *Collision zones to be avoided during the swing phase*

In Figure 2.22, all points P_k of the sole median AB can be pushed back beyond the dotted curve by fulfilling the condition:

$$t \in I_{SA},\ h_k^{nce}(q(t)) := f(x_{P_k}(q(t))) - y_{P_k}(q(t)) < 0 \qquad [2.147]$$

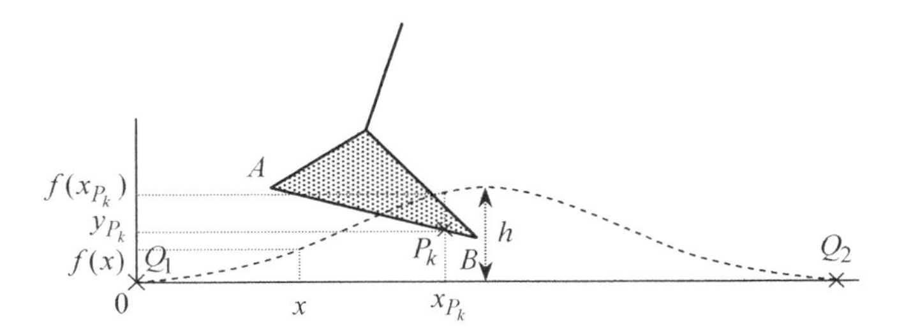

Figure 2.22. *The foot must clear the collision zone defined by the dotted curve*

In this case, a very limited set of points can be used, limited, for example, to the foot extremities A and B and possibly to the middle point of AB. Height h of the curve must be adjusted depending on a maximal desired security margin.

If, as in Figure 2.21, two avoidance curves are superimposed, the constraint function [2.147] can be redefined as:

$$t \in I_{SA},\ h_k^{nce}(q(t)) := Max[f_1(x_{P_k}(q(t))), f_2(x_{P_k}(q(t)))] - y_{P_k}(q(t)) < 0 \qquad [2.148]$$

where each of the functions f_1 and f_2 define one of the curves under consideration.

Finally, defining the vector function:

$$h^{nce} = \left(h_1^{nce}, \ldots, h_{n_{P_k}}^{nce}\right)^T$$

where n_{P_k} represents the number of points P_k under consideration, the conditions [2.147] or [2.148] can be summarized as:

$$t \in I_{SA},\ h^{nce}(q(t)) < 0\ (\in \Re^{n_{P_k}}) \qquad [2.149]$$

The preceding constraints are characteristic of walking to be generated and of the locomotion system which performs the movement created. They must be fulfilled by any solution extracted from the dynamics equations which govern the movement. These constraints together with the models described in this chapter will be studied further in Chapter 4, which presents a general construction technique of solutions provided by optimization methods.

2.6. Conclusion

In this chapter we studied the main aspects of kinematic and dynamic modeling for walking robots. The models which have been developed here are based on splitting up walking steps into successive phases which are characterized by the way the feet are in contact with the ground.

A major determinant of walking gaits to be generated and controlled is therefore the way the feet make contact with the ground during each step. This statement has resulted in the formulation of specific constraints which have enabled us to model the dynamics of the step sub-phase by sub-phase.

The importance of formulating contact conditions which guarantee well-balanced ground footholds must be noted while allowing for an appropriate propulsive effect. In particular, the notion of *CoP* associates the dynamics of movement with support contact locations. This has led to the formulation of simple conditions which enable us to prescribe footholds appropriately centered on assigned footprints. These conditions will play an essential role in the construction of optimal steps and the development of control laws.

The notions, models and constraints which have been formulated here will be developed in the following chapters for numerical synthesis and control of walking.

2.7. Bibliography

[ARM 94] ARMSTRONG-HÉLOUVRY B., DUPONT P., CANUDAS DE WIT C., "A survey of models, analysis tools and compensation methods for the control of machines with friction", *Automatica*, vol. 30, no. 7, p. 1083–1138, 1994.

[BAM 81] BAMBERGER Y., *Mécanique de l'ingénieur I. Système de corps rigides*, Hermann, Paris, 1981.

[BAU 90] BAUSSET M., *Mécanique des systèmes de solides*, Masson, Paris, 1990.

[BES 04] BESSONNET G., CHESSE S., SARDAIN P., "Optimal gait synthesis of a seven-link planar biped", *International Journal of Robotics Research*, vol. 33, p. 1059–1073, 2004.

[BES 05] BESSONNET G., SEGUIN P., SARDAIN P., "A parametric optimization approach to walking pattern synthesis", *International Journal of Robotics Research*, vol. 24, p. 523–536, 2005.

[BOU 04] BOURGEOT J.M., Contribution à la commande de systèmes mécaniques non réguliers, PhD Thesis, Institut national polytechnique de Grenoble, 2004.

[BRA 89] BRACH R.M., "Rigid body collisions", *Transactions of the ASME, Journal of Applied Mechanics*, vol. 56, p. 133–138, 1989.

[CAN 95] CANUDAS DE WIT C., OLSSON H., ASTRÖM K.J., LISCHINSKY P., "A new model for control of systems with frictions", *IEEE Transactions on Automatic Control*, vol. 40, no. 3, p. 419–425, 1995.

[CHA 88] CHATELIN F., *Valeurs propres de matrices*, Masson, Paris, 1988.

[CHE 01] CHEVALLEREAU C., AOUSTIN Y., "Optimal reference trajectories for walking and running of a biped robot", *Robotica*, vol. 19, p. 557–569, 2001.

[DEN 55] DENAVIT J., HARTENBERG R.S., "A kinematic notation for lower-pair mechanisms based on matrices", *Journal of Applied Mechanics*, vol. 77, p. 215–221, 1955.

[FRE 91] FREEMAN P.S., ORIN D.E., "Efficient dynamic simulation of a quadruped using a decoupled tree-structure approach", *The International Journal of Robotics Research*, vol. 10, no. 6, p. 619-627, 1991.

[FUR 95] FURUSHO J., SANO A., SAKAGUCHI M., KOIZUMI E., "Realization of bounce gait in a quadruped robot with articular-joint-type legs", *Proceedings of the IEEE International Conference on Robotics and Automation*, p. 697–702, 1995.

[GAL 01] GALLIER J., *Geometric Methods and Applications, For Computer Science and Engineering, Series: Texts in Applied Mathematics*, vol. 38, Springer-Verlag, New York, 2001.

[GAR 94] GARCIA DE JALON J., BAYO E., *Kinematic and Dynamic Simulation of Multibody Systems*, Springer-Verlag, New York, 1994.

[GEN 98] GENOT F., Contributions à la modélisation et à la commande des systèmes mécaniques de corps rigides avec contraintes unilatérales, PhD Thesis, Institut national polytechnique de Grenoble, 1998.

[HIR 98] HIRAI K., HIROSE M., HAIKAWA Y., TAKENAKA T., "The development of Honda humanoid robot", *Proceedings of IEEE International Conference on Robotics And Automation*, Leuven, Belgium, p. 1321–1326, 1998.

[HIR 03] HIRUKAWA H., KANEHIRO F., KAJITA S., FUJIWARA K., YOKOI K., KANEKO K., HARADA K., "Experimental evaluation of the dynamics simulation of biped walking of humanoid robots", *Proceedings of IEEE International Conference on Robotics And Automation*, Taipei, China, p. 14–19, 2003.

[HUN 75] HUNT K.H., CROSSLEY F.R.E., "Coefficient of restitution interpreted as damping in vibroimpact", *ASME Journal of Applied Mechanics*, p. 440–445, 1975.

[HUR 92] HURMUZLU Y., CHANG T.-H., "Rigid body collisions of a special class of planar kinematic chains", *IEEE Transactions on Systems, Man, and Cybernetics*, vol. 22, no. 5, p. 964–97, 1992.

[HUS 90] HUSTON R.L., *Multibody Dynamics*, Butterwoth-Heinemann, Stoneham, Massachusetts, 1990.

[KAN 85] KANE T.R., LEVENSON D.A., *Dynamics: Theory and Applications*, McGraw-Hill, New York, p. 348, 1985.

[KAN 04] KANEHIRO F., HIRUKAWA H., KAJITA S., "OpenHRP: open architecture humanoid robotics platform", *The International Journal of Robotic Research*, vol. 23, p. 155–165, 2004.

[KEL 86] KELLER J.B., "Impact with friction", *Transaction of the ASME, Journal of Applied Mechanics*, vol. 53, p. 1-4, 1986.

[KHA 86] KHALIL W., KLEINFINGER J.F., "A new geometric notation for open and closed loop robots", *Proceedings of IEEE International Conference on Robotics And Automation*, San Francisco, Etats-Unis, p. 1174–1179, 1986.

[KHA 87] KHALIL W., KLEINFINGER J.-F., "Minimum operations and minimum parameters of the dynamic model of tree structure robots", *IEEE Transactions of Robotics and Automation*, vol. 3, no. 6, p. 517–526, 1987.

[KHA 02] KHALIL W., DOMBRE E., *Modélisation, identification et commande des robots*, Hermes, Paris, 2002.

[LÖF 03] LÖFFLER K., GIENGER M., PFEIFFER F., "Sensors and control concept of walking "Johnnie", *The International Journal of Robotics Research*, vol. 22, p. 229–239, 2003.

[LUH 80] LUH J.Y.S., WALKER M.W., PAUL R.P.C., "On-line computational scheme for mechanical manipulators", *Transaction ASME Journal of Dynamic Systems, Measurement and Control*, vol. 102, no. 2, p. 69–76, 1980.

[MIO 04] MIOSSEC S., Contribution à la marche d'un bipède, PhD Thesis, University of Nantes and l'Ecole centrale de Nantes, 2004.

[MOR 88] MOREAU J.J., PANAGIOTOPOULUS P.D., "Unilateral contact and dry friction in finite freedom dynamics", in J.J. MOREAU, P.D. PANAGIOTOPOULUS (ed.), *CISM Courses and Lectures*, Springer-Verlag, Vienn, New York, vol. 302, p. 1–82, 1988.

[MUR 03] MURARO A., CHEVALLEREAU C., AOUSTIN Y., "Optimal trajectories of a quadruped robot with trot, amble and curvet gaits for two energetic criteria", *Multibody System Dynamics*, vol. 9, p. 39–62, 2003.

[OGU 06] OGURA Y., AIKAWA H., SHIMOMURA K., KONDO H., MORISHIMA A., LIM H., TAKANISHI A., "Development of a new humanoid robot WABIAN-2", *Proceedings of IEEE International Conference on Robotics and Automation*, Orlando, USA, p. 76–81, 2006.

[ORH 94] ORHANT P., Contribution à la Manipulation Fine. Étude de la Phase d'Impact, PhD Thesis, l'Institut national polytechnique de Grenoble, Laboratoire d'automatique de Grenoble, 1994.

[OUE 03] OUEZDOU F.B., KONNO A., SELLAOUTI R., GRAVEZ F., MOHAMED B., BRUNEAU O., "ROBIAN biped project – a tool for the analysis of the human-being locomotion system", *Proceedings of 6th International Conference on Climbing and Walking Robots*, Catania, Italy, p. 375–382, 2003.

[PFE 96] PFEIFFER F., GLOCKER C., *Multibody Dynamics with Unilateral Contacts*, Wiley Series in Non Linear Sciences, Wiley-WCH, Berlin, 1996.

[PFE 03] PFEIFFER F., LÖFFLER K., GIENGER M., "Humanoid robots", *Proceedings of 6th International Conference on Climbing and Walking Robots*, Catania, Italy, p. 505–516, 2003.

[RAD 94] RADCLIFFE C.W., "Prosthetics", in J. ROSE, G. JAMES (ed.), *Human Walking*, Williams and Wilkins, Baltimore, USA, p. 165–199, 1994.

[ROS 94] ROSE J., GAMBLE J., *Human Walking* (2nd Edition), William and Wilkins, Baltimore, 1994,.

[SAI 03] SAIDOUNI T., BESSONNET G., "Generating globally optimised sagittal gait cycles of a biped robot", *Robotica*, vol. 21, p. 199–210, 2003.

[SAK 02] SAKAGAMI Y., WATANABE R., AOYAMA C., MATSUNAGA S., HIGAKI N., FUJIMURA K., "The intelligent ASIMO: system overview and integration", *IEEE/RSJ International Conference on Intelligent Robots and Systems*, p. 2478–2483, 2002.

[SAR 98] SARDAIN P., ROSTAMI M., BESSONNET G., "An anthropomorphic biped robot: dynamic concepts and technological design", *IEEE Transactions on Systems Man and Cybernetics*, vol. 28a, p. 823–838, 1998.

[SEG 05] SEGUIN P., BESSONNET G., "Generating optimal walking cycles using spline-based state-parameterization", *International Journal of Humanoid Robotics*, vol. 2, p. 47–80, 2005.

[STR 80] STRANG G., *Linear Algebra and its Applications*, 2nd edition, Academic Press, New York, 1980.

[SUT 94] SUTHERLAND D.-H., KAUFMAN K.-R., MOITOZA J.-R., "Kinematics of normal human gait", in J. ROSE, G. JAMES (ed.), *Human Walking*, Williams and Wilkins, Baltimore, USA, p. 23–44, 1994.

[VUK 72] VUKOBRATOVIC M., STEPANENKO J., "On the stability of anthropomorphic systems", *Mathematical Biosciences*, vol. 15, no. 1, p. 1–37, 1972.

[VUK 90] VUKOBRATOVIC M., BOROVAC B., SURLA D., STOKIC D., *Biped Locomotion: Dynamics, Stability, Control and Applications*, Springer-Verlag, Berlin, 1990.

[VUK 04] VUKOBRATOVIC M., BOROVAC B., "Zero-moment point – thirty five years of its life", *International Journal of Humanoid Robotics*, vol. 1, no. 1, p. 157–173, 2004.

[WAL 82] WALKER M.W., ORIN D.E., "Efficient dynamic computer simulation of robotics mechanisms", *Journal of Dynamic Systems, Measurement, and Control*, vol. 104, p. 205–211, 1982.

[WHI 04] WHITTAKER E.T., *A Treatise on the Analytical Dynamics of Particules and Rigid Links*, Cambridge University Press, p. 232, 1904.

[WIT 77] Wittenberg J., *Dynamics of Systems of Rigid Bodies*, Teubner, Stuttgart, 1977.

[YOK 04] YOKOI K., KANEHIRO F., KANEKO K., KAJITA S., FUJIWARA K., HIRUKAWA H., "Experimental study for humanoid robot HRP1-S", *The International Journal of Robotics Research*, vol. 23, p. 351–362, 2004.

Chapter 3

Design Tools for Making Bipedal Robots

3.1. Introduction

Since the end of the 1980s, numerous walking and humanoid robot prototypes have been created by higher education [AOU 06, CHE 03, CHEV 03, SAR 99, ZON 02] and industrial [HIR 98] research laboratories. Current day studies in how these bipedal robots perform are working towards improving agility, environmental adaptation and walking gait celerity.

When creating an autonomous walker robot, there are many choice criteria to take into consideration: structure, actuator type (in order to optimize the weight/power ratio), sensors, dynamics and production cost of the task it has to carry out. The robot designer then has to make difficult compromises.

Numerous theoretical studies and simulations are needed to harmonize the desired characteristics of existing technological solutions. This chapter aims to give study possibilities and choice criteria. We will also give a brief overview of current technologies concerning multibodied robotic systems, actuated chains and sensors.

Section 3.2 is dedicated to the problem of mass distribution for a three-link or a five-link planar robot. The purposely limited number of degrees of freedom (DoF) of these two robots enables us to highlight the laws of mass distribution and size design. In this way, the energy consumption criteria guide us to achieve the most reliable leg mass. Spherical feet also have a favorable effect on energy consumption. A robot reaches higher forward velocities when it has spherical feet. The joints, as well as the motorized articulations between body-links, determine the robot's mobilities and influence its dynamics.

Section 3.3 presents previously created or possible structures for planar walking robots and then for 3D robots. We then consider the problem of force or torque transmission, and the principles and characteristics of different commercialized gear reducers. As bipedal robot walkers are expected to perform better all the time, the choice of actuator is becoming crucial.

Section 3.4 deals with the possible technological solutions for the actuators. An inventory of the different types of actuator is made according to its power supply.

Section 3.4.2 mainly deals with the detailed characterization of electric actuators and their velocity or torque controls. Particular attention is given to the DC switchmode converter principle and to the current loop. The important notion of the global efficiency of the actuator or gear drive must not be forgotten when designing a robot walker. This problem will be dealt with in section 3.4.5. It is necessary to locate a bipedal robot in relation to its environment and to add proprioceptive or exteroceptive sensors in order to establish effective control laws.

In section 3.5, we focus on the sensors which help us to locate bipedal robots: accelerometers, gyrometers, gyroscopes, GPS and force sensors. The latter measure the forces which result from the reaction between the ground and the stance foot, and detect impact.

3.2. Study of influence of robot body masses

The distribution of a bipedal robot's different body masses, as well as the moments of inertia of these same bodies and the positions of their center of gravity in relation to the articulations of the robot, have a noticeable influence on its dynamic performances and on its energy consumption during walking or running. The global mass of a body is determined by three main functions: (1) the shape, the material and the rigidity of the body's structure itself; (2) the kinematic transmission of the movements to the articulations; and (3) the actuating. The technological choices have a significant influence during this step of the mechanical design.

The displacement of one mass (that of the motor, for example) requires the utilization of a different or supplementary driving mechanism which increases the general mass. It is therefore necessary to compare the advantages and the disadvantages for each solution.

The energy consumption is highly influenced by the actuator control laws. Certain authors suggest control laws based on the zero moment point (ZMP) for robot walkers [BES 04, CHEV 01, ONO 02], or based on an internal oscillator developed with a bio-mimetic approach [AOI 05, RIG 06], or in a more traditional way, in control of robust trajectory tracking control [AOU 06, TZA 96, ZON 02].

These control laws are very different from each other and there are numerous parameterization settings. The energy consumption depends on both the structure and the control, and it is very difficult to separate the influence of the control law from the mechanical structure that was chosen at the design step.

This chapter is focused on the three control laws put forward by [GRI 01] and which enables us to prove walking stability.

For their practical application, these control laws must nevertheless be adapted to real robot cases. The same control laws will be used for the stabilization of walking for a three-link as well as for a five-link robot.

This study enables us to highlight how the mechanical design step influences the globally obtained performances.

This chapter deals with the complex problem of a bipedal robot's mechanical design by trying to establish simple and, if possible, general design rules.

In this way, the study of three-link or five-link robot structures enables us to limit the number of design and control parameters. This in turn enables us to easily solve these simple cases.

3.2.1. *Case 1: the three-link robot*

This planar bipedal robot is made up of a trunk and two rigid legs joined at hip level. This robot does not have any feet. The extremity of the legs is in direct contact with the ground. This extremity can be of any shape and size but, in general, we use two generic forms as is shown in Figures 3.1 and 3.2.

Point contact with the ground is interesting as it enables us to simplify the dynamic equations and to understand the fundamental principles of the design and control of such robots. A spherical foot, which is characterized by its $R = CO_2$ radius, shows a behavior model which is close to the roll of a human foot during a walking movement.

Two actuators, which create a torque between each leg and the trunk, bring about this type of robot movement. The joint coordinates define the orientation of each of these bodies. The geometric and dynamic models of this robot are given in the appendix (see section 3.7).

To obtain the dynamic model, each of the robots' bodies is modeled by its global mass m_i, $i=\{1, 2, 3\}$ (placed at the center of gravity) and by an inertia matrix. For the legs, we consider the thigh mass and tibia mass separately. By supposing that the knee articulation is rigid, we can calculate the equivalent mass and inertia matrix for the whole leg. The following simplified hypotheses have been used for the calculations:

– each body's center of gravity, placed along the leg's symmetric axis, is defined by its distance $s_i = O_1G_i$, $i=\{1, 2, 3\}$ from the hip's rotational axis; and

– each body's inertia matrix is the diagonal $I_{Ci} = diag\left(I_{xi} \quad I_{ri} \quad I_{zi}\right)$, $i=\{1, 2, 3\}$ and the single parameter I_{ri} (which intervenes in the dynamic model of this planar robot) is therefore relative to the axis y_0.

The angles q_i, $i=\{1,2,3\}$ identify the absolute orientation of the bodies in relation to the vertical terrestrial field of gravity, and the angles θ_i, $i=\{1,2\}$ give the position of the leg relative to the trunk. The angle $\alpha = q_1 + \pi/2$, which is very important for control, identifies the inclination of the robot's stance leg in relation to the supposed horizontal ground. The actuating torques Γ_i, $i=\{1,2\}$ are applied between each leg and the trunk.

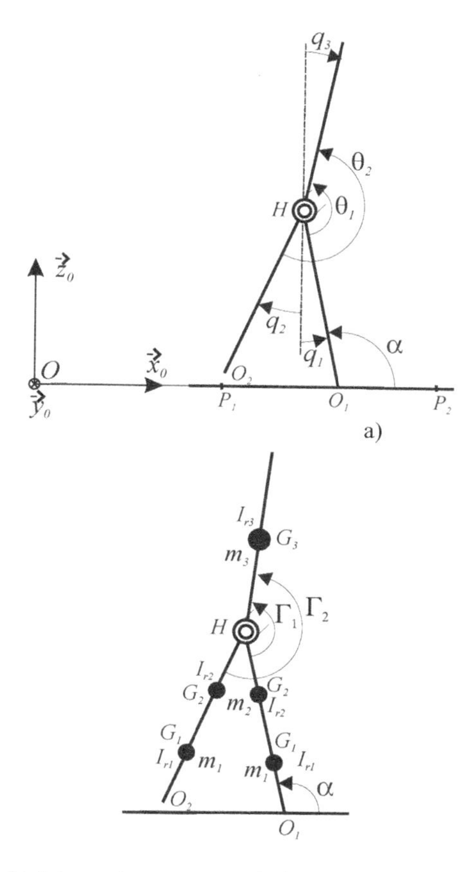

Figure 3.1. *Robot which has point contact with the ground: (a) frames and coordinates; (b) masses, inertias, centers of gravity and torques*

When designing a robot we need to choose the actuators and give a shape to the robot's main bodies. This enables us to determine the inertias, the masses and the centers of gravity of the different bodies. The total mass of the actuators could represent up to 40% of the robot's total mass; it is very important to locate the position of the actuators. This choice influences or imposes the kinematics of the transmission elements, and modifies the different body masses, inertias and gravity centers. The robot under consideration for this example has a hip height of about 80 cm from the ground. The leg and trunk dimensions and the physical limits for this robot are given in the appendix (see section 3.7).

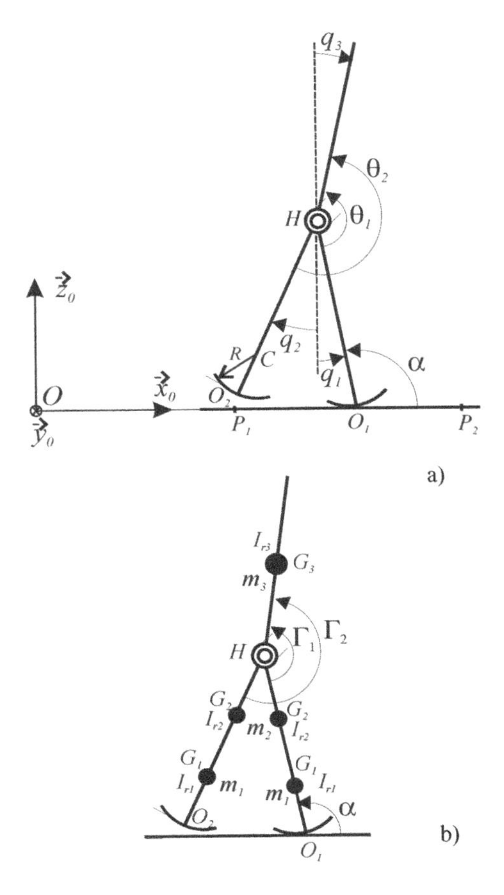

Figure 3.2. *Robot with spherical feet: (a) frames and coordinates; (b) masses, inertias, centers of gravity and torques*

In the case of these robots, the mass distribution and the inertias of the mechanical structure need to be known approximately for the control law and the optimization of the walking gaits.

When this distribution influences the energy consumption and the efficiency of the robot, it is then necessary to optimize the mass distribution at the same time as that of the walking gaits.

We are then ready to simulate the behavior of a three-link robot depending on the forward velocity and on the distribution of the different masses.

Research studies [GRI 01] have shown that in the hypothesis of a rigid impact with the ground, supposing that the stance leg does not slide or bounce, this robot is controllable and the zero dynamics of the controlled system is stable. The PD control law, which has been chosen for what follows, requires a definition of reference trajectories.

Walking gaits are defined by two reference trajectories:

$$q_{3d} = \text{constant}; \quad q_{2d} = -\alpha - \pi/2 = -q_1 \qquad [3.1]$$

These trajectories enable us to continuously synchronize the robot's trunk and movements of its balancing leg with the rotation (defined by angle α) of the whole robot on its stance leg. The movement of angle α is not controlled and depends solely on the angular momentum along the fixed point and its profile [AOU 03]. The initial value of angle α is obtained by imposing a distance between the two legs which corresponds to the approximate length of the forward step. The final value of α is calculated when the balancing leg touches the ground. The chosen control law is defined as follows:

$$\Gamma_1 = K_{p1}(q_{3d} - q_3) - K_{v1}\dot{q}_3 \qquad [3.2]$$

$$\Gamma_2 = \Gamma_1 + K_{p2}(q_1 + q_2) + K_{v2}(\dot{q}_1 + \dot{q}_2) \qquad [3.3]$$

The Γ_1 torque is applied to the stance leg and stabilizes the position of the trunk. The Γ_2 torque is applied to the mobile leg and allows for the forward movement of the balancing leg. As the robot is under-actuated, the rotation of the stance leg is not under control; its profile depends on the biped's dynamics. The coefficients K_{p1}, K_{p2}, K_{v1} and K_{v2} are adjusted so that its behavior resembles that of an equivalent passive robot. Figures 3.3 to 3.6 show, in the nominal case, the behavior of the robot under this control law for the first 40 walking steps. The robot progresses from an initial configuration to a stable walking movement and a regular gait. The step gaits become stabilized very rapidly and the average forward velocity per step becomes constant after a few steps.

We also notice that the steps are symmetric (same gait for the left and right leg). Figure 3.7 shows the torque curves for the stance and balancing legs. The motor mass, which is capable of producing the simulated torques, is totally compatible with the body masses under consideration during the simulation.

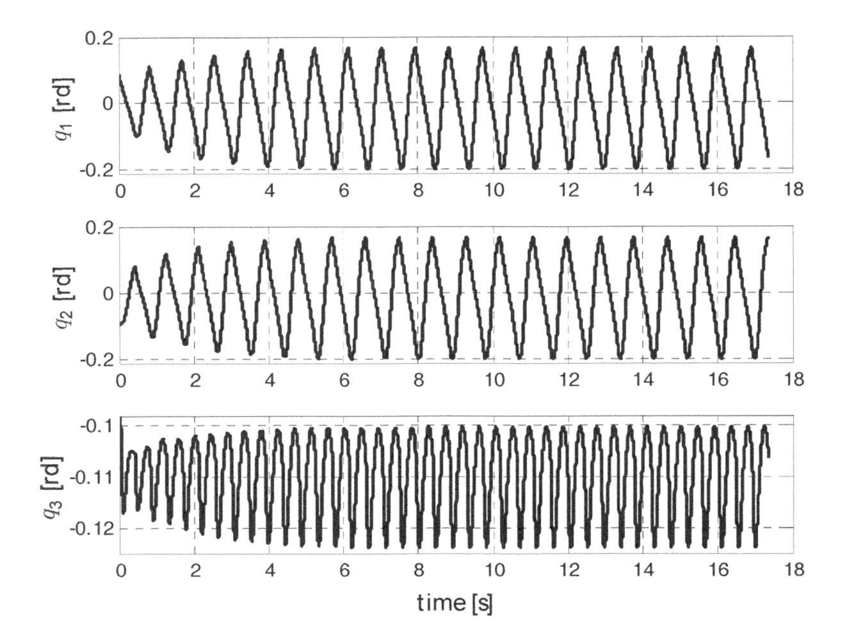

Figure 3.3. *Profile of absolute angles*

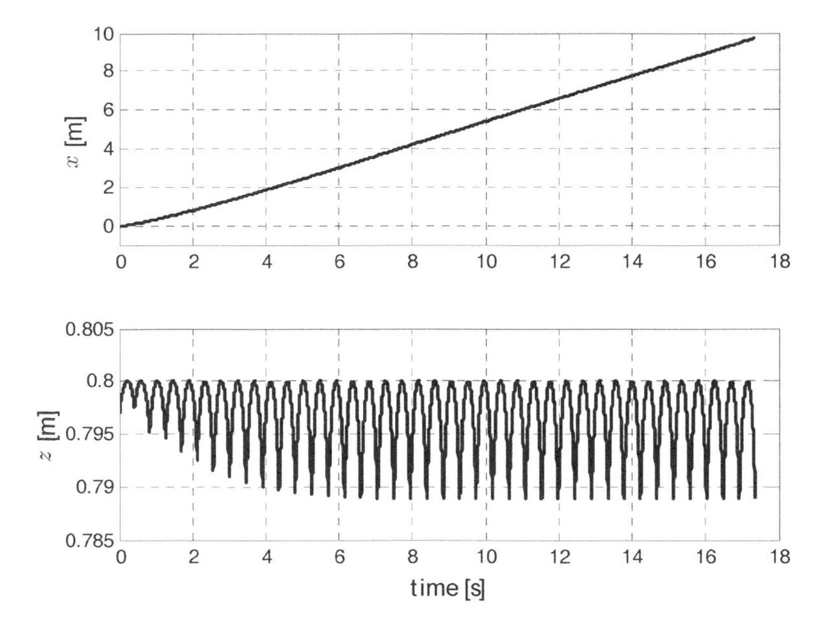

Figure 3.4. *Profile of Cartesian coordinates*

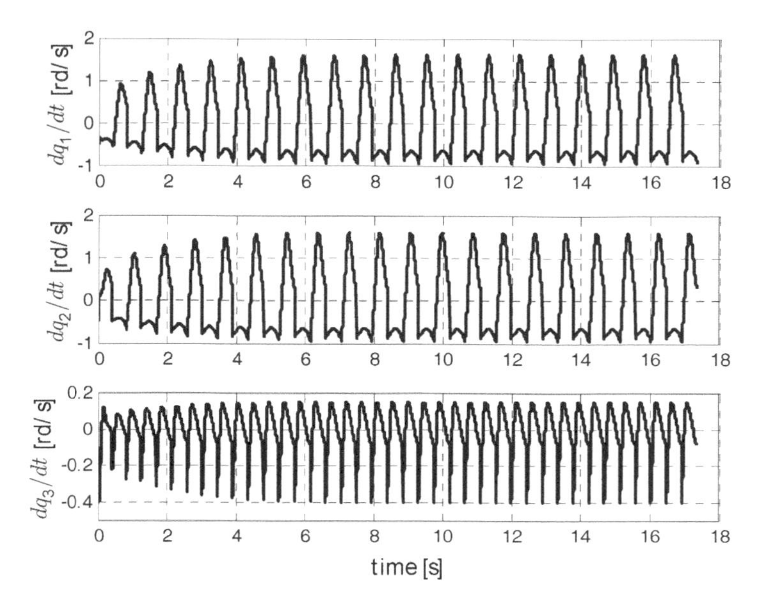

Figure 3.5. *Profile of absolute revolute velocities*

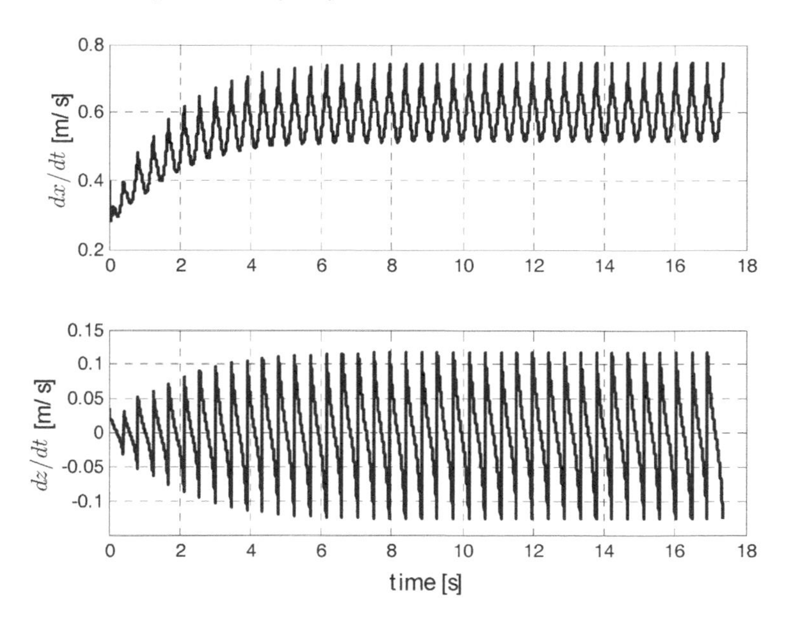

Figure 3.6. *Profile of Cartesian velocities*

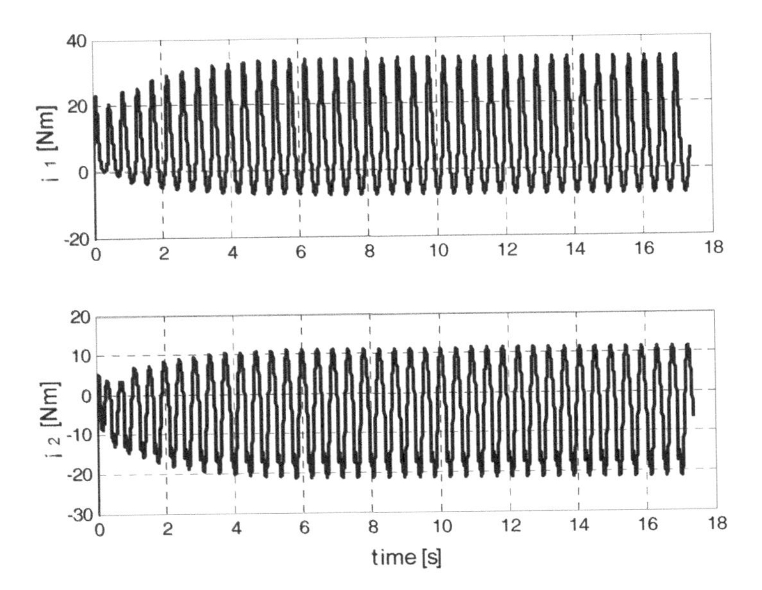

Figure 3.7. *Profile of joint torques*

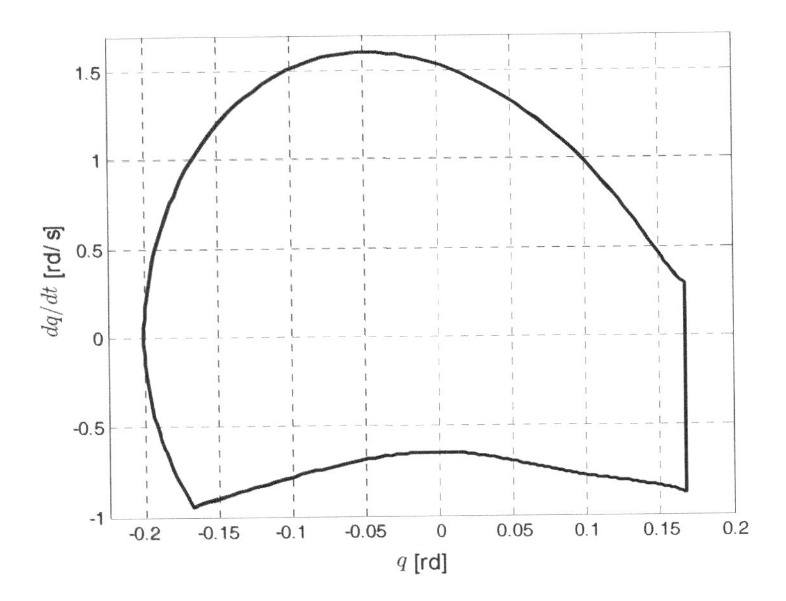

Figure 3.8. *Phase plane profile for angles* q_1 *and* q_2

Figure 3.8 shows the phase plane of the angular variable q_1 and q_2 for about 20 steps after the walk has become stable. It must be noted that there is a strong similarity of this phase plane with that of passive robots of analogous structure [GOS 96, GOS 97, MCG 90].

The distribution of the previously chosen masses can be modified in order to optimize the global behavior of the robot and to improve its energy autonomy. The study shows how the masses and the position of the mass centers of each of the robot's three bodies can influence the energy characteristics and the choice of motor.

The control law is the same for each simulation and its optimization can be improved and adapted for each robot model. The proposed law nevertheless remains close to the passive behavior of the robot's profile.

The motors' average power curves enable us to predict the nominal temperature and the losses. The maximum power curves enable us to choose the optimal motor and deduce the necessary mass for the actuators as a whole.

Table 3.1 shows the different case studies of variations of mass (kg) and the relative position of the centers of gravity (m). The mass and position reference values which make up the elements of case 1 are chosen in accordance with the design of the Rabbit[1] robot.

The inertia moments are constant and have a value of $I_{r1} = 0.05$ kg m^2, $I_{r2} = 0.07$ kg m^2 for the tibias and thighs respectively and $I_{r3} = 1.56$ kg m^2 for the trunk. These simulations aim to find a compromise between the distributions of mass on the thighs, tibias and trunk.

In this way, case 2 is a hypothesis of a transfer of 1 kg (motor mass for example) from the tibia to the thigh. Case 3 envisages a 4 kg transfer of mass from the trunk to the thighs, which raises the mass of each thigh by 2 kg.

Cases 4 and 5 attempt to bring the gravity centers of the leg and trunk closer to their respective rotational axes. This reduces the inertia moments of these bodies.

Cases 6 and 7 consider the mass reductions (divided by 2) for both the trunk and the legs.

---

1. See http://robot-rabbit.lag.ensieg.inpg.fr/.

Figures 3.9 to 3.15 show the results obtained for the different robot prototypes which correspond to the aforementioned cases.

Case	m_1	m_2	m_3	s_1	s_2	s_3
1	3.2	6.8	16.5	0.527	0.163	0.2
2	2.2	7.8	16.5	0.527	0.163	0.2
3	3.2	10.8	12.5	0.527	0.163	0.2
4	3.2	6.8	16.5	0.468	0.082	0.2
5	3.2	6.8	16.5	0.527	0.163	0.1
6	3.2	6.8	8.25	0.527	0.163	0.2
7	1.6	3.4	16.5	0.527	0.163	0.2

Table 3.1. *Parameter variations for the three-link robot simulation model*

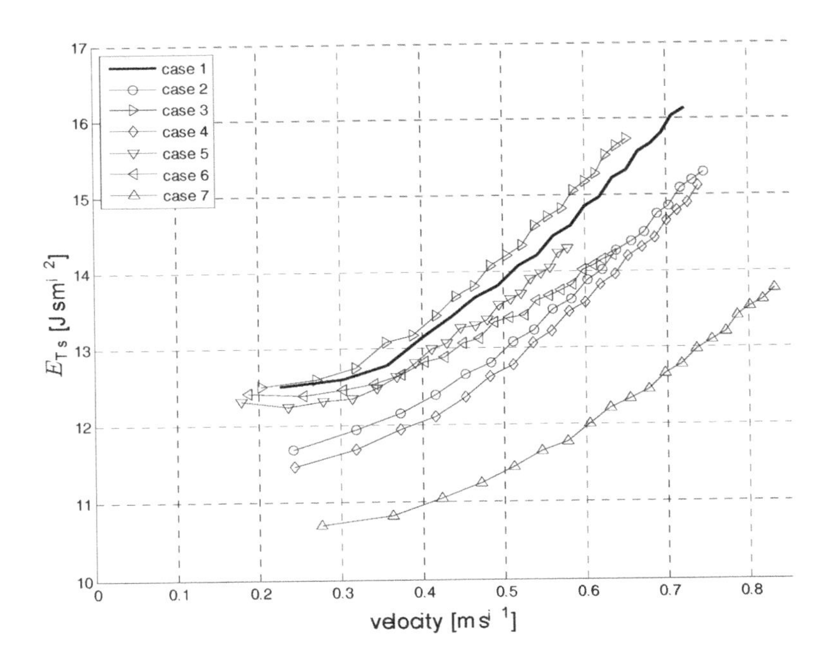

Figure 3.9. *Total energy versus velocity*

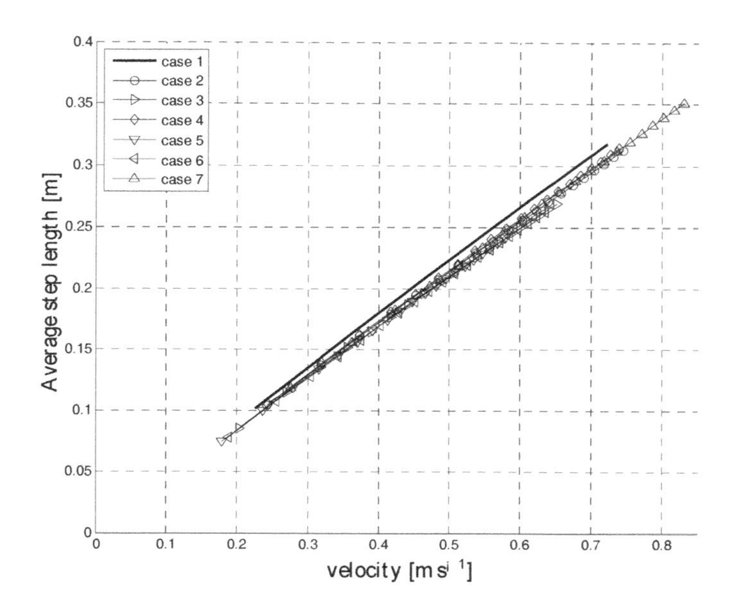

Figure 3.10. *Step lengths versus velocity*

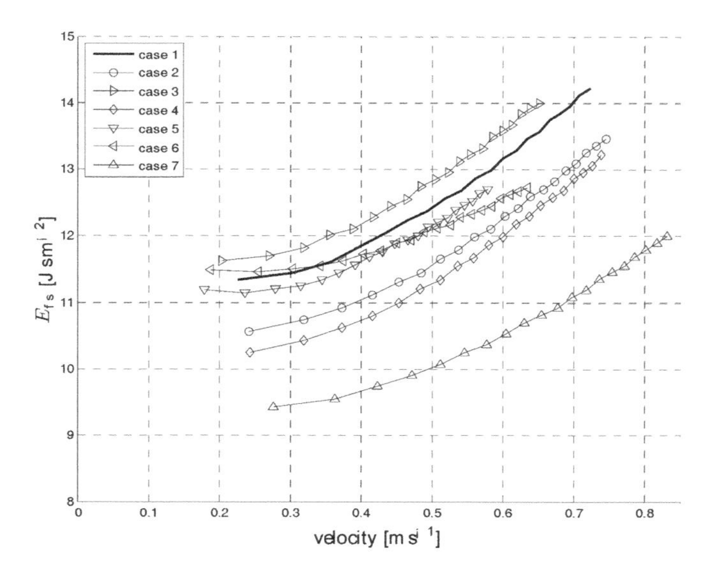

Figure 3.11. *Stance leg energy versus velocity*

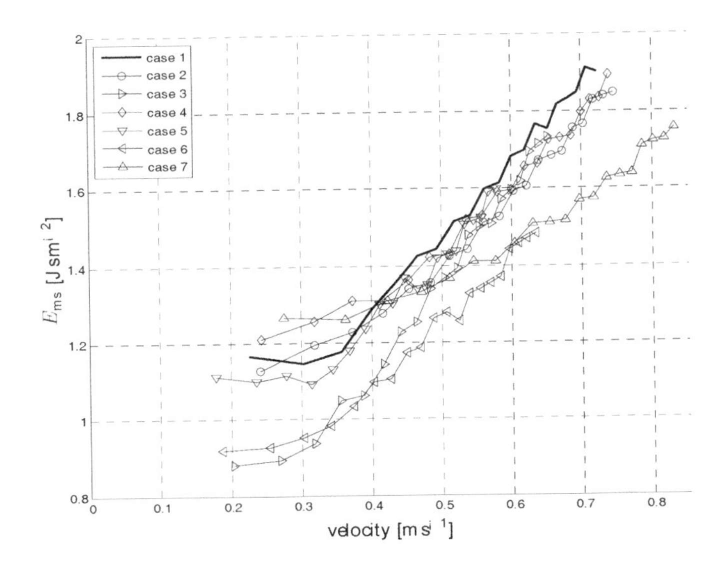

Figure 3.12. *Swing leg energy versus velocity*

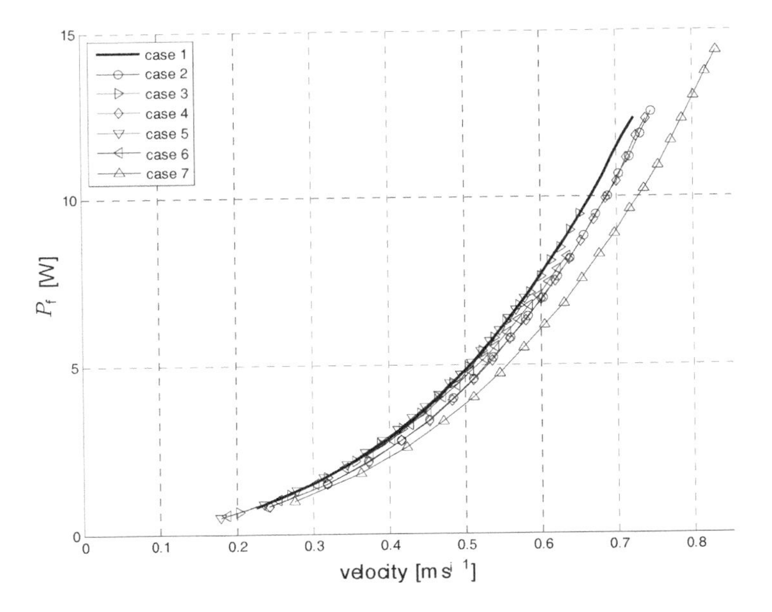

Figure 3.13. *Average power for stance leg*

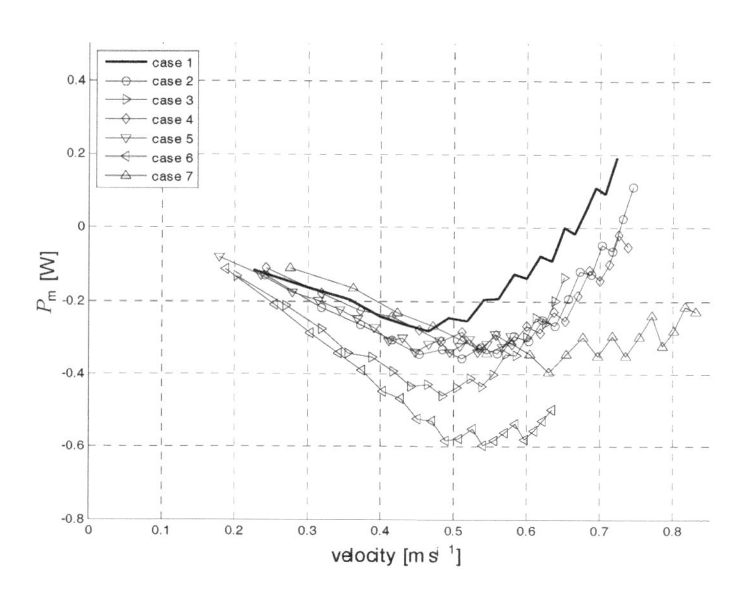

Figure 3.14. *Average power for swing leg*

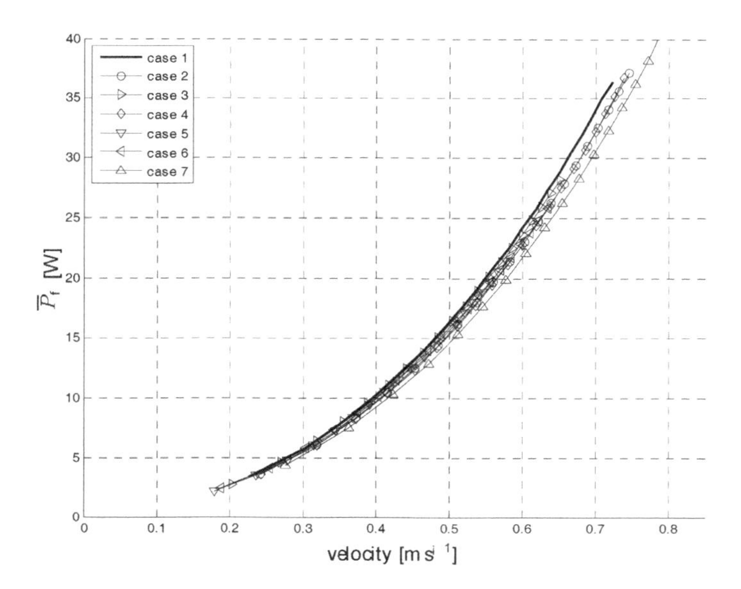

Figure 3.15. *Maximum power for stance leg*

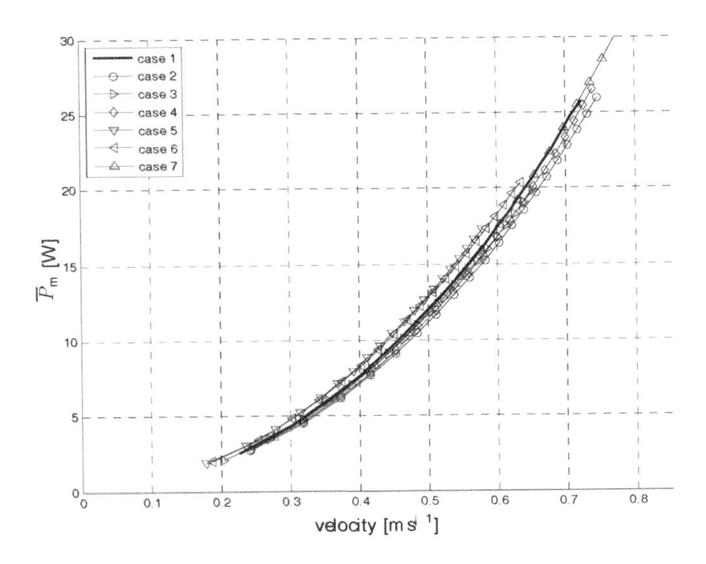

Figure 3.16. *Maximum power for swing leg*

Figures 3.9, 3.11 and 3.12 give (depending on the average forward velocity indicated in the abscissa) the energy consumption per meter traveled, divided by the robot's average forward velocity when the gait has been stabilized. The length of the steps, depending on the average forward velocity, is shown in Figure 3.10. Figures 3.13–3.16 give the average and maximum necessary powers at actuator level, depending on the forward velocity, the Joule losses and the mechanical friction losses which are not taken into consideration in these simulations. We will now comment on the above curves.

The chosen control law essentially leads to lengthening of the steps when a higher forward velocity is requested. A step period is close to 0.45 s. The length of steps is not greatly influenced by the distribution of the body masses. In Figure 3.9, we see that cases 2, 4 and 7 result in a lower energy consumption, and we gain up to 13% in case 7. For the same total mass for motors installed on the robot, the power curves show that we gain 5% maximum velocity, or for an imposed velocity of 0.7 m s^{-1}, we lower the necessary mass for the motors by 10%. In short, when designing a robot, it is interesting to lower the maximum leg mass. The trunk mass (see case 6), however, has a much smaller influence on the energy performances.

In the case of the robot with spherical feet (see Figure 3.2), it is also important to define the feet's curvature radius (which is taken as constant). In order not to weigh down this study case, the circle center (which defines the foot) is placed along the leg axis. For designing and dimensioning this robot, we need to determine each body mass, the center of gravity positions and the spherical foot radius. Figures 3.17–3.24 show the results obtained for a robot prototype with three different radius values for spherical feet.

We notice that for all the previous curves, the increase in the foot radius is always advantageous to robot walking. Indeed, for the same amount of energy consumed, the robot attains greater forward velocities. The length of step is not actually modified. For the same maximum forward velocity, we can choose motors that are 10% less powerful and therefore the actuator mass is reduced by 9%. Similarly, when we have a given forward velocity and spherical feet, we notice that the energy consumption is reduced by 12% to 16%.

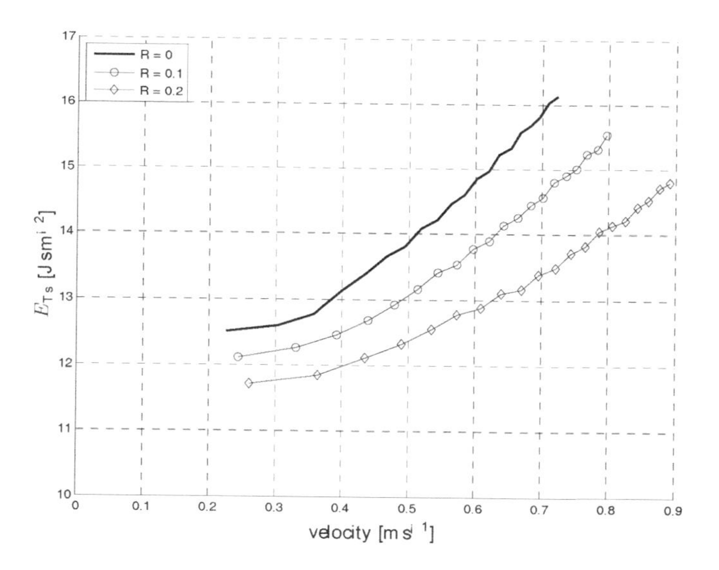

Figure 3.17. *Total energy versus velocity*

All these simulations show that in the case of a three-link robot, it is better to build a robot with spherical feet, reduce the total leg masses and choose driving mechanisms which enable us to place the actuators on the trunk. The shape of the legs must ensure that the leg mass center is as close as possible to the hip axis, and that their inertia moment is reduced in relation to hip revolute axis. When creating a robot with spherical feet, it is therefore extremely important to build feet which have the smallest possible mass.

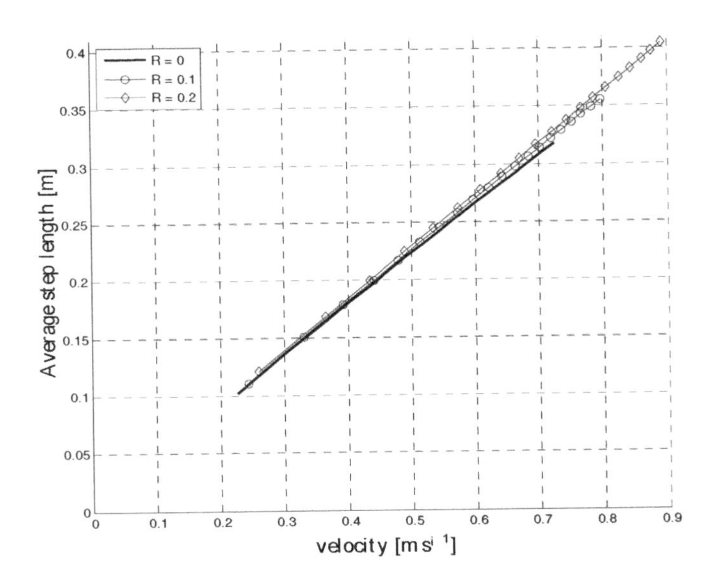

Figure 3.18. *Step lengths versus velocity*

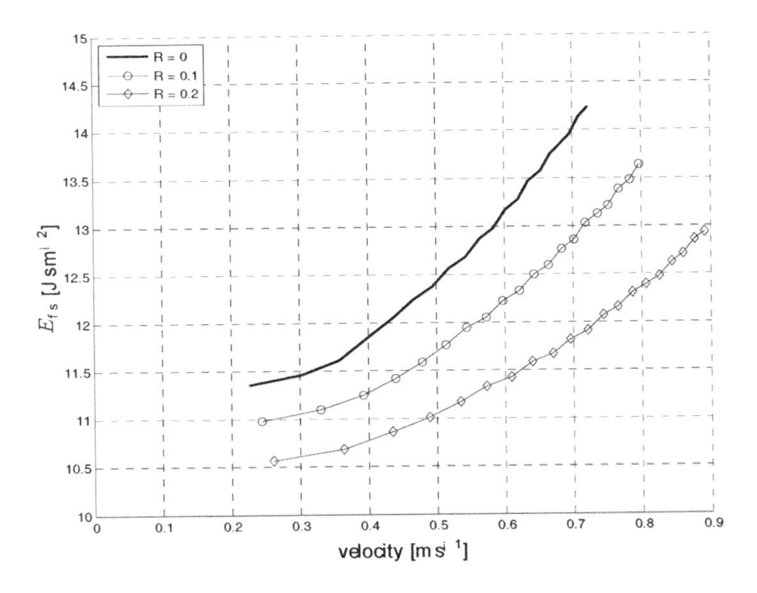

Figure 3.19. *Stance leg energy versus velocity*

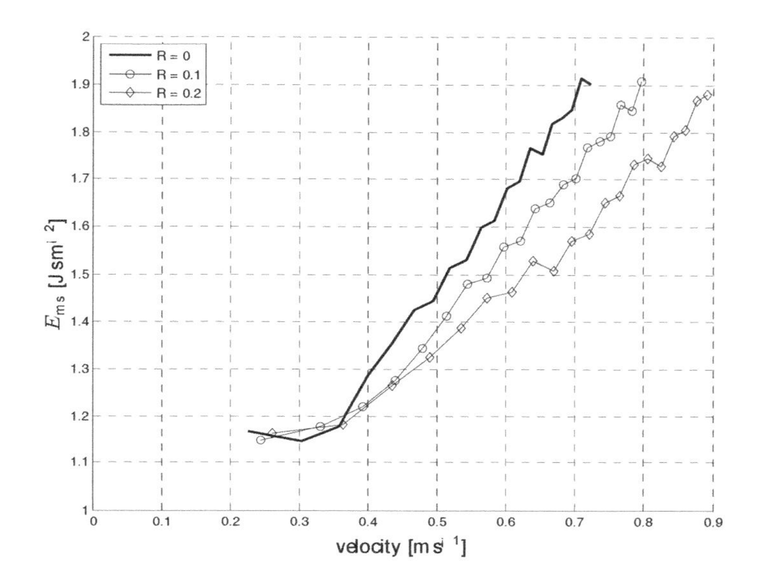

Figure 3.20. *Swing leg energy versus velocity*

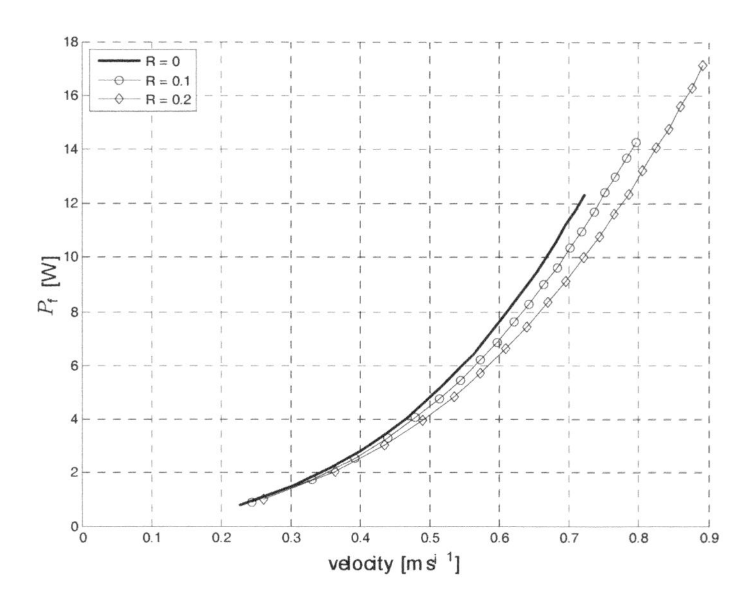

Figure 3.21. *Average power for stance leg*

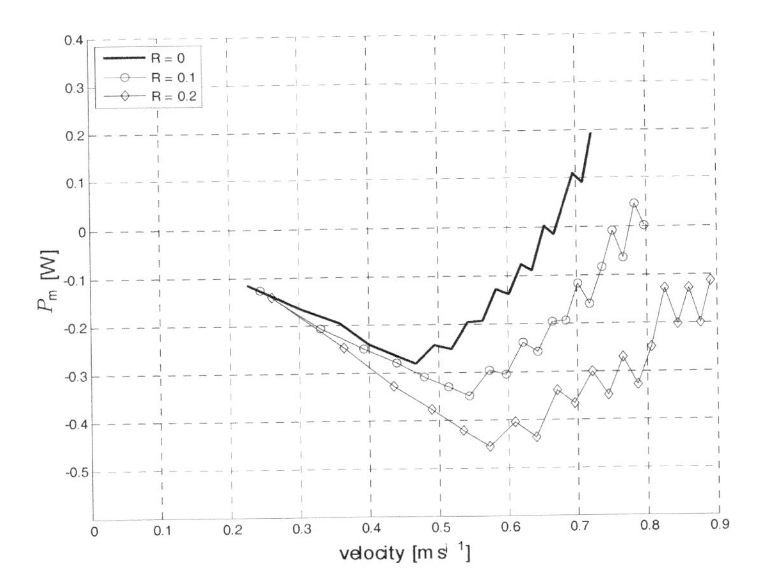

Figure 3.22. *Average power for swing leg*

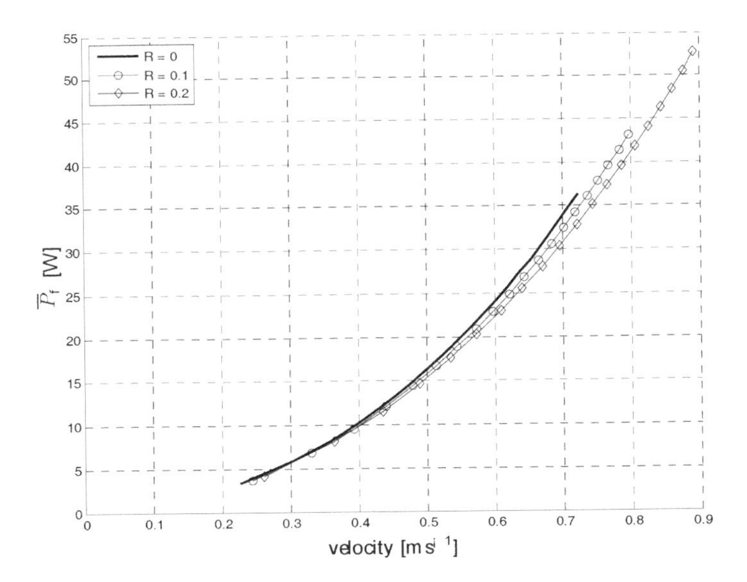

Figure 3.23. *Maximum power for stance leg*

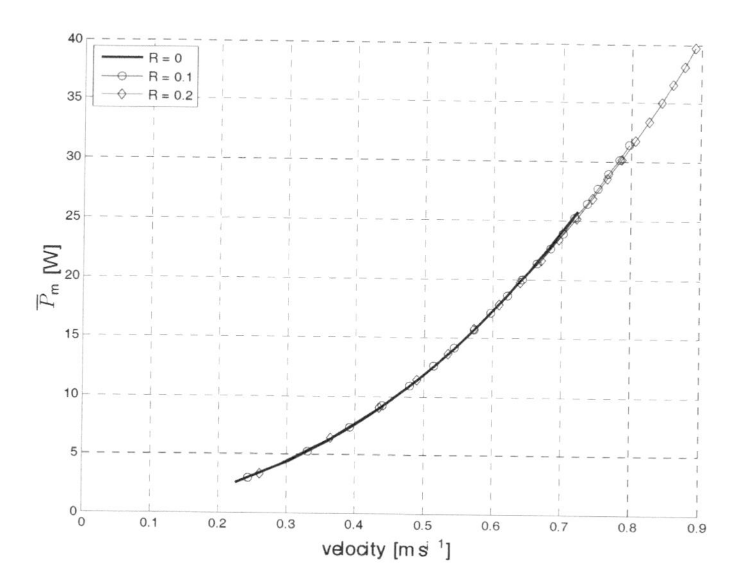

Figure 3.24. *Maximum power for swing leg*

3.2.2. *Case 2: the five-link robot*

A five-link bipedal robot is made up of a trunk and legs articulated at knee and hip level. This type of robot does not have feet as such. The tibia extremities (bodies S1 and S5) are in direct contact with the ground. In this case we will also take into consideration the cases which were discussed in the three-link robot: that of point contact and spherical feet (see Figures 3.25 and 3.26). The spherical foot is characterized by a radius of $R = CO_5$, which is a design parameter.

The angles q_i, $i = \{1 \cdots 5\}$ mark the bodies' absolute orientation in relation to the vertical terrestrial gravity field. The four actuators, which create a torque between each thigh and the trunk and at the two knees, bring about this robot's movement. The torque of the stance and swing legs are denoted Γ_2 and Γ_3, respectively The knee torque of the stance and swing legs are denoted Γ_1 and Γ_4, respectively. The absolute joint coordinates define the orientation for each body. To obtain a dynamic model, we make the same simplified hypotheses which were used for the three-link robot. The body S_i, $I = \{1,\ldots,5\}$ is therefore characterized by three parameters: its mass m_i, the position of its center of gravity s_i and its inertia moment I_{ri} in relation to the body's rotational axis.

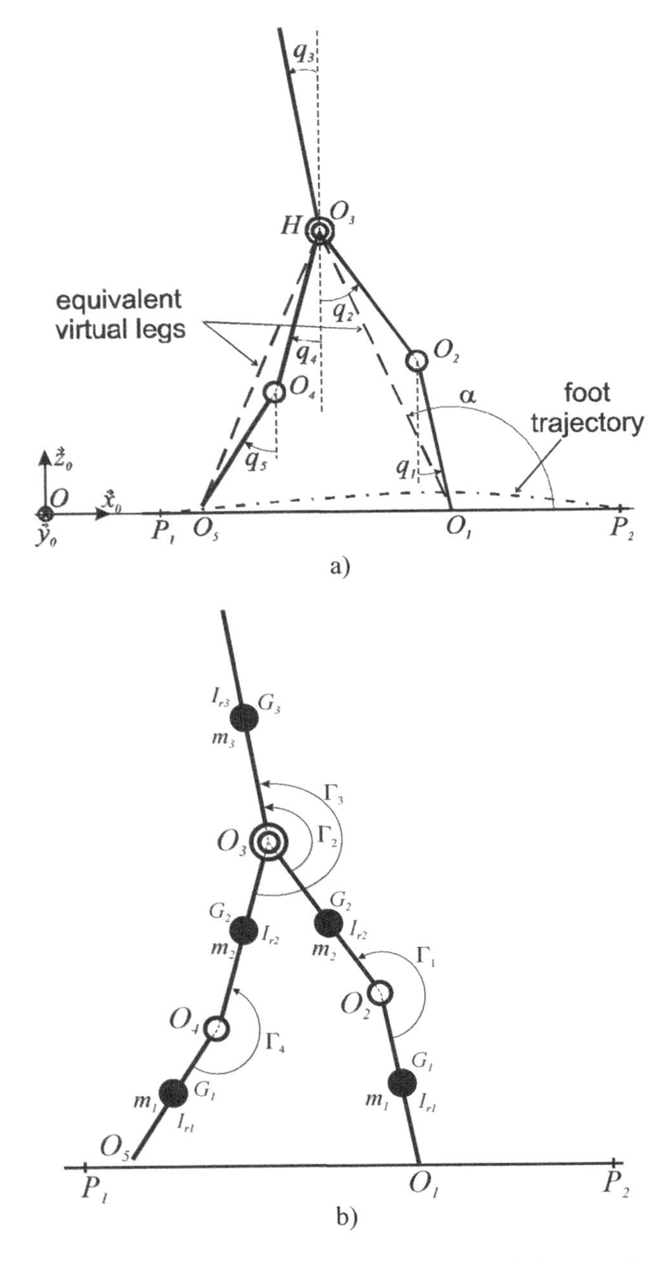

Figure 3.25. *Five-link robot: point contact with the ground*

When controlling a robot of this type, it is interesting to introduce the notion of virtual legs which are equivalent to the fixed leg and swing leg.

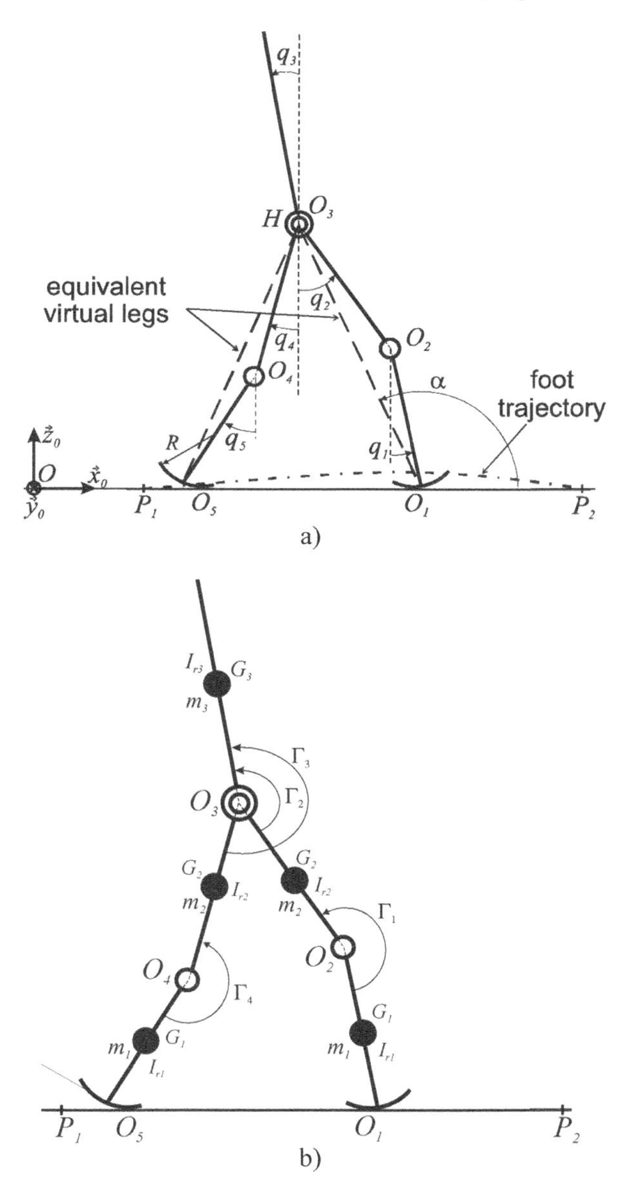

Figure 3.26. *Five-link robot with spherical feet*

Angle α marks the inclination of the virtual stance leg in relation to the ground.

Simulations enable us to find the optimal locations for the actuators and provide indications of the geometry and ideal shape of the different bodies which make up the robot.

Before any simulation, it is important to ensure that we can obtain dynamically stable walking for the robot. The dynamic model for walking (with a fixed leg which is always in contact with the ground) has five DoF.

This robot only has four control torques, and is therefore under-actuated. As in the case of the three-link robot, it has been proved [PLE 03] that the robot can be controlled and that the controlled system's zero dynamics is stable (see Chapter 5).

Studies have shown that it is also possible to stabilize the zero dynamics of this five-link robot with proportional and derived controller [MOR 05]. Consequently, the chosen control law has proportional and derived terms. The angular motorized variables are chosen as outputs θ_i $i = \{1, 2, 3, 4\}$ as in

$$\begin{aligned} \theta_1 &= q_2 - q_1 + \pi, \quad \theta_2 = q_3 - q_2 + \pi \\ \theta_3 &= q_3 - q_4 + \pi, \quad \theta_4 = q_4 - q_5 + \pi \end{aligned} \qquad [3.4]$$

Fourth-order polynomials are defined as reference trajectories θ_{id} $i = \{1, 2, 3, 4\}$ for these outputs. These polynomials are functions of the virtual stance leg's orientation variable α.

The α variable behaves monotonously, depending on the time during a half-step in single support (see sections 4.5.3.1 and 5.5).

For each reference polynomial trajectory $\theta_{id}(\alpha)$, the polynomial's five coefficients are calculated by determining the initial and final configurations and velocities, as well as the intermediary configuration.

The intermediary configuration is defined by specifying the Cartesian positions of the hip articulation, the swing leg's foot and the trunk orientation.

By using the inverse kinematic model, we can deduce the relative angular variables for each leg.

By deriving the reference trajectories $\theta_{id}(\alpha)$, we obtain the desired joint velocities. These are also used by the proportional and derived control laws, as in the following:

$$\Gamma_i = K_{pi}\left(\theta_{id} - \theta_i\right) + K_{vi}\left(\dot{\theta}_{id} - \dot{\theta}_i\right), \quad i = \{1,2,3,4\} \tag{3.5}$$

with $\dot{\theta}_1 = \dot{q}_2 - \dot{q}_1$, $\dot{\theta}_2 = \dot{q}_3 - \dot{q}_2$, $\dot{\theta}_3 = \dot{q}_3 - \dot{q}_4$ and $\dot{\theta}_4 = \dot{q}_4 - \dot{q}_5$.

The simulation approach for this five-link robot becomes more complicated because of the high number of possible variable settings. Indeed, when designing the robot, we can chose an initial configuration defined by (L_p, z_h). The set parameters for the reference trajectories are defined by (z_{h1}, z_{h2}, q_{30}), the coefficients K_{pi}, $i = \{1,\dots,4\}$ and K_{vi}, $i = \{1,\dots,4\}$ for the control law, as well as the mass values and the position of the center of gravity. Figures 3.27 to 3.31 show the robot's behavior during the first 11 steps of the walking (in the nominal case) when it is controlled by this law. The robot progresses from the initial configuration to a stable walking movement and has a regular gait.

After a few steps, the gait becomes rapidly stable and the average forward velocity per step also becomes constant (in the 0.59 m s^{-1} example). We also notice that the steps are symmetric (same gait for the left and right leg). Figure 3.30 shows the torque curves for both legs. The maximum torque of 80 Nm is reached at knee level (stance leg) during the first step. For the movement as a whole, the torque stays clearly under the maximum value of the admissible 200 Nm for motors and reducers. Figure 3.31 shows the phase plane for the joint variables q_1, q_2 and q_3. The knee and hip behavior is influenced by the way the feet hit the ground. The figure which shows the knee joint variables resembles those of the hips for the three-link robot.

The simulations then enable us to show the importance of mass allocation (parameters s_1, s_2 and s_3) and how it is distributed along the trunk, thighs and tibias. The cases under consideration here use the same values as those listed in Table 3.1.

The aim of these simulations is to find a compromise when allocating mass to the thighs, tibias and trunk. Case 1 is the reference configuration, and the other cases are the same as those studied in the three-link robot.

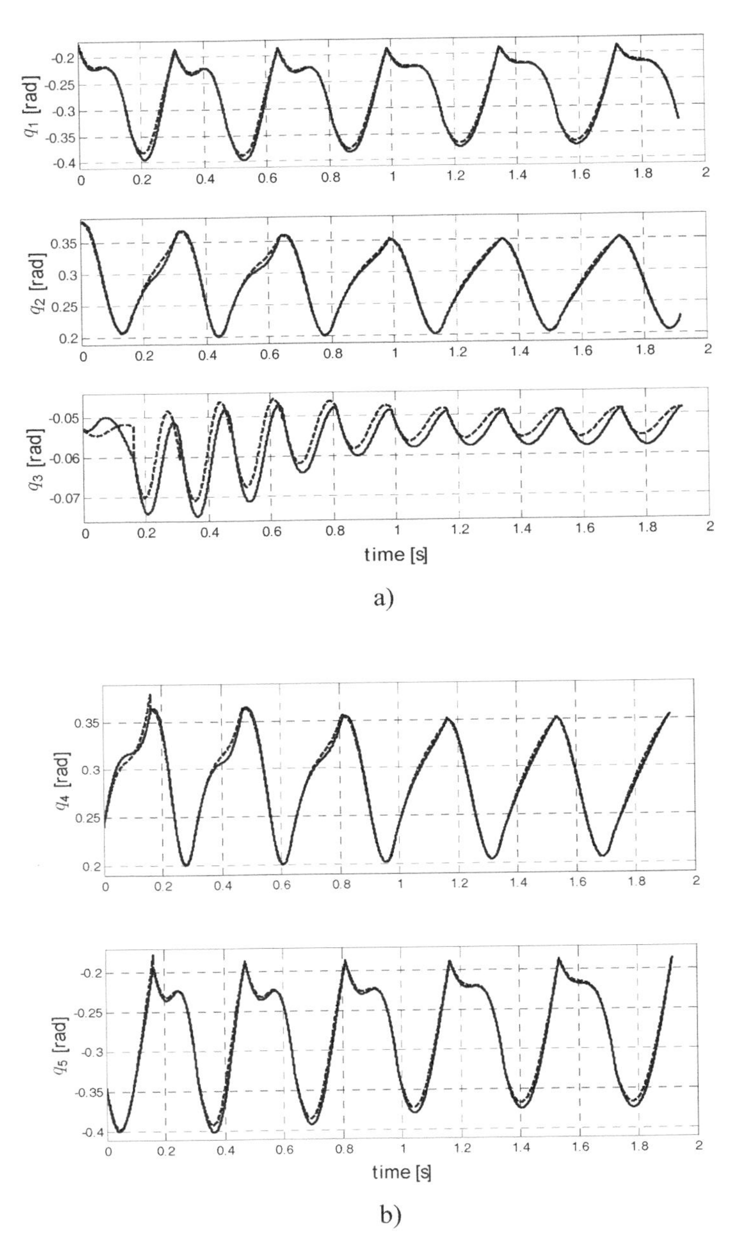

Figure 3.27. *Profile of absolute joint angles*

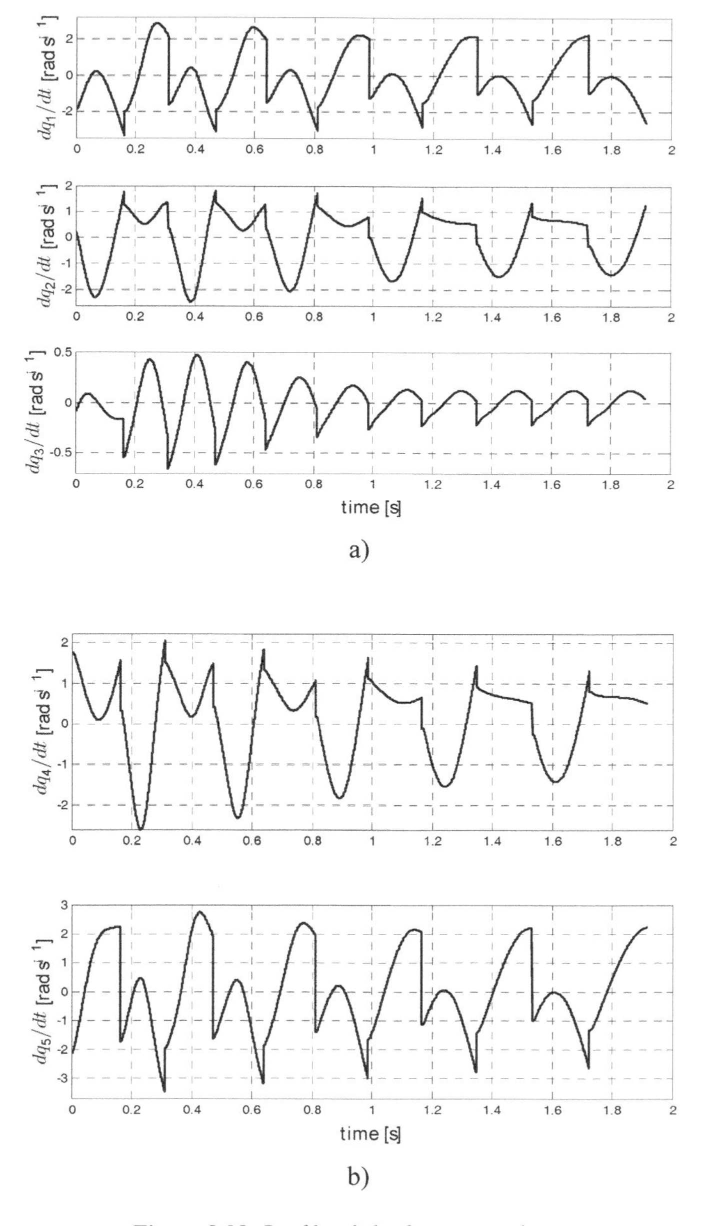

Figure 3.28. *Profile of absolute joint velocities*

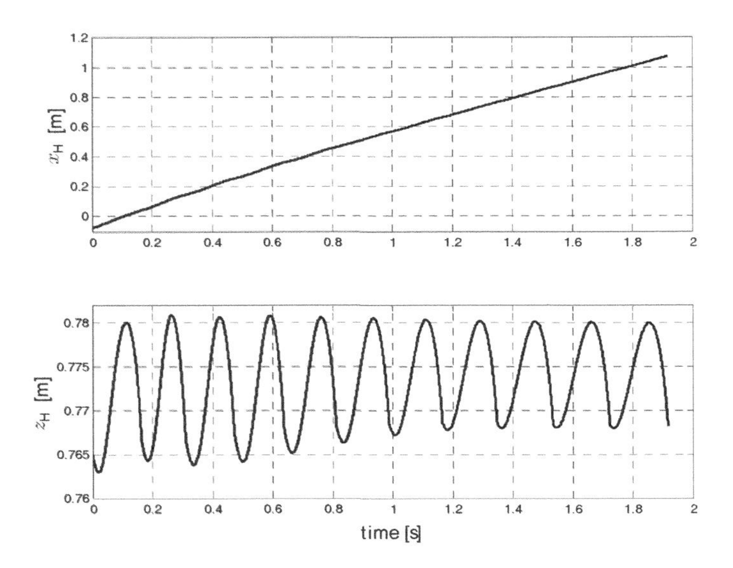

Figure 3.29. *Profile of the hip's Cartesian coordinates*

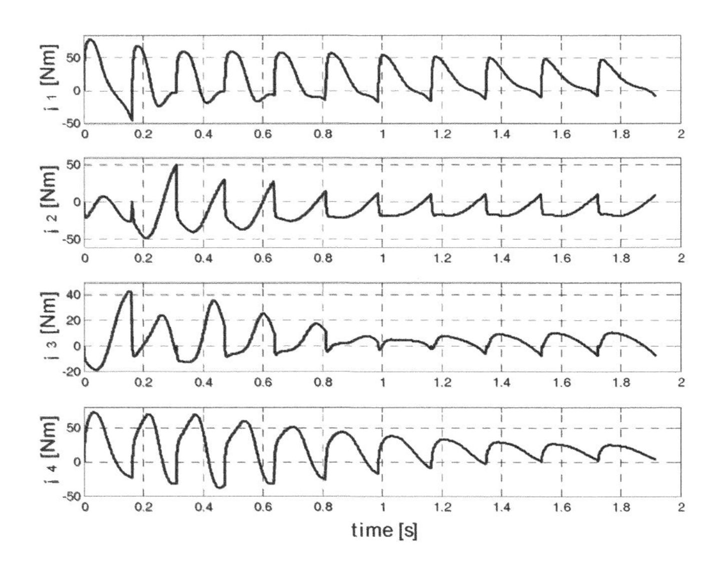

Figure 3.30. *Profile of joint torques*

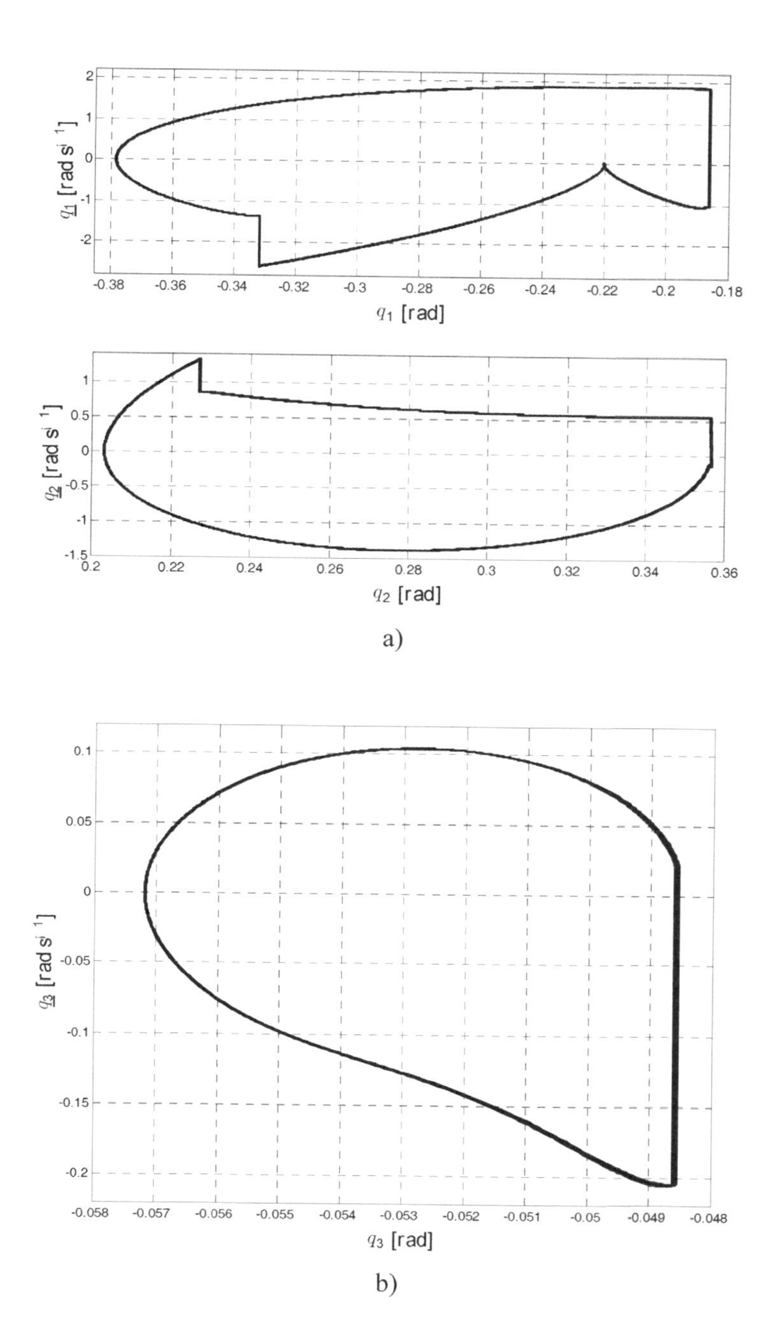

a)

b)

Figure 3.31. *Profile of angles q_1, q_2 and q_3 phase plane*

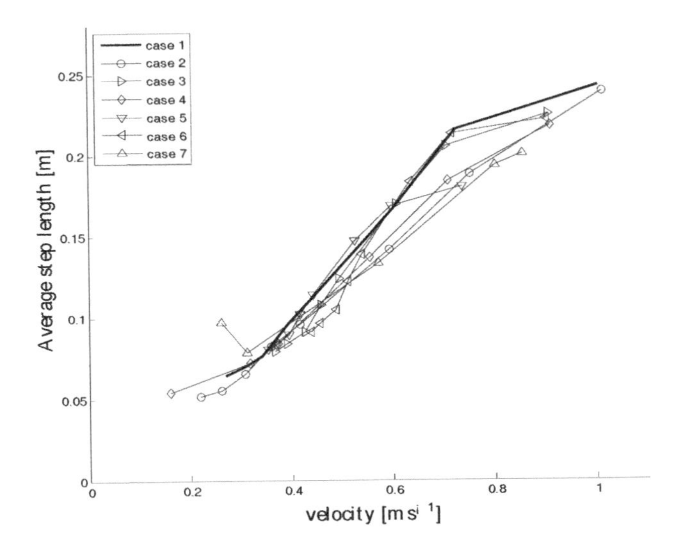

Figure 3.32. *Length of steps versus velocity*

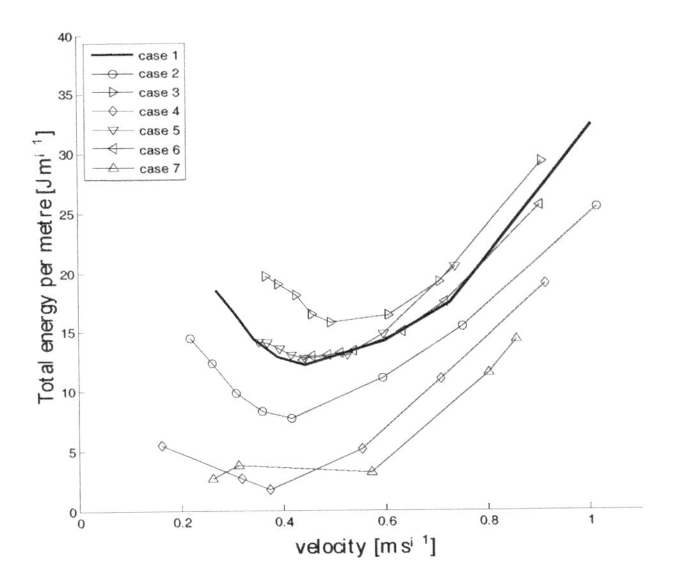

Figure 3.33. *Total energy versus velocity*

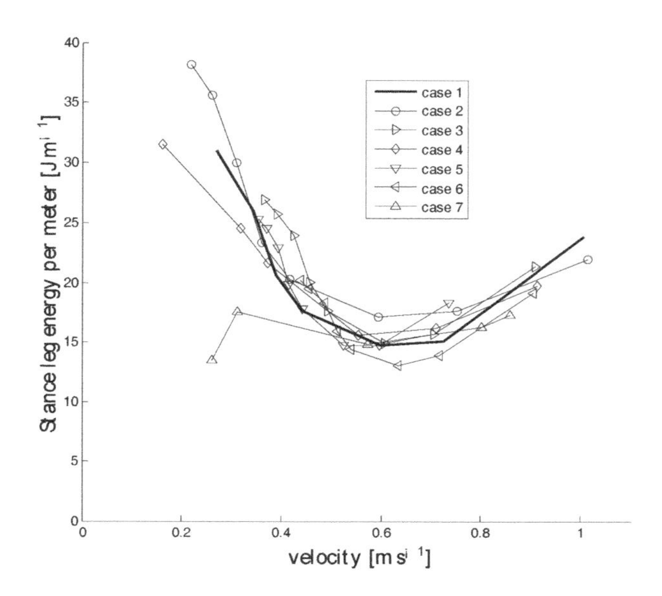

Figure 3.34. *Stance leg energy versus velocity*

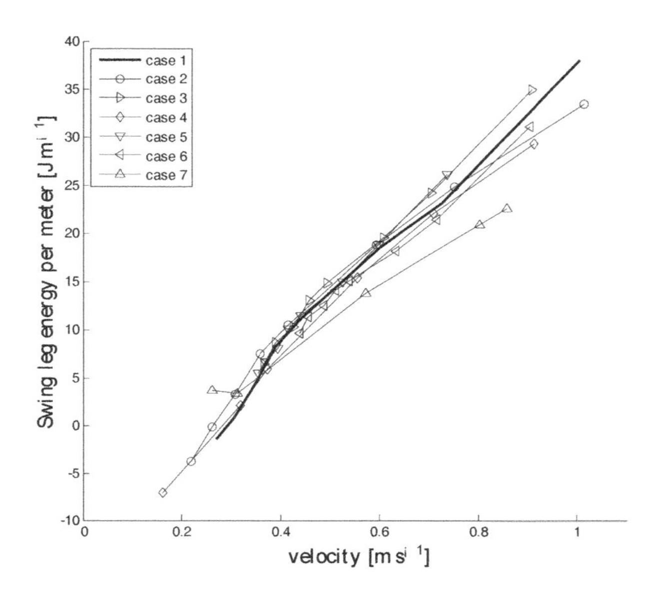

Figure 3.35. *Swing leg energy versus velocity*

We notice that the proportional and derived control actions (Figure 3.32) lead to a variation in the length of steps which is close to proportional velocity functions.

Figure 3.33 clearly shows that the total energy necessary to cover a meter is reduced when we bring the leg masses closer to the rotational axes or when we reduce the robot's total mass. The tibia masses have the greatest influence on the total amount of energy consumed.

For a velocity of 1 m s^{-1} for example, the transfer of 1 kg of tibia mass (Case 2) to that of the thigh helps to reduce the total required energy by 22%. Dividing the leg mass by 2 (Case 7) results in an energy economy of 42%.

Figure 3.34 shows that there is an optimal gait velocity and that it leads to minimum energy consumption.

Figure 3.35 shows that this type of proportional and derived controller slows down the swing leg's forward progression and results in a partial recuperation of the energy received by the stance leg.

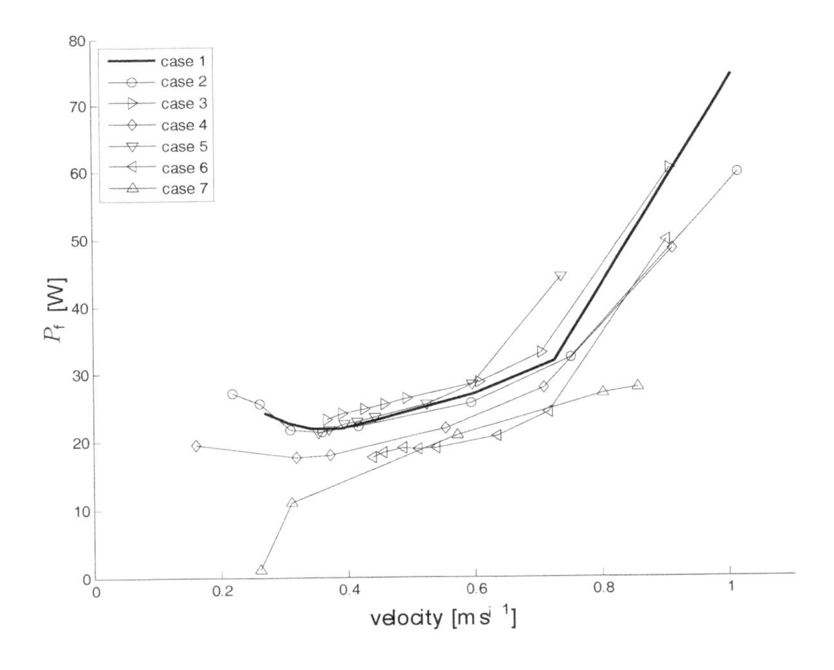

Figure 3.36. *Average power for stance leg*

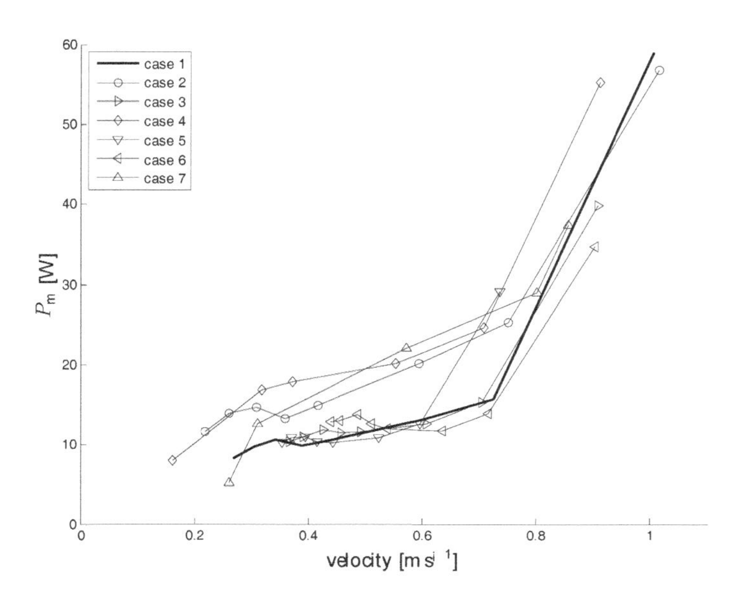

Figure 3.37. *Average power for swing leg*

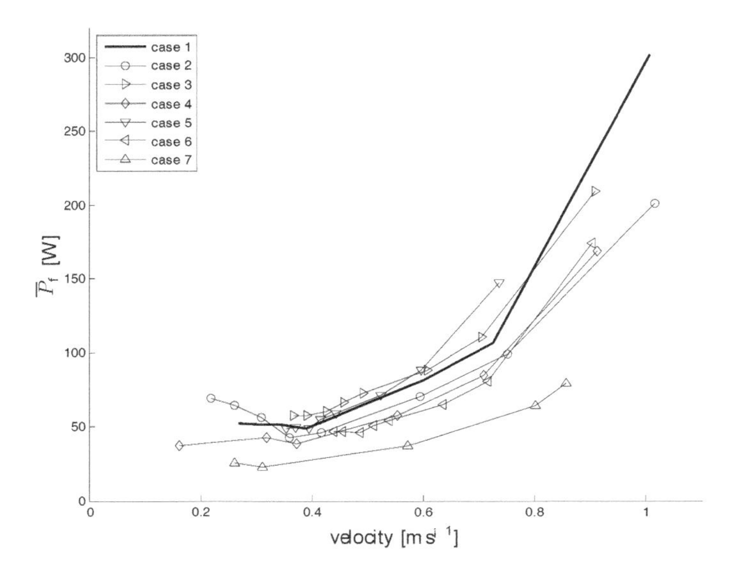

Figure 3.38. *Maximum power for stance leg*

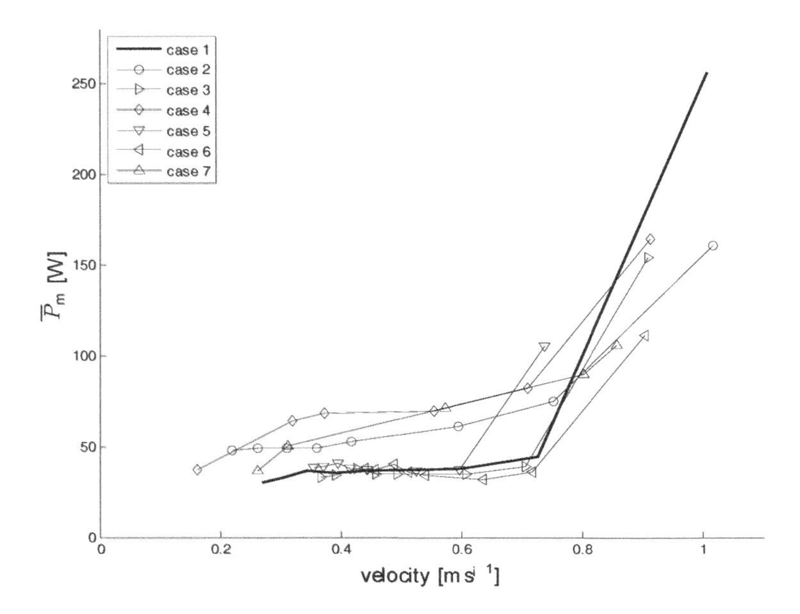

Figure 3.39. *Maximum power for swing leg*

Figures 3.36 to 3.39 give the average and maximum powers necessary for each leg at its actuator. The most unfavorable cases (Cases 3, 5) correspond to making the trunk lighter or reducing its inertia.

The most favorable case (Case 7) corresponds to a reduction of the leg mass. In relation to the nominal values (Case 1) and for a maximum velocity of 1 m s^{-1}, the reduction of tibia mass in order to raise thigh mass (Case 2) reduces the average power by 17% and maximum power by 34%. In turn, the motor mass can be reduced by 40%.

Equally, dividing the leg mass by 2 (Case 7) enables us to divide the average motor mass by 2 and its maximum power by 3. We therefore reduce motor mass by 56% (for power to weight ratio relations, see section 3.3.4).

The model for the five-link robot with spherical feet (see Figure 3.26) is also of interest from an energy consumption point of view.

Figures 3.40 to 3.47 give the results for three different values (2 cm, 5 cm and 10 cm) of foot radius and compare them to the nominal case with a point contact.

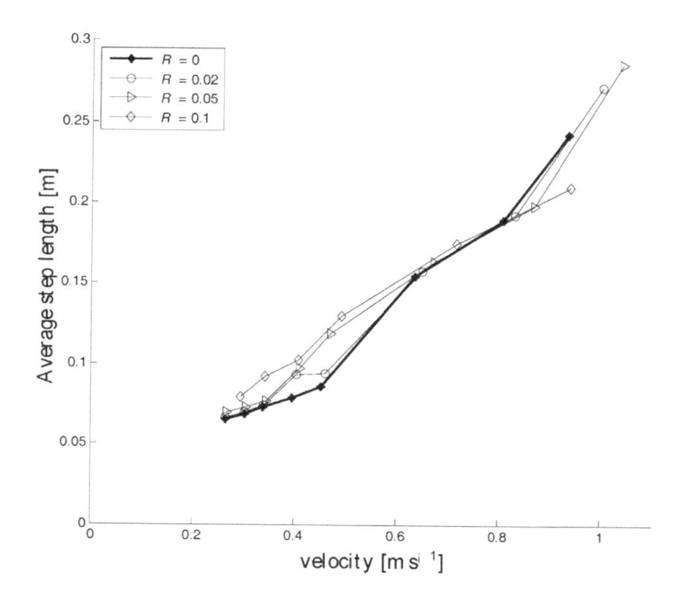

Figure 3.40. *Step lengths versus velocity*

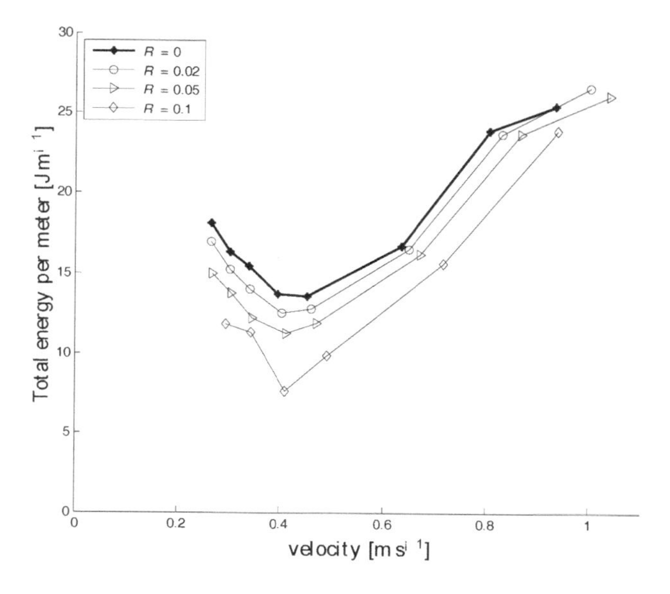

Figure 3.41. *Total energy versus velocity*

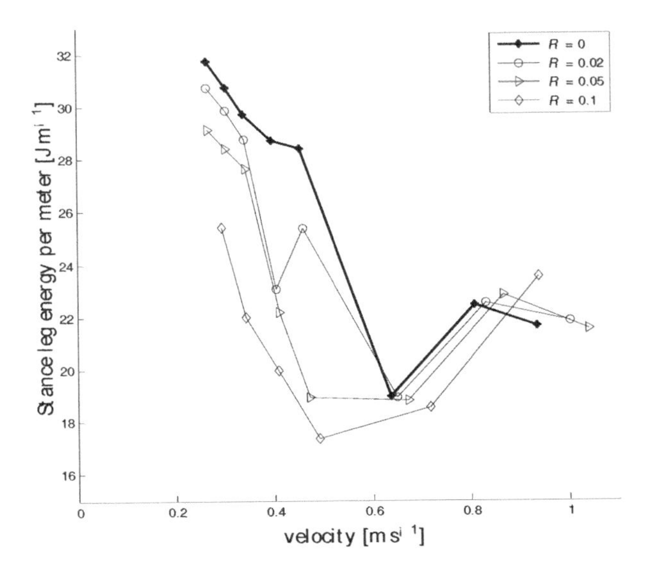

Figure 3.42. *Stance leg energy versus velocity*

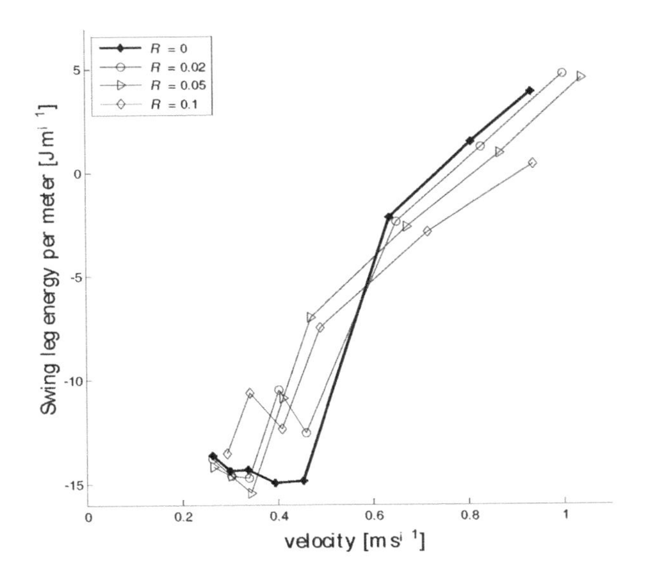

Figure 3.43. *Swing leg energy versus velocity*

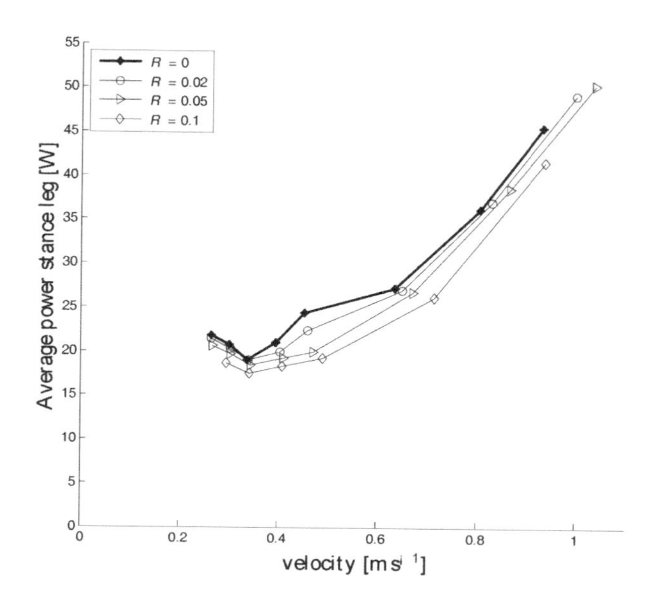

Figure 3.44. *Average power for stance leg*

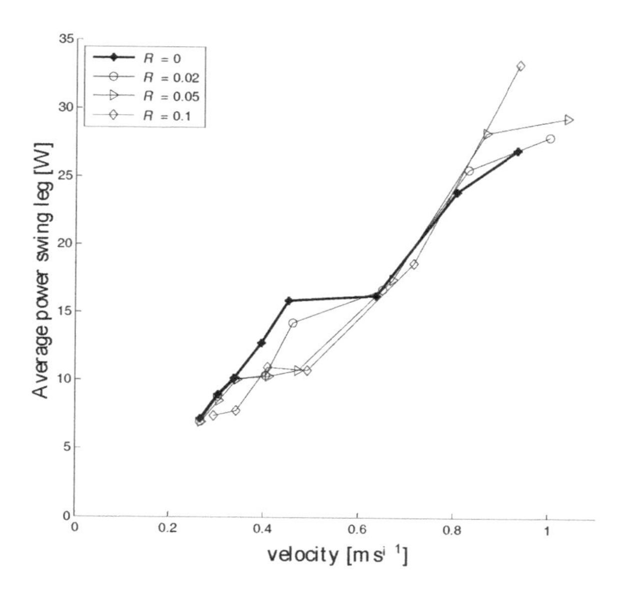

Figure 3.45. *Average power for swing leg*

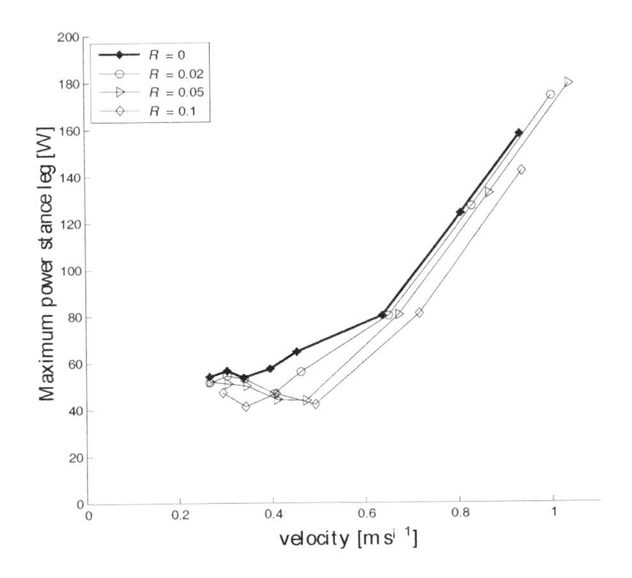

Figure 3.46. *Maximum power for stance leg*

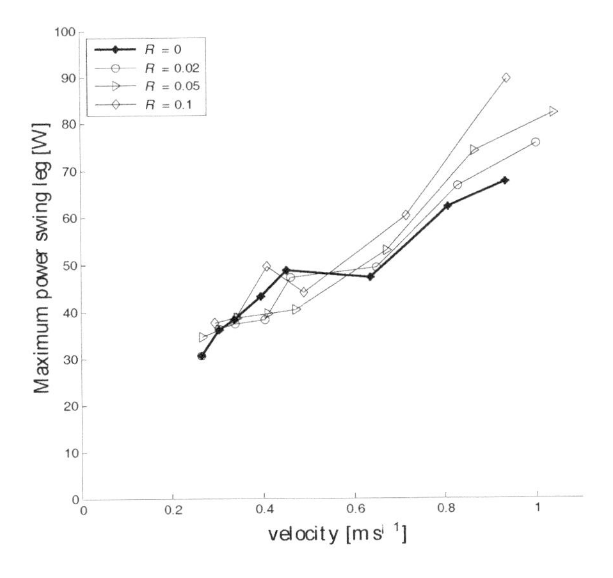

Figure 3.47. *Maximum power for swing leg*

The greatest influence of the foot's radius concerns the total energy consumed per meter traveled, i.e. reduced significantly with radius R. The step length is not greatly affected by varying the radius. Figures 3.42 and 3.43 show that the stance leg always acts as a motor (positive absorbed energy), whereas the swing leg acts as a break for the low velocities (negative absorbed energy). The average power (Figures 3.44 and 3.45) and the maximum power (Figures 3.46 and 3.47) curves show that the increase in radius R enables us to reduce the size of the motors slightly. Nevertheless, it is to be noted here that these results depend on the trajectory profile set-up $\theta_{id}(\alpha)$, $\dot{\theta}_{id}(\alpha)$, i = {1, 2, 3, 4}. Note also that the trajectories and the control laws must be optimized in conjunction with the energy study.

3.3. Mechanical design: the architectures carried out

3.3.1. *The structure of planar robots*

A bipedal robot's main movement is performed in the sagittal plane. In this plane, angular leg deflection has the greatest magnitude. The forward propulsion of the robot's mass center is also important. As a result of this, planar robots which only move in the sagittal plane were created. Biomechanical studies (see Chapter 1) as well as control and stability studies have been the inspiration for the simplest structures. The following kinematic descriptions give an overview of the main structures currently under study. Their fundamental properties and the solutions retained for the creation of the mechanical structure will also be presented.

Planar robots often have a high number of revolute joints with orthogonal axes to the sagittal plane. There are also planar bipeds which have a certain number of DoF with prismatic joints. For the revolute joints, the mechanical design must take all the creation solutions into account in conjunction with the constraints studied in section 3.1. Let us reiterate them briefly:

– a revolute joint between two bodies has to have significant stiffness for all DoF except for the DoF which is actuated;

– the joint mass must be minimal;

– the actuator must be placed as close as possible to the trunk;

– the driving mechanisms must be simple and lightweight.

For robots of planar structure, all the joints with one single DoF have an actuator and driving mechanism at the revolute joint axis. Three main structures result from this:

(a) the actuator axis and the revolute joint axis merge;

(b) the actuator axis is collinear to that of the revolute joint axis; and

(c) the actuator axis is perpendicular to the revolute joint axis.

Figure 3.48 shows, schematically, the three structural solutions listed above. For the knee articulation, solution (b) enables us to shift the motor mass nearer to the trunk. This therefore means that no significant problems are generated.

The third solution is best used for the ankle articulation as it helps to avoid oversized leg profiles. The bevel gear transmission, however, must be carefully carried out.

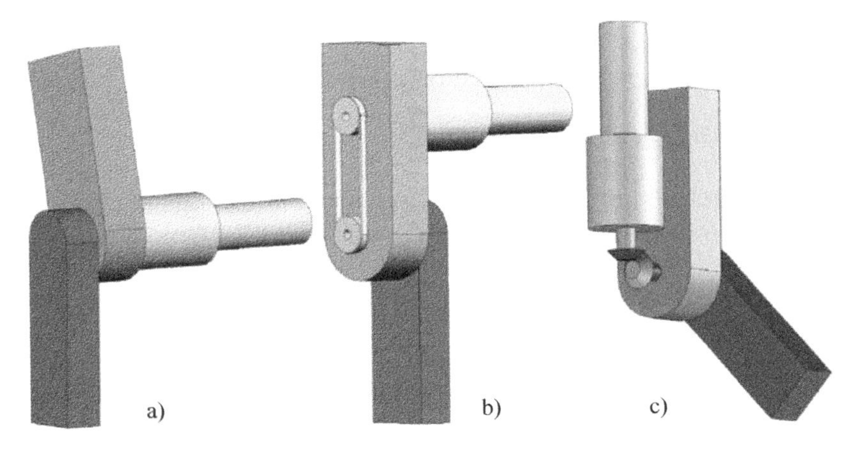

Figure 3.48. *Diagram of the revolute joint principle for planar structures with one DoF*

There are also planar bipedal robots with driving mechanisms which transform the actuator's rotational motion into prismatic motion. Four main structures of this type can be found.

– The actuator axis is collinear to the prismatic joint axis (nut and screw drive).

– The actuator axis is perpendicular to the prismatic joint axis and there is a passive intermediary cylindrical joint (oblong hole).

– The actuator axis is perpendicular to the prismatic joint axis (belt pulley drive or rack and pinion transmission).

– The actuator axis is perpendicular to the prismatic joint axis and there is a passive intermediary revolute joint (crank mechanism).

Figure 3.49 shows, schematically, the four solutions mentioned above. The nut and screw transmission often helps to reduce the size and the bulk of the reducer, and can even eliminate it altogether. This reduces general mass and friction at the joint.

The rack and pinion solution (with no backlash) is difficult to make and is therefore not often used. The last two solutions are interesting when the maximum stroke is small but need to look to limiting the frictional forces.

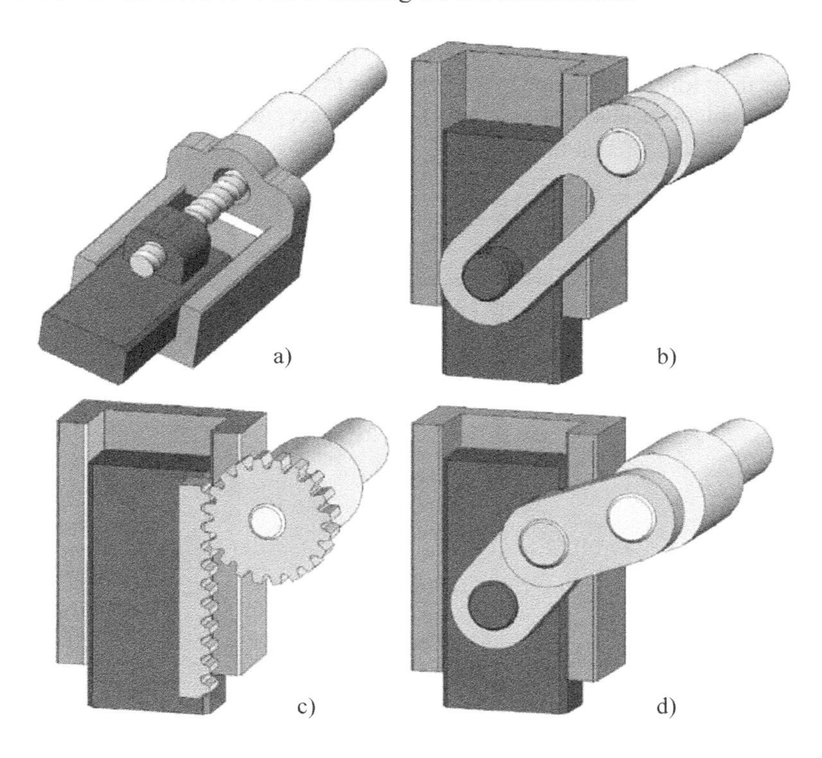

Figure 3.49. *Diagram of the prismatic joint principle for planar structures with one DoF*

In the structures presented in Figure 3.49 we have considered movement transmission without any intermediary parts, apart from the particular case of the solution called the crank mechanism where there is a non-motorized revolute joint between the two parts. There are also bipedal robots which transform the linear actuator movement into a rotational movement. There are also many possible basic structures:

– the actuator axis is mobile in relation to the revolute joint axis;

– the actuator axis is perpendicular to the revolute joint axis;

– the actuator axis is perpendicular to the revolute joint axis and there is a passive intermediary revolute joint (crank mechanism).

These constructions have the advantage of having actuators with a thrust force axis which is collinear to the supporting body's axis. This limits the joint obstruction in volume. An appropriate choice of thrust center for the mobile part enables us to obtain an optimum reduction ratio. Figure 3.50 shows, schematically, the three solutions mentioned above. These solutions allow for bio-mimetic approaches to the design, as their kinematic action mode is similar to that of muscles. It is nevertheless difficult to obtain mechanical solutions with an acceptable efficiency. This means that they are of less interest to us.

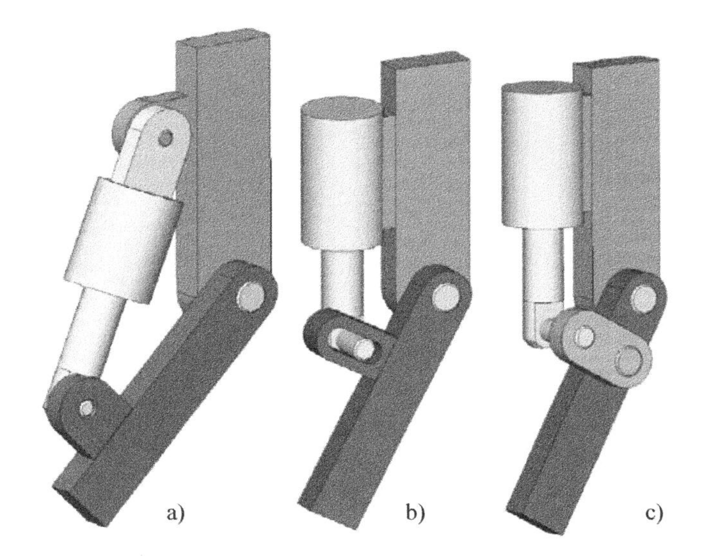

Figure 3.50. *Diagram of the revolute joint principle for planar structures with one DoF and driven by a linear actuator*

3.3.2. *3D robot structures*

For so-called 3D bipedal robots, the main movements are executed in the sagittal and frontal planes. In the sagittal plane, the angular leg deflections are the most significant movements, but there are also movements in the frontal plane. Frontal plane movements mainly serve to keep the robot laterally balanced. Nevertheless, the ankle articulation in the frontal plane often offers a better adaptation to the terrain. The hip articulations ensure possible lateral displacements or ensure that balance is maintained when there are high centrifugal forces (going round bends at high velocities). The structures which have been built are based on biomechanical studies (see Chapter 1). These studies were based on the control and stability studies of these same structures and by a desire to create anthropomorphic structures. The following kinematic descriptions give an overview of the main study cases, giving their fundamental properties, the obtained solutions and the final creation of their mechanical structures.

Figure 3.51 shows three solutions for a 2-DoF spherical joint which can be used for 3D robot structures. Figure 3.51a uses traditional motor reducers which are mounted with their axis in prolongation of the joint axis. This solution often has the disadvantage of a voluminous articulation and a high mass at the articulation levels. Figure 3.51b places the two motors on the higher body which lightens the legs (or reduced the inertia of the leg bodies). This improves the energy consumption, as was previously observed. The main disadvantage of this solution is linked to the difficulty in making the conical gear transmission. Figure 3.51c shows a sophisticated and compact solution, although the inertia of the lower part is not minimal.

Bipedal robots often need DoF which have relatively low angular deflection, especially in the frontal plane. Figure 3.52 gives a solution which is well adapted to this case.

Transmission via an eccentric shaft and oblong hole allows for a reduction of velocity. This facilitates the choice of output reducer for the drive motor. The eccentric shaft and oblong hole mechanism must be adjusted with care, in order to avoid friction. The eccentric shaft and oblong hole mechanism can also be replaced by a crank mechanism.

The most sophisticated bipedal robots need articulations with three DoF between each body. The hip liaison often needs this type of structure. Figure 3.53 shows a traditional solution for this case. The disadvantages of this solution are the relatively high dimension and inertia at joint level.

The combination of the solutions found in Figures 3.51 and 3.52 can be very advantageous for a hip articulation. This is because the DoF in the sagittal plane needs a lot of deflection whereas the rotation of the leg in the frontal plane requires only little deflection.

When the same articulation needs two or three DoF, it is sometimes useful to study a parallel structure. The most pertinent choice criteria are the rigidity of the articulation, the global size and the total mass of the mechanical structure.

Numerous studies on these promising issues for future bipedal and humanoid robots are currently underway [CHE 03, MOR 00, SUG 04].

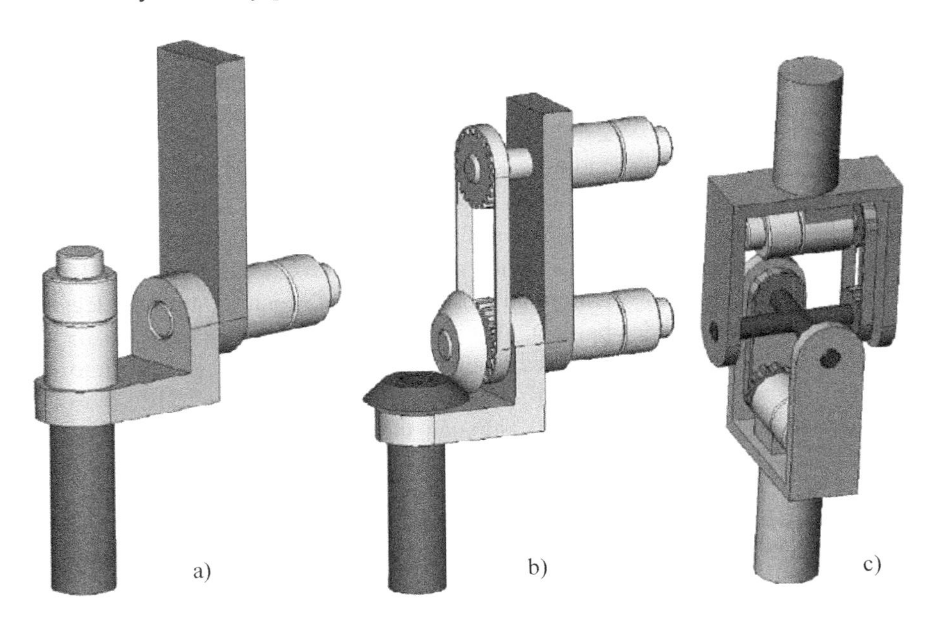

Figure 3.51. *Principle of articulation with a 2-DoF spherical joint and for 3D structures*

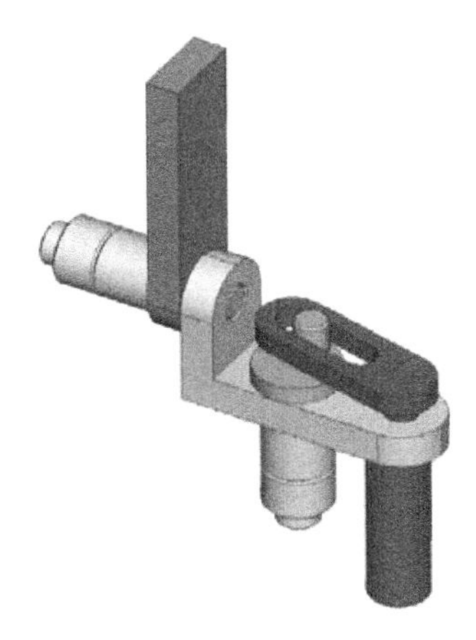

Figure 3.52. *Principle of eccentric articulation for a 2-DoF spherical joint*

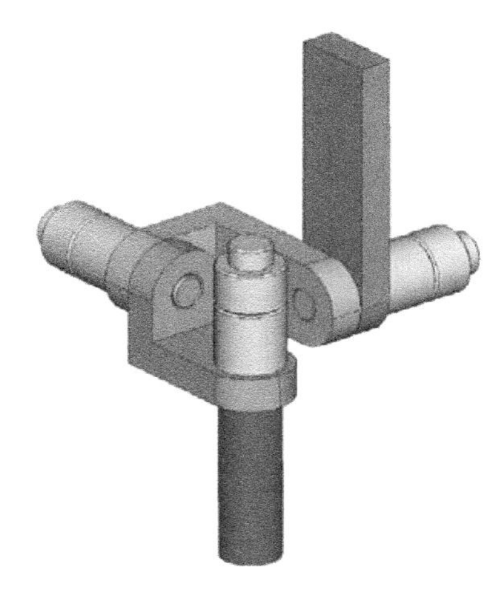

Figure 3.53. *Principle of articulation with a 3-DoF spherical joint for 3D structures*

Figure 3.54 shows a more elaborate solution for a spherical joint with three motorized DoF. This is the result of studies carried out on parallel robot structures. The advantage of this solution is that all the motors on the body are placed above the spherical joint.

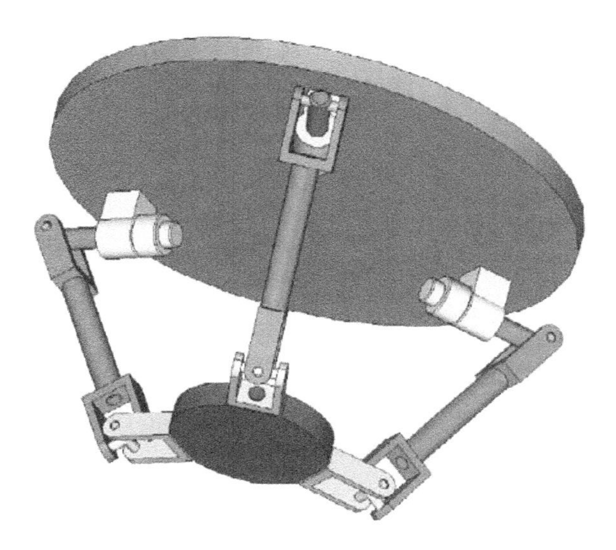

Figure 3.54. *Principle of articulation with a spherical joint and a parallel structure*

The structures with linear actuators can also be used advantageously in the case of an articulation with two or three revolute joints. Figure 3.55 gives an example of this type.

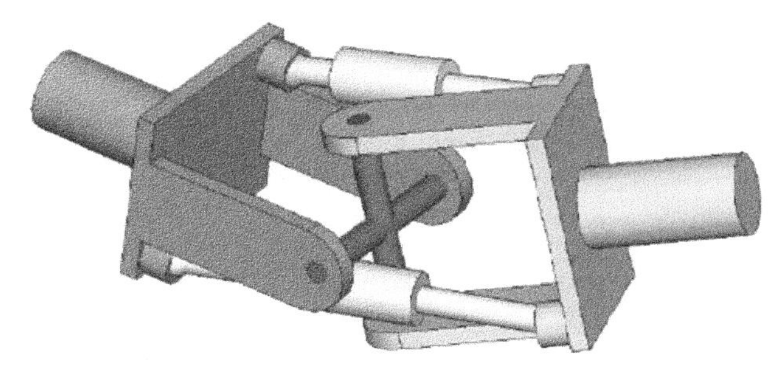

Figure 3.55. *Principle of articulation with a 2-DoF spherical joint and a parallel structure*

The latter solution is used for the BIP robot's ankle articulations. Figure 3.56 shows a photograph of this successful structure.

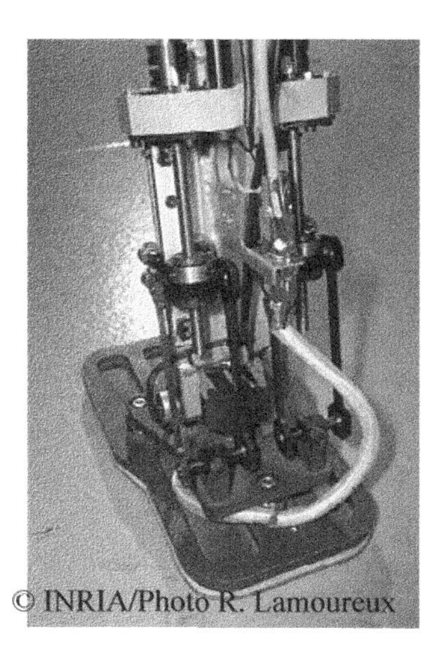

Figure 3.56. *BIP robot's ankle articulation, INRIA Rhône Alpes and LMS Poitiers [SAR 99]*

3.3.3. *Technology of inter-body joints*

Robot inter-body joint technology is vast and depends on many constraint factors which vary for each study case. The most common cases will be discussed here, as will their advantages and disadvantages.

A sleeve bearing is the simplest solution for creating a revolute joint. The two joint bodies are directly in contact via the intermediary of a bushing which reduces friction. This solution has the advantage of having a low mass, especially if the bushing can be made out of a light material. It is nevertheless limited due to the five constrained DoF.

A more sophisticated solution which is commonly used for creating a revolute joint is a ball bearing. Considering the effort level of bipedal robot joints, the use of two four-point-contact radial ball bearings, or two angular-contact O-arrangement ball bearings are very well adapted. The mass is nevertheless often higher than in a sleeve bearing. This is due to the impact efforts of the feet which necessitate greater ball bearing size.

The spherical joint is made either by combining revolute joints mentioned above, or by sliding a spherical extremity into another hollow spherical part. Movement between the two parts is executed by relative slide, which results in a great deal of friction especially when there is also a great amount of joint effort. The advantage of this type of technology is that it is simple; only two parts are needed for the revolute joint with concurrent axes. However, the motorizing of the three DoF is complex.

The prismatic joints are made up of two elements: a trolley and linear unit with one or many rails. The trolley moves prismatically along the linear guides. To reduce friction between the parts, we add a bushing made of material with a low friction coefficient. This is a simple solution but it has many disadvantages: the guiding rectitude is difficult to obtain and the frictional forces remain high, especially when the normal force is also high. In addition, joints with many contact areas often have overconstrained structures, which also increase the frictional forces.

In order to reduce the friction effects, we use prismatic joints with recirculating ball slide linear guide. These linear ball slides have the same advantages as the radial ball bearings in the revolute joints. However, once complete, this solution is bulkier and heavier.

There is one last possible joint type which is based on the theory of contact between solids. The elementary joint between solids is a one-point contact with a plane. By using this idea, we achieve the two joints types shown in Figure 3.57. This type of joint is based on how one body rolls on the surface of another. Many institutes are currently studying how to design robots in this way [SCA 04]. This solution should reduce the joint friction, but further studies are needed to understand how the movements can be transmitted from an actuator placed between these two bodies.

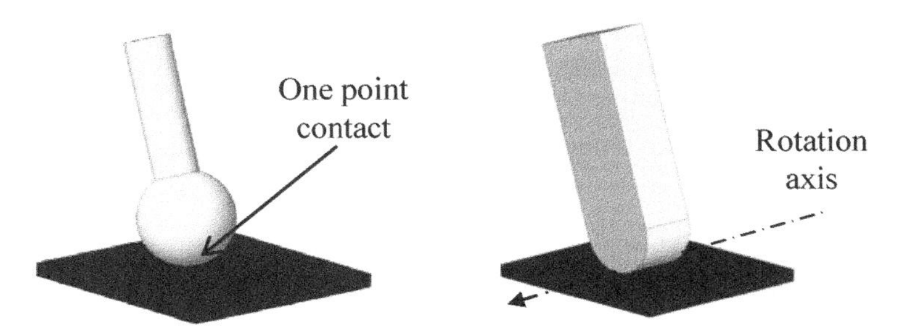

Figure 3.57. *One-point contact and linear contact between two solids*

3.3.4. *Drive technology*

Drive technology is vast and depends on many constraint factors which vary for each study case.

The main objective of a driving system is to modify the kinematic relation either by modifying the type of motion itself (rotation, prismatic or helicoids) or by modifying its support entity (prismatic or rotational axes).

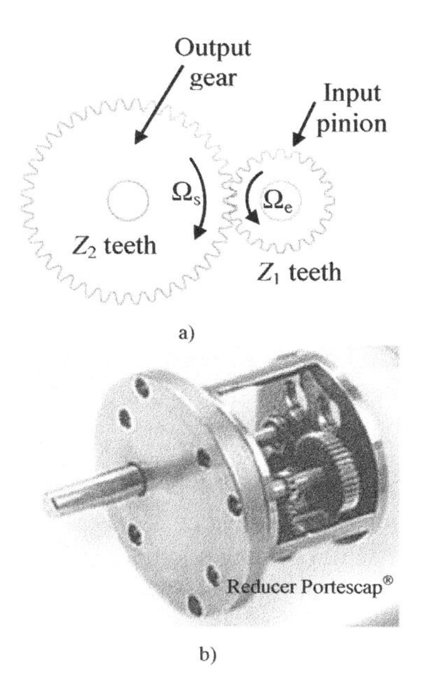

Figure 3.58. *(a) Principle of spur gear reducer [LEB 92]; b) example of a K series reducer made by Portescap, Switzerland*

A bipedal robot's movements are mainly made up of rotational and prismatic movements. We will consider the most common cases here, as well as their advantages and disadvantages.

A spur gear reducer modifies the maximum rotational velocity of a body driven by its joint, and adapts it to the actuator's optimal performance, placed at its input. These spur gear reducers have very good energy efficiency but their reduction ratios are limited to values of between 1/5 and 1.

Figure 3.58 shows the principle of these reducers and gives an example. The input pinion has Z_1 teeth and drives the output gear with Z_2 teeth in the opposite direction. The gear ratio and the relation between input and output velocities are given in equations [3.6] and [3.7]. There are also reducers with many reduction stages (3 to 6 stages). This can yield gear ratios of up to 1/1000.

Spur gear reducers only have a single tooth contact line (see Figure 3.58) per stage, with only one slide contact area with kinematic and Coulomb friction. The efficiency is about 0.9 per stage. As there is only one contact per stage, spur gear reducers have limited output torques. Compared to other types of reducer, one stage occupies a lot of space and also increases the total mass of these reducers. We avoid using them when we require high gear ratios, i.e.

$$\Omega_S = \frac{\Omega_e}{\mathrm{N}} \qquad [3.6]$$

$$\frac{1}{N} = \frac{Z_1}{Z_2} \qquad [3.7]$$

Table 3.2 summarizes the main data collected by most manufacturers.

No. of stages	2	3	4	5
Gear ratio $1/N$	$5 \le N \le 11$	$12 \le N \le 30$	$20 \le N \le 150$	$90 \le N \le 250$
Efficiency	$\cong 0.81$	$\cong 0.73$	$\cong 0.65$	$\cong 0.59$
Maximum input speed (rpm)	8000	8000	8000	8000

Table 3.2. *Main characteristics of spur gear reducers*

Reducers with epicyclic gear trains, which are also called planetary gears, enable us to obtain gear ratios which are almost identical to the spur gear reducers.

However, they are more compact and their principle allows for naturally aligned input and output axes.

Figure 3.59a shows the working principle and Figure 3.59b is a CAD representation of this type of reducer.

There are also other versions of this type of reducer with many reduction stages (1 to 5). This enables us to obtain ratios of up to 1/1000. Figure 3.59c is an example of one of these achievements or realizations.

For this type of reducer, the gear ratio of $1/N$ is obtained with the following formula (application of the Willis formula, fixed crown, and output via the planet gear carrier):

$$\frac{1}{N} = \frac{\Omega_s}{\Omega_e} = -\frac{Z_1}{Z_3} \qquad [3.8]$$

For reducers with many stages, their gear ratio of $1/N$ is obtained by the multiplication of each gear ratio of all stages.

For a smaller occupied volume, we obtain high gear ratios which will enable us to increase the robots' articulation performances (as will be seen in section 3.3).

The torque-to-weight ratio or specific torque for these reducers is high, especially for high gear ratios.

The efficiency is a little lower for reducers with only one stage. If a high gear ratio is necessary, these reducers have a better efficiency than spur gear reducers.

Table 3.3 gives a summary of the main data collected by most manufacturers.

The maximum input velocities mainly depend on the size of the reducer and therefore also on the transmitted torque.

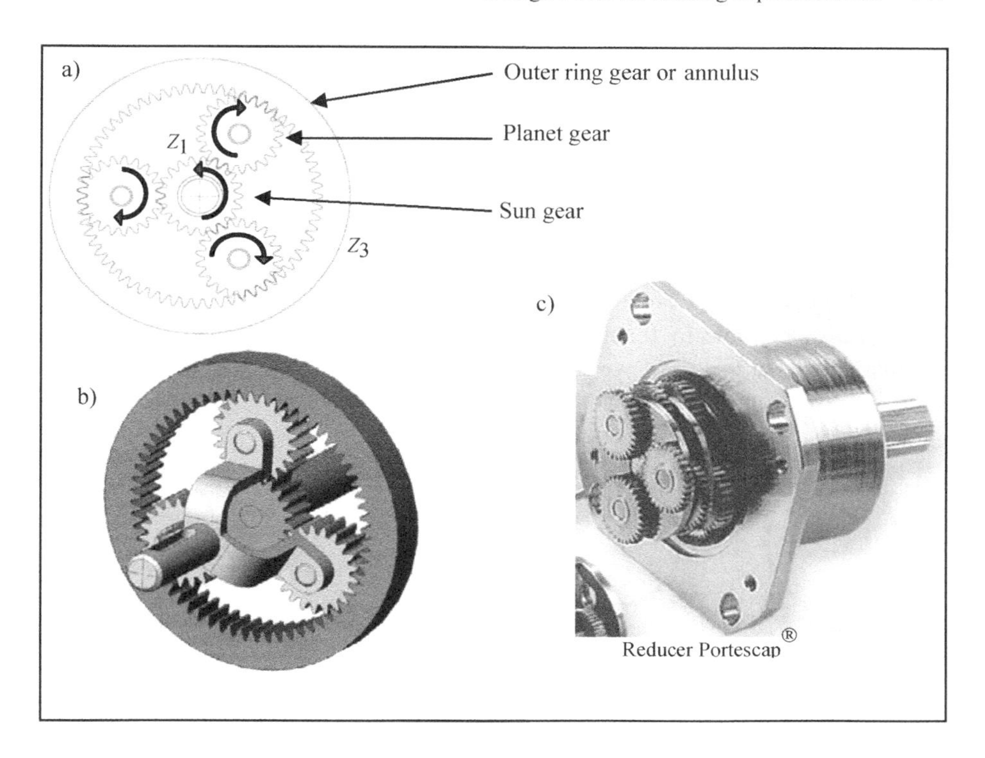

Figure 3.59. *Epicyclic gear reducer: (a) working principle; (b) CAD scheme; (c) example of a creation made by Portescap, Switzerland*

No. of stages	1	2	3	4
Gear ratio $1/N$	$3.5 \leq N \leq 6$	$13 \leq N \leq 33$	$49 \leq N \leq 236$	$180 \leq N \leq 1000$
Efficiency	$\cong 0.85$	$\cong 0.77$	$\cong 0.67$	$\cong 0.59$
Maximum input speed (rpm)	10000	10000	10000	10000

Table 3.3. *Main characteristics of epicyclic or planet gear reducers*

Walter Musser invented the Harmonic Drive reducer in 1955. He used a flexible cup named flexspline which is placed between a circular spline (which is usually fixed) and an elliptical wave generator which is driven by an input shaft. Figure 3.60 shows the working principle for this type of reducer. The flexspline has a tooth number of Z_F which is slightly inferior to that of the circular spline Z_C. When the wave generator makes a complete rotation, the flexspline shifts by the difference in tooth number. We therefore obtain a reduction ratio, which can be very high, from the formula:

$$\frac{1}{N} = \frac{\Omega_s}{\Omega_e} = \frac{Z_C - Z_F}{Z_C} \qquad [3.9]$$

It is also possible to fix the flexspline and to attach the output shaft to the circular spline. In this case, we obtain an inverse rotation with the same reduction ratio. Table 3.4 summarizes the main data concerning this type of reducer.

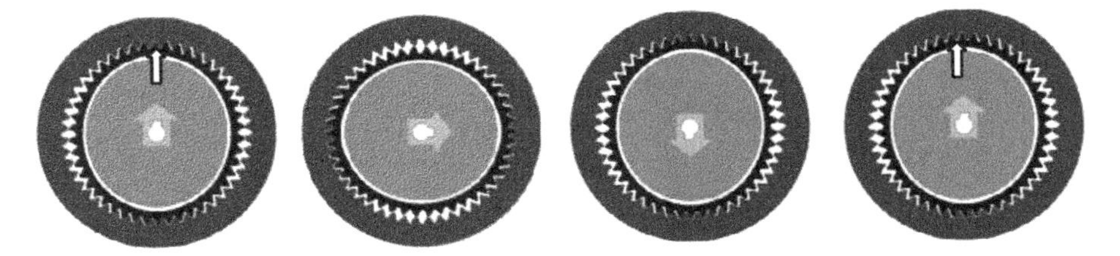

Figure 3.60. *Working principle of a Harmonic Drive® reducer*

Size	11	17	25	40
Reduction ratio $1/N$	$30 \leq N \leq 100$	$30 \leq N \leq 120$	$30 \leq N \leq 160$	$50 \leq N \leq 160$
Efficiency	0.72–0.75	0.70–0.75	0.65–0.75	0.65–0.75
Maximum input speed (rpm)	14000	10000	7500	5600
Maximum repetitive torque (Nm)	4.5–11	16–54	50–176	402–647

Table 3.4. *Main characteristics of Harmonic Drive® reducers*

There are numerous advantages of Harmonic Drive® reducers. They are small in size, they are precise, their backlash is low, they have excellent repeatability, a long operating life, high efficiency, high torsional stiffness for high reduction ratios, a high reduction ratio per stage and a high available torque capacity. Their main disadvantage is their high financial cost.

Figure 3.61 is a photograph of a reducer of this type. These reducers are available in cartridge versions and can be directly integrated into the mechanical robot design structure. There is also a version within a housing which has a lower specific torque.

Figure 3.61. *Example of a Harmonic Drive® reducer*

Choosing a reduction element associated with the drive motor is an essential stage of designing a bipedal robot.

The specific torque enables us to choose the best-adapted technology and, in turn, reduce the amount of mass given over to driving mechanisms.

The gear ratio is chosen depending on the criteria linked to the motor (see section 3.3) and also enables us to choose the reducer which has the best energy efficiency.

Figure 3.62 provides relevant information concerning these two choices. The reducers presented here are the spur gear reducers (labeled "spur"), epicyclic gear reducers made of different materials (labeled "plastic", "metallic", and "ceramic") and the Harmonic Drive® reducers with or without a housing (labeled "harmonic" and "harmonic carter").

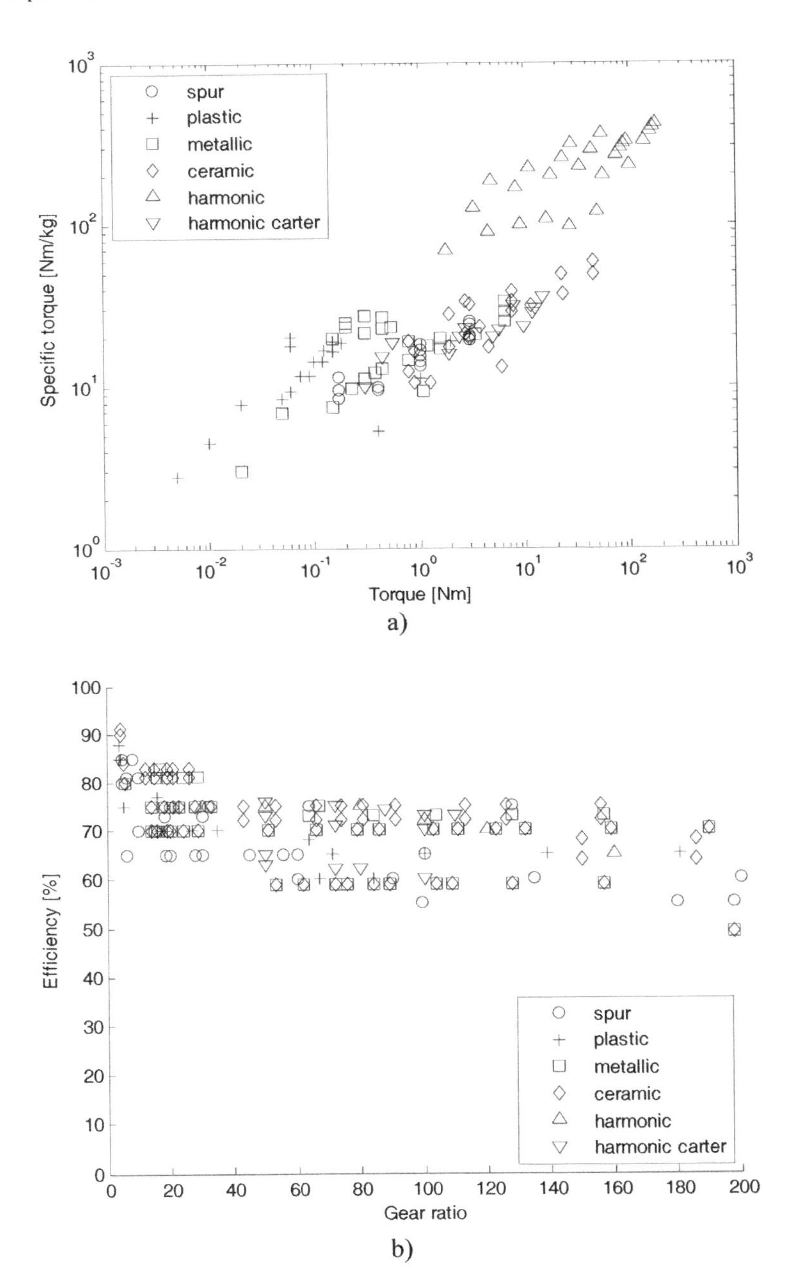

Figure 3.62. *Specific torque in relation to the torque and efficiency depending on the gear ratio for the different reducer technologies*

The non-housing Harmonic Drive® reducers benefit from the reduction in mass due to the absence of a housing. They therefore have the highest specific torque.

We then have epicyclic gear reducers which are very interesting and we chose plastic, metallic (tempered steel) or ceramic materials depending on the output torque required.

Spur gear reducers generally have a lower specific torque because of the housing. They also have multiple idler shafts with guidance. The reducer efficiency is linked to the number of friction contacts against each other.

Spur gear reducers with a low gear ratio have an efficiency of 0.9. Epicyclic gear reducers are well suited for average and high gear ratios.

In the ratio range of 30–160, the Harmonic Drive® reducers have very good efficiencies, especially if the robot design can incorporate models with no housing.

For a certain number of bipedal robot articulation designs, other types of driving mechanisms are necessary. In particular, those with miter gears, crank mechanisms, rack-pinion drives, belt pulley or nut-screw systems as well as Cardan arrangements can be considered.

3.4. Actuators

3.4.1. *Actuator types*

The kinematic descriptions mentioned above show the necessity of motorizing the DoF, either in rotation or prismatically. We therefore need actuators which are capable of generating a torque to drive the revolute joints, or a force for the prismatic displacements. The majority of actuators, which are generally electric, produce a torque and we then use an element of transmission to carry out a kinematic rotation/prismatic change. In the same way, the articulations which have many DoF and, more precisely, revolute joints and two or three axis spherical joints, also need specific driving elements for torque transmission.

Choosing an actuator type is one of the most important stages in legged robot design [CHEV 00]. Actuators can be categorized depending on the source of power supply fluid (electric, hydraulic, pneumatic) and the kinematic motion which is directly generated by the actuator. Table 3.5 shows the main actuator types.

		Power supply fluid		
		Electric	Hydraulic	Pneumatic
Motion	Rotation motion	DC motor brushless motor	Rotating hydraulic cylinder	Pneumatic rotating cylinder
	Prismatic motion	Linear brushless motor	Hydraulic cylinder	Pneumatic cylinder, artificial muscle

Table 3.5. *Actuator types*

There are also other actuator types but their usage in the bipedal robot domain is still restricted. These actuator types have stepper, switched reluctance, synchronous reluctance, AC motors and reluctance linear motors [LAC 99].

Within the electric actuator group, DC brushed motors with permanent magnets and DC brushless motors with permanent magnets are of most interest.

Indeed, they all have a linear relationship between the control variable I_C and the torque (or force) produced along the motor axis. This is an interesting property for the calculated torque controls or for the linearization techniques which are used for walking stabilization.

In the hydraulic actuator group, we can make a distinction between the hydraulic motors and linear cylinders. Hydraulic motors often have limited angular motion (this differentiates them from electric actuators).

Hydraulic actuating control is carried out by an electro-hydraulic element which is called a servo valve.

In hydraulic control, the servo valve is often a sensitive and onerous component. The performances of the servo controls also depend on the quality of the fluid used and which must be purified regularly.

The idea of using an actuator which has the same characteristics and a similar shape to that of muscles has led to the development of pneumatic actuators which are often called artificial muscles.

This type of pneumatic actuator has a behavior which is very different to that of actuators and their usage for bipedal robot control has lead to numerous research studies [PLA 05, SAG 02, TAK 04, TON 00, VAN 05].

Figures 3.63–3.71 show examples of actuators made by different manufacturers.

Figure 3.63. *DC motor with Samarium-Cobalt magnets*

Figure 3.64. *Flat motor with Alnico magnets manufactured by Parker®*

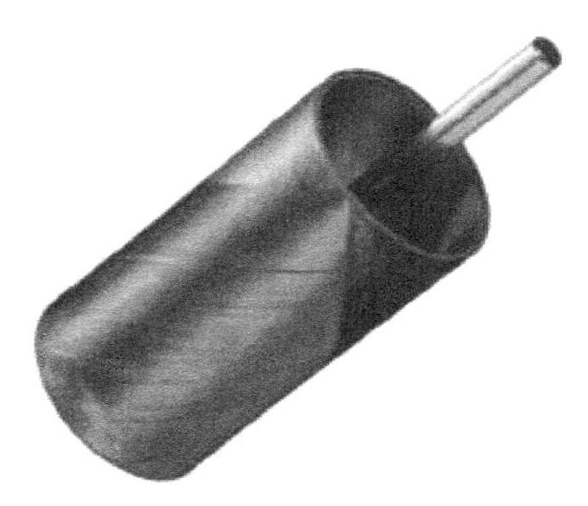

Figure 3.65. *Armature of a non-iron motor*

Figure 3.66. *DC brushless motor with Neodyme-Iron-Bore magnets, made by Maxon®*

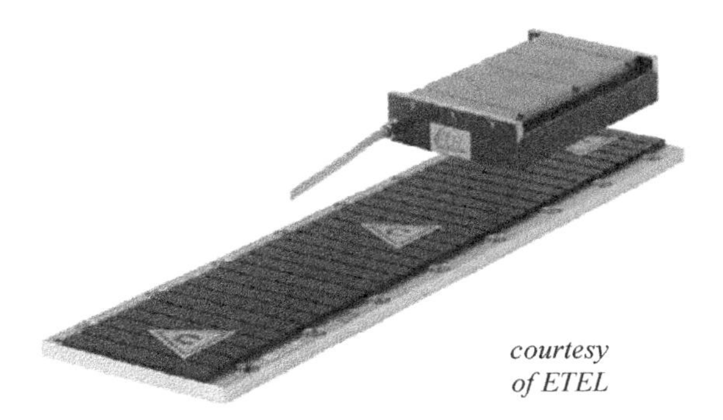
courtesy of ETEL

Figure 3.67. *Linear synchronous motor made by Etel®*

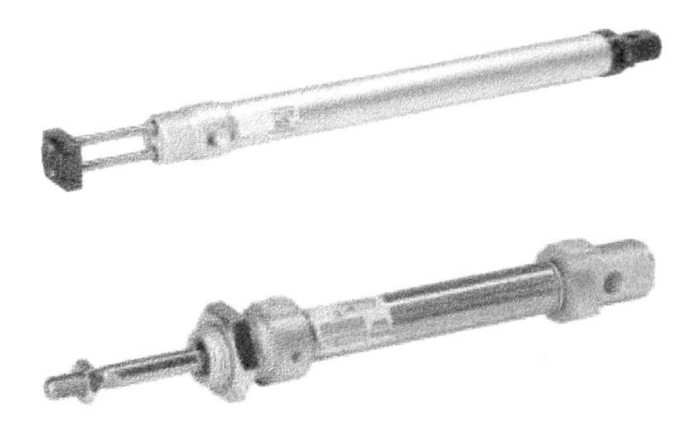

Figure 3.68. *Pneumatic cylinders made by Hoerbiger-Origa®*

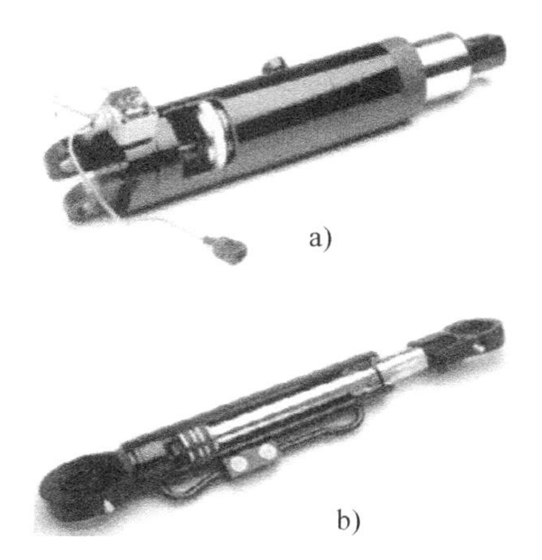

Figure 3.69. *Hydraulic cylinders made by Serta®*

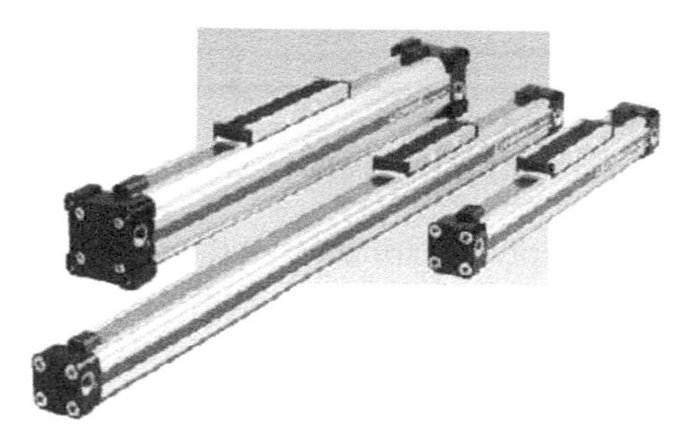

Figure 3.70. *Pneumatic plunger cylinders made by Hoerbiger-Origa®*

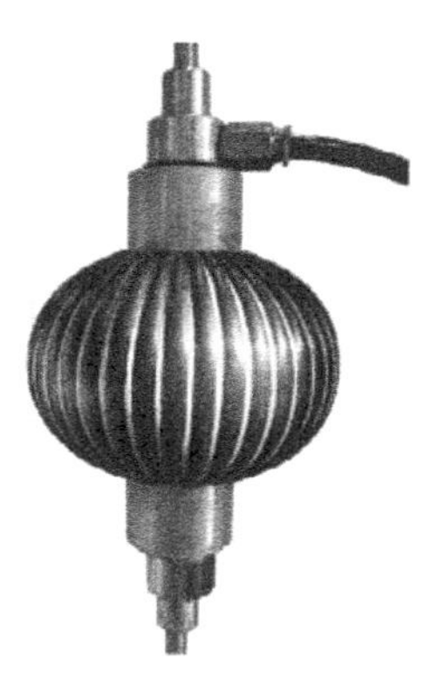

Figure 3.71. *Artificial muscle*

3.4.2. *Characteristics of electric actuators*

Electric actuators are often used in robotics. This is because this solves the problem of energy storage on an autonomous mobile system, i.e. a bipedal robot.

Figure 3.72 shows the general organization for the distribution of energy for such a robot.

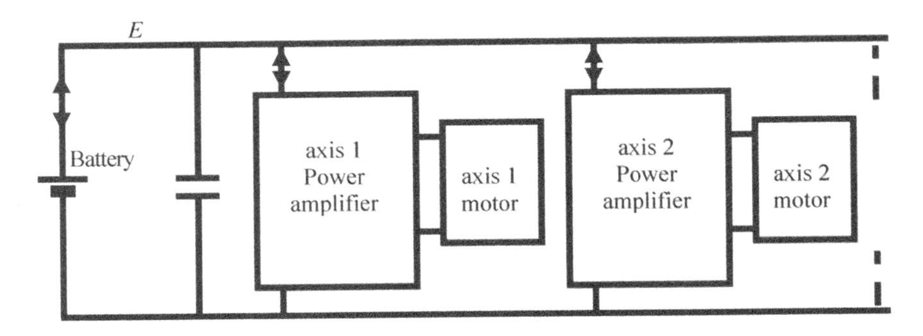

Figure 3.72. *Distribution of electric energy*

The power converters supply the armature of the machine (or each phase for the synchronous motors). They must be bi-directional for the current and the voltage. This encourages the use of a full bridge structure which has four power switches controlled by PWM (pulse-width modulation). Figure 3.73 shows the traditional structures used for DC motors (DC) and for synchronous motors (SM).

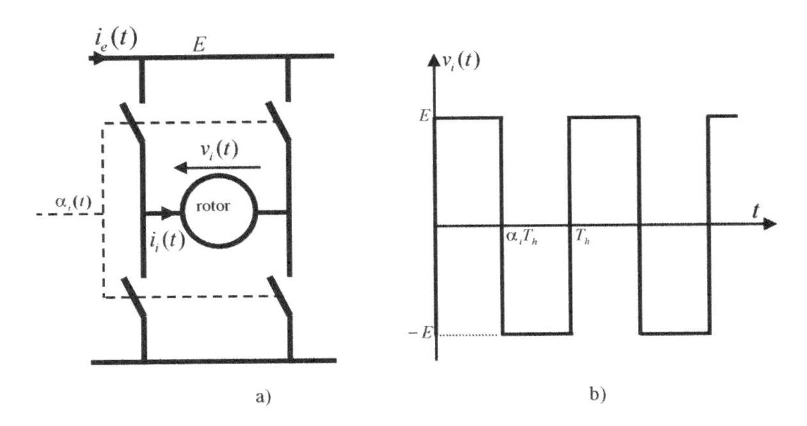

Figure 3.73. *Full bridge converter for supply a DC motor and armature voltage gait for a cyclic ratio of 0.5*

The switches (usually power transistors) alternate to supply the armature with positive voltage E and negative voltage $-E$ (E being a constant power voltage). The commutation occurs at instants $\alpha_i T_h$ and T_h. The switching frequency is defined by $f_h = 1/T_h$ and the cyclic ratio is defined by α_i ($0 < \alpha_i < 1$). The DC motor produces a torque which is proportional to the average value of the armature current. The modification of the cyclic ratio enables us to modify the average voltage at the armature and in turn, control the velocity of the motor. When the profile of the cyclic ratio $\alpha_i(t)$ is slow in relation to the switching period T_h (bandwidth of $\alpha_i(t)$ which is significantly lower than f_h), we can obtain average voltage values of $\overline{v}_i(t)$ and of current $\overline{i}_e(t)$ per switching period, with the following relations:

$$\overline{v}_i(t) = \left(2\alpha_i(t) - 1\right) E \qquad [3.10]$$

$$\overline{i}_e(t) = \left(2\alpha_i(t) - 1\right)\, i_i(t) \qquad [3.11]$$

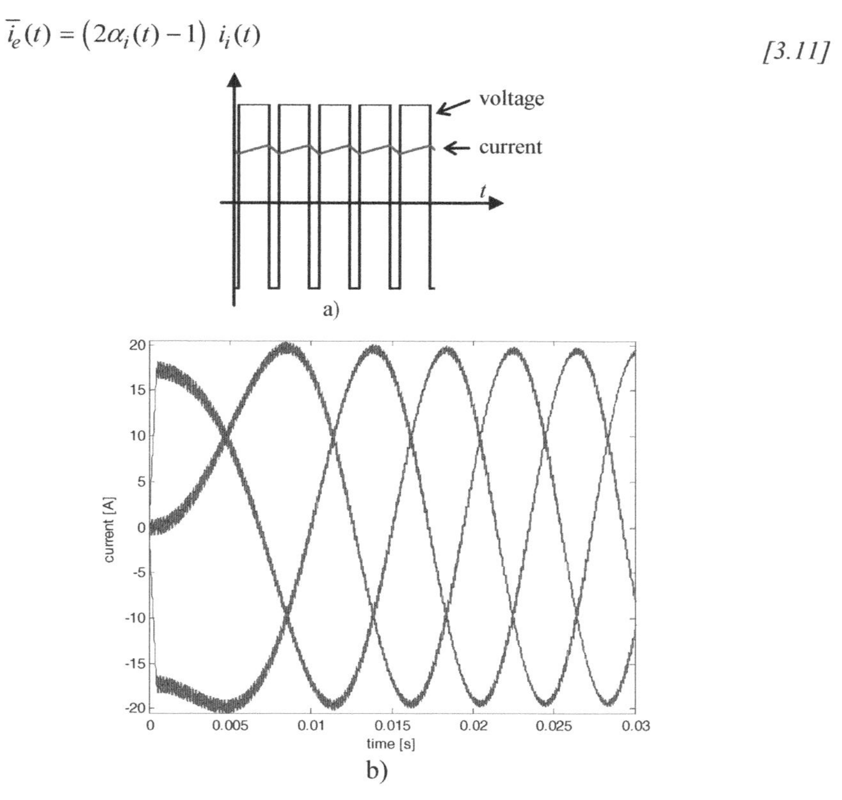

Figure 3.74. *(a) Voltage and current gait of a DC motor armature and (b) shape of tri-phases currents during the start-up of a synchronous motor*

For a synchronous motor, the plotted profile of $\alpha_i(t)$ must enable us to create phase supply currents of sinusoidal shape.

Figure 3.74 gives the typical current shapes found in DC and SM motors. A phase-current control loop enables us to control the armature current shape for the DC motor and the phase-currents for the SM motor. Figures 3.75 and 3.76 show the typical structures for these control loops.

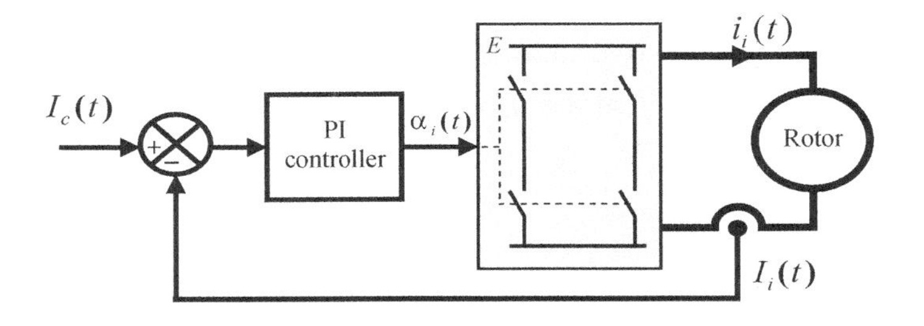

Figure 3.75. *Control loop structure of a DC motor*

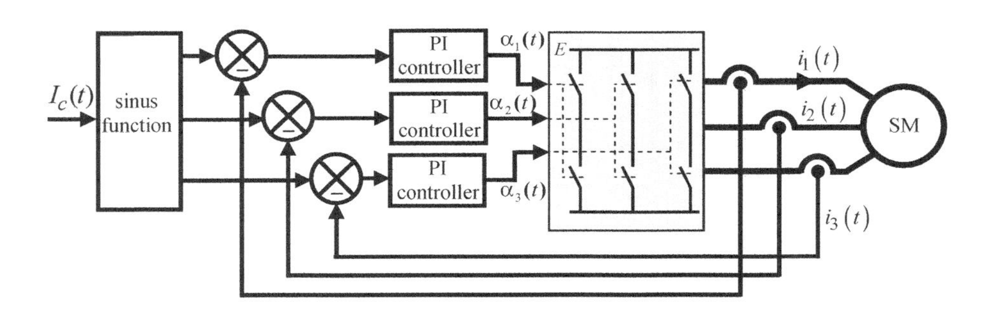

Figure 3.76. *Control loop structure of SM phase currents*

The proportional and integral current control loop are adjusted with high gains, which enables us to obtain high bandwidths. For the DC motor and for the variations of the desired $I_c(t)$ currents, which are slow in relation to the frequency f_h, the average armature current value $\overline{i_i}(t)$ is therefore equal to:

$$I_i(t) = \overline{i_i}(t) = I_c(t) \quad [3.12]$$

For the synchronous motor and for the variations of the reference value $I_c(t)$ which are slow in relation to the frequency f_h, the effective value of the fundamental phase-current $I_i(t)$ is therefore equal to:

$$I_i(t) = I_c(t), \quad i = \{ 1, 2, 3 \} \qquad [3.13]$$

Equations [3.12] and [3.13] enable us to deduce the fundamental property of the pulse width modulation (PWM) control for these electric motors, in short:

$$C_m(t) = k_e I_c(t) \qquad [3.14]$$

or

$$F_m(t) = k_e I_c(t) \qquad [3.15]$$

with C_m as the motor torque and k_e the torque constant for the global servomotor system. The driving force $F_m(t)$ is defined in the case of linear motors.

For a DC motor, the coefficient k_e is given by the manufacturer. It is guaranteed to be precise and is stable over time. In certain cases, it can be dependent on the motor's working temperature and can therefore vary (by a variation of up to 10%). For the SM motor, the coefficient k_e depends on the adjustment of the control loop and the motor's parameters, such as the armature's magnets. As for a DC motor, the coefficient k_e depends on the SM's working temperature and can therefore vary. The phase-current control loop enables the synchronous motor to have a torque characteristic which is identical to that of the DC motor. This property is the origin of the name given to the *brushless* DC motor.

Current control loop precision performances are obtained as long as there is no control saturation. This condition requires that the maximum armature winding voltage values remain lower than the converter supply E. We therefore have to fulfill the following inequality for the DC motor:

$$E_i(t) = k_e \Omega(t) < E \qquad [3.16]$$

For an SM motor, the condition of the electromotive forces becomes:

$$E_i(t)\sqrt{2} = \frac{2}{3} k_e \; \Omega(t) < E \qquad [3.17]$$

The PWM control inevitably leads to slight current variations of a triangular shape around the reference value of the control loop. These variations induce supplementary Joule losses in the windings. We therefore introduce a shape factor coefficient to take into account this extra heat and a downgrading coefficient when choosing the motor type.

3.4.3. *Elements of choice for robotic actuators*

The choice elements for electric actuators are linked to the particular usage of these actuators within the context of legged robotics and to the specific conditions of how electric motors work. We must first establish limited working conditions for the actuators and, in particular, the four main limited static working values which are: the peak torque, the instantaneous power limit, the maximum velocity limit and the limited working temperature.

An electric actuator has four main parts: the rotor, the stator, the airgap and the rotating or prismatic guidance. The principle of conversion of an electric actuator is based on the interaction of the fixed magnetic field created by the currents circulating in the windings.

In the case of DC motors, the fixed field is linked to the stator and the currents circulate in the rotor's armature windings. The current must be alternating in the armature's windings and have a frequency which is in phase with the rotational frequency. These alternative currents are obtained by the mechanical split ring commutator.

In the case of synchronous motors, the fixed field is linked to the rotor which, by turning, induces alternative voltages in the stator windings. Alternative currents must therefore supply the stator winding.

For actuators used in robotics and which require the best dynamic characteristics, the magnetic field of the rotor is created by permanent magnets. The magnetic curve of DC motor magnets is represented in Figure 3.77 and shows the plotted profile of the magnets' induction B in relation to the magnetic field H. When putting the motor together, the provider establishes a normal profile zone for the magnets' magnetic field. The magnetic induction of the magnets therefore always remains close to the nominal value B_N. When the motor's working range remains in this working zone, the characteristics of the motor are those which have been envisaged by the provider. If we then impose too great a current $I > I_{\lim}$ in the armature windings, the working range shifts to the zone marked D in Figure 3.77. This is called demagnetizing the machine. Indeed, when the current returns to its nominal value, the magnets' working range then becomes point E and the curve shows that the residual induction of the magnets B_E is lower than its nominal value B_N. This demagnetizing effect is permanent and the motor is irremediably damaged. There is an identical phenomenon with the synchronous motors.

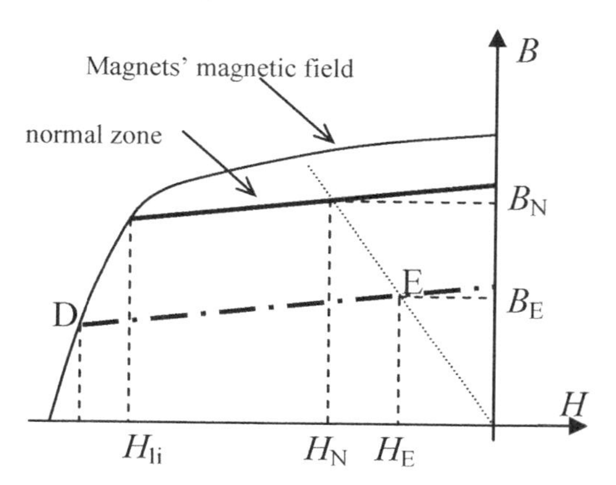

Figure 3.77. *Magnetic characteristics of the magnets of a DC motor*

Limitations in velocity for electric motors are mainly due to mechanical considerations (or phenomena). The first phenomenon is due to the centrifugal forces which are exerted on the rotor's mechanical parts. These centrifugal forces tend to break up the rotor and therefore exert deformation and traction constraints on all the rotor parts, and especially on the parts mounted on the surface. A DC motor's biggest constraint is exerted on the split ring of the commutator, which is mounted on the surface. For an SM motor, the constraint is exerted on the magnets. The machine's velocity must therefore remain lower than a limited value of Ω_{lim}.

There are two types of power limitation. The first type is linked to thermal phenomena. Indeed, the mechanical and electrical losses must be removed through the external surfaces of the machine (convection) and through the circulation of air (transportation of calories). This condition of thermal equilibrium leads to the characteristic of nominal power which is expressed by a curve of a maximum admissible permanent torque. The circulation of air around the rotors greatly depends on the rotational velocity and the machine design. The permanent admissible torque, therefore, has an expression which varies with the velocity and the motor range. For DC motors, the second limitation is linked to the problems of current commutation at the commutator, which imposes a maximum power P_{lim} admissible to the motor. In the velocity-torque plane of Figure 3.78, this limitation is expressed by a limited hyperbole which is framed by P_{lim}.

Figure 3.78 shows all of these limited values for a DC motor in a single graph.

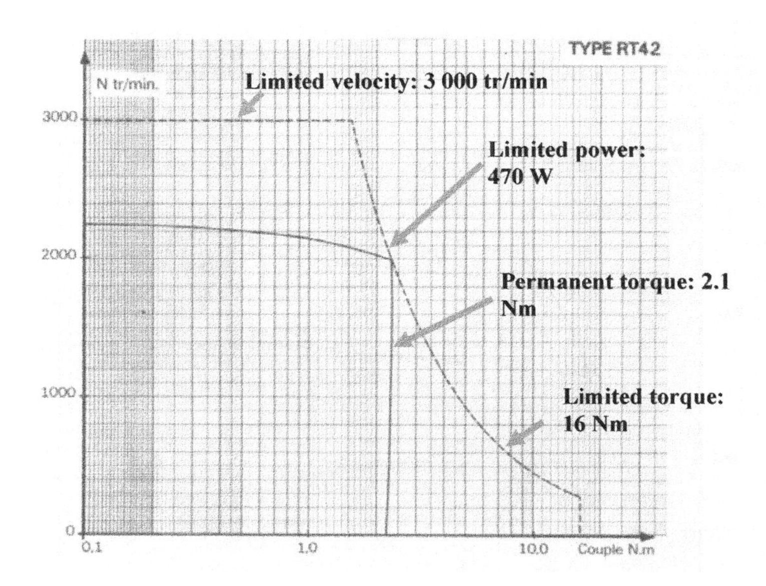

Figure 3.78. *Static and intermittent working limits of a DC motor: the numerical values correspond to an RT42 motor with a SmCo magnet made by Parker®*

For a synchronous motor, the same physical principles are involved in the limitations of the static values. There are, however, no problems linked to the current commutations. The instantaneous maximum power motor limitation is now linked to the phenomenon of the demagnetization of the magnets and to the limitation of the maximum voltage per phase. Figure 3.79 is a typical diagram of velocity–torque. It is necessary to check, when choosing an actuator, that the extreme working area/point for the robot's walking gait (or any other gait) respects at all instants the three limited conditions of velocity, power and torque and that the permanent working function respects the thermal limits imposed by the permanent admissible torque curve.

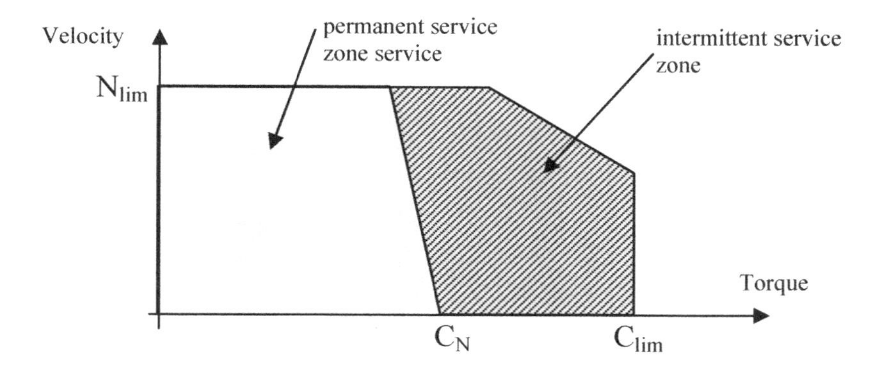

Figure 3.79. *Permanent (united zone) and intermittent (chopped zone) working limits of a synchronous motor*

3.4.4. *Comparing actuator performances*

A robot designer is often confronted with a multitude of choices linked to the desired performances and to the technological constraints of these choices. It is therefore useful to have general design rules to help guide the designer. Performances of actuators with no driving load are often compared. This approach is practical and enables manufacturers to put forward comparative elements without knowing any mechanical characteristics of the system to be controlled. The main comparative elements are the mass performances (specific torque and power to mass ratio), energy performances (efficiency and operating life), dynamic performances (inertia, impulsive acceleration) and the performances of electromagnetic design (time constant and quality coefficient).

Figures 3.80–3.85 give the general comparative elements of numerous manufacturers and enable us to select the most interesting motor technologies. These figures show the category types for DC motors: motors with ferrite magnets, flat motors, motors with rare earth magnets (samarium-cobalt, neodymium-iron-boron) and motors with no iron at the rotor. For brushless motors there are synchronous motors with rare earth magnets (samarium-cobalt, neodymium-iron-boron).

Also to be found in Figures 3.80–3.85 are DC motors called pancake motors. They have a great number of pole pairs and are meant to work on a slow rotational velocity, therefore with no reducer.

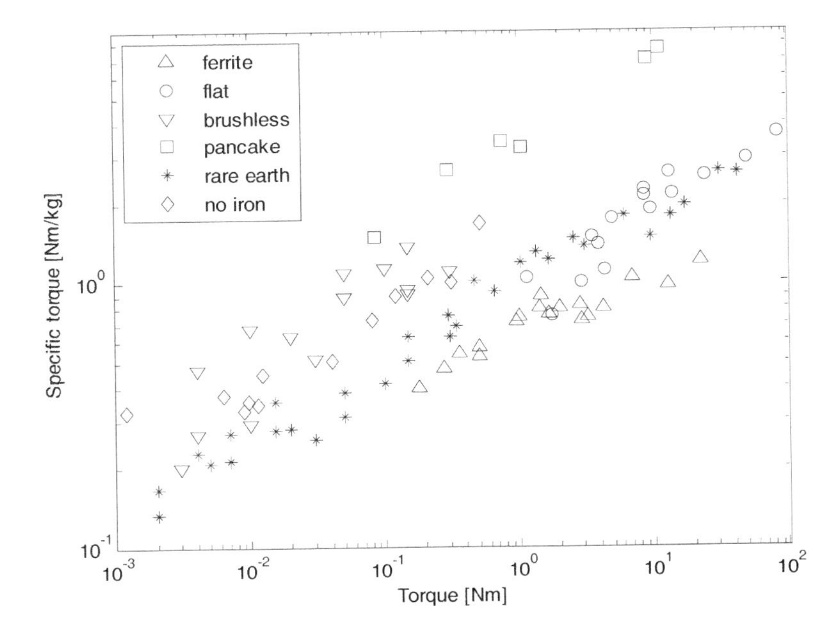

Figure 3.80. *Specific torque in relation to torque*

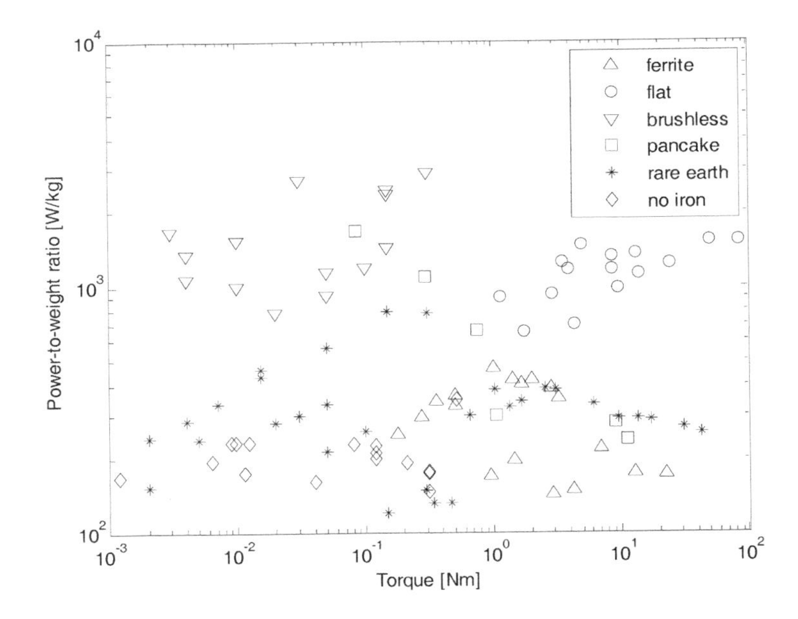

Figure 3.81. *Power-to-weight ratio in relation to torque*

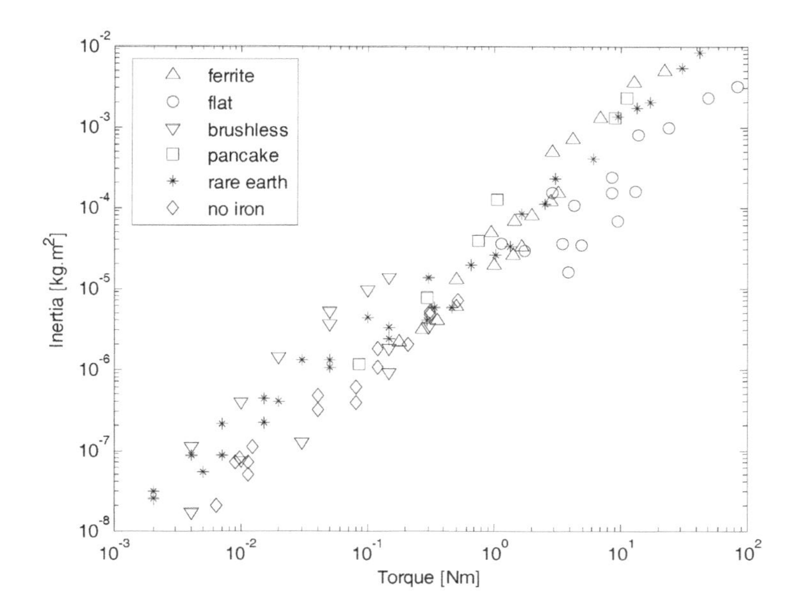

Figure 3.82. *Rotor inertia in relation to torque*

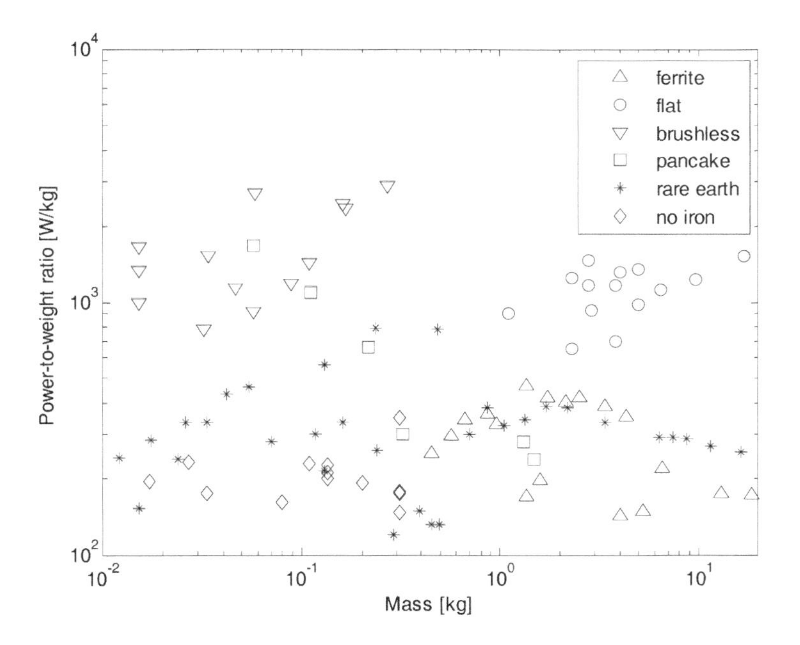

Figure 3.83. *Power-to-weight ratio in relation to mass*

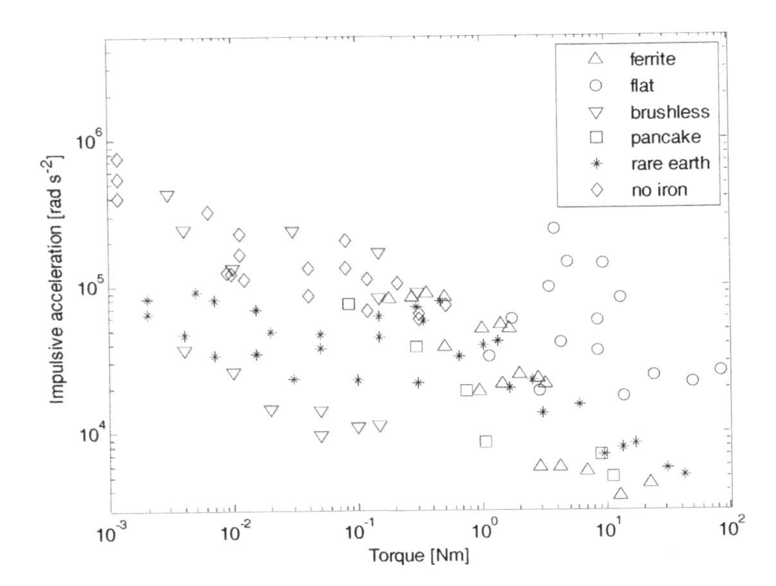

Figure 3.84. *Impulsive acceleration in relation to torque*

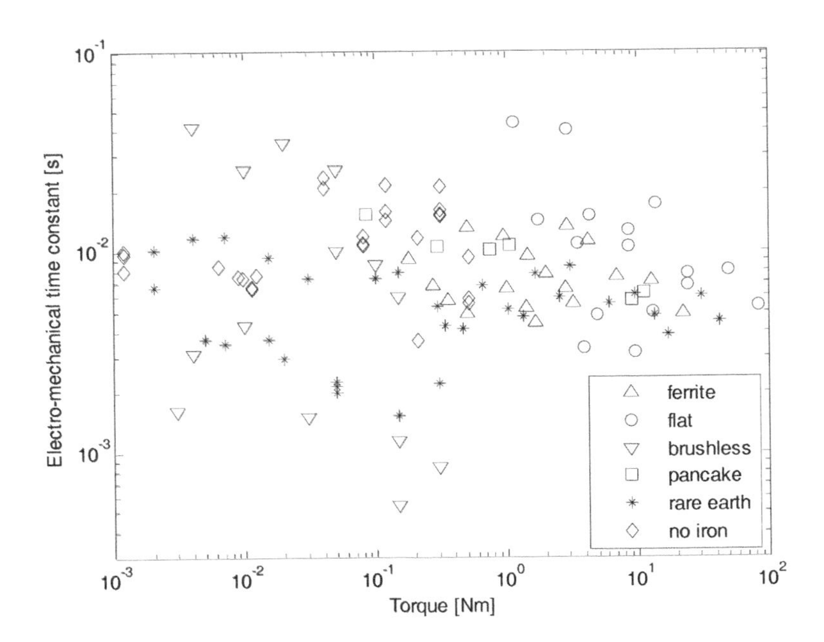

Figure 3.85. *Electro-mechanical time constant, dependent on torque*

Figure 3.80 gives the specific torque in relation to the nominal torque for different types of motor technology. We notice that, for a given nominal torque, the pancake and the synchronous motors have the best specific torques. In second place, we have the motors with no iron at the rotor (but these are limited to relatively low nominal torques). Next best are DC motors with rare earth magnets, the flat DC motors, and finally the DC motors with ferrite magnets. Figure 3.81 gives power-to-weight ratio in relation to the nominal torque. Synchronous and flat DC motors have the highest power-to-weight ratios. This means that when they are associated with a well-adapted reducer, these motors enable us to obtain the lowest actuator mass for a given application. The motors with rare earth magnets and the pancake motors also have, for certain applications, interesting power-to-weight ratios. Figure 3.83 also gives the power-to-weight ratio expressed relative to the motor mass.

Figures 3.82 and 3.84 give the dynamic performances for the different types of motor technology relative to the nominal torque. Here we aim to have the lowest inertia possible in order to accelerate rapidly and to stock as little kinematic energy as possible in the motor. The non-iron and the flat motors (just the motors) have the best dynamic performances. Next best are rare earth magnets and the pancake motors. The performances of synchronous motors vary greatly depending on the manufacturer. Motors with ferrite magnets have the highest rotor inertia.

Figure 3.85 gives the electro-mechanical time constant for just the motor. Good control loop performances can be reached when the motor has a low electro-mechanical time constant. With a view to this end, it is therefore interesting to choose certain synchronous motors, DC motors with rare earth magnets and certain motors with no iron at the rotor. Flat motors should not be used, however.

When building a bipedal robot, we must take into consideration the fact that it must have autonomous energy. For a given battery mass, it must also be capable of covering the greatest distance possible without requiring to be recharged. The motor efficiency and the driving elements are therefore very important criteria.

The motor efficiency is defined in the formula:

$$\eta = \frac{P_m}{P_a} \quad [3.18]$$

where P_m is the mechanical power given to the motor shaft and P_a is the absorbed electric power.

It is easier to determine the efficiency by expressing the power in relation to the electro-magnetic power P_e transmitted in the airgap. We therefore obtain:

$$\eta = \frac{P_e - C_f \Omega_m}{P_e + R I_i^2} \qquad [3.19]$$

where C_f is the mechanical friction torque, Ω_m is the rotational velocity of the motor, R is the armature resistance and I_i is the absorbed current.

The efficiency can only be expressed according to the velocity and the torque motor C_m through the formula:

$$\eta = \frac{\left(C_m - C_f\right) \Omega_m}{C_m \Omega_m + \frac{R}{k_e^2} C_m^2} \qquad [3.20]$$

This formula enables us to draw the constant efficiency curves seen in Figure 3.86. We notice that the maximum efficiency is situated in the surrounding area of the nominal torque and the maximum motor velocity.

This observation is very important as it shows that it is necessary to associate a mechanical transmission with an electric motor (and in particular a reducer) when the maximum velocity is not adapted to that desired for the robot's articulation.

Figure 3.87 shows the efficiency values for the nominal working range obtained for the different types of motor technologies that we have seen. The synchronous motors and DC motors with ferrite or rare earth magnets have the best efficiencies.

Flat and pancake motors have smaller efficiencies. This is mainly because of the rotor shape which raises the winding resistance and therefore leads to Joule losses.

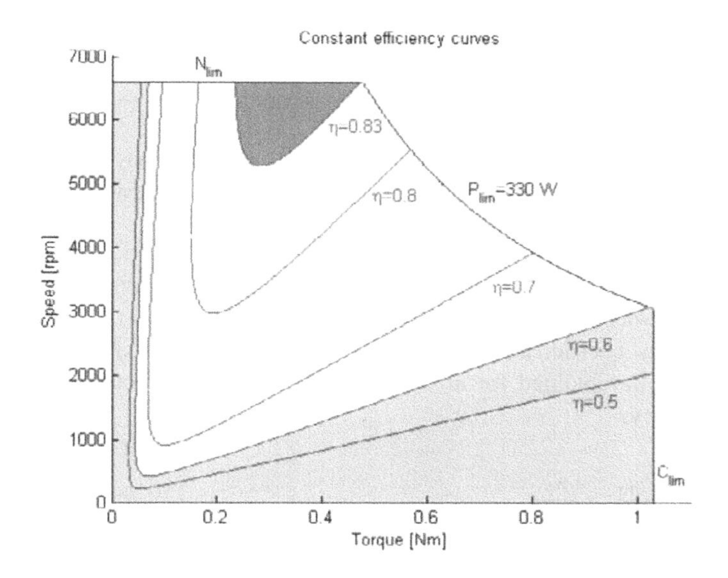

Figure 3.86. *Motor efficiency for a DC motor based on the working range*

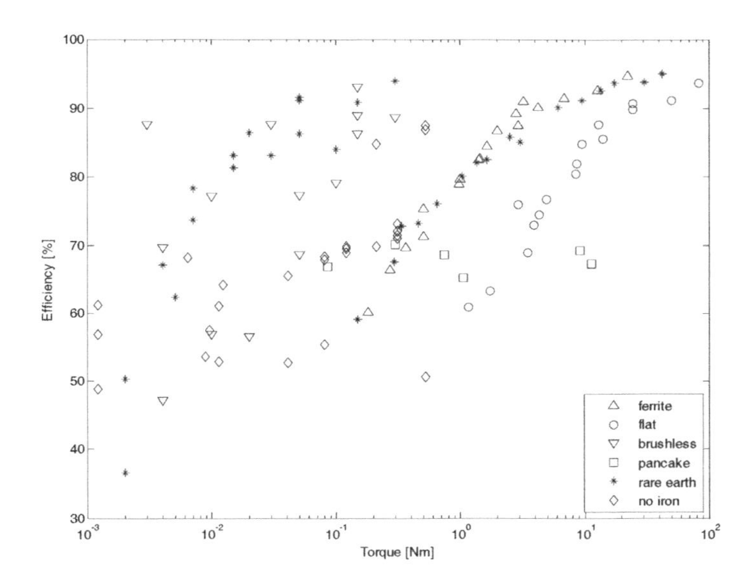

Figure 3.87. *Nominal working point efficiency for different types of motor technology*

Choosing the motors is a capital moment in designing legged robots [CHEV 03]. Nevertheless, the comparative elements presented above are not sufficient for determining the motor choice for certain robot designs.

When the total robot mass is constant, a rise in the payload of the robot necessitates a mass reduction of the mechanical elements, the transmissions and the motors.

Studies on dimensioning electric motors [WAL 95] have shown that the determining element for choosing a motor for a bipedal robot is its effectiveness coefficient. This coefficient is defined by referring to DC motors, but a similar coefficient can be obtained for all electrical motors. The effectiveness coefficient K_m is defined by:

$$K_m = \frac{C_{\lim}}{\sqrt{P_J}} = \frac{k_e}{\sqrt{R}} \qquad [3.21]$$

where $C_{\lim}$ is the limited instantaneous motor torque, P_J the corresponding Joule losses, k_e the torque constant and R the armature resistance. The effectiveness coefficient is proportional to the motor mass with a power of 5/6. When it is necessary to compare different power motors or torque motors, it is useful to refer to the motor's quality coefficient Q_m as defined by:

$$Q_m = \frac{K_m}{m^{5/6}} = \frac{C_{\lim}}{m^{5/6}\sqrt{P_J}} \qquad [3.22]$$

where m is the motor mass.

The motors' quality coefficient is based on scaling analysis during construction. The quality coefficient is constant when using the same motor type (homothetic size) and the same constitutive materials.

Indeed, an analysis of the magnetic state of a machine with permanent magnets enables us to demonstrate that the torque is proportional to the rotor mass with a power of 7/6:

$$C_{\lim} \propto A_L B_M V_e \qquad [3.23]$$

where A_L is the linear current density, B_M the maximum induction in the airgap and V_e the airgap volume. Since $A_L \propto l^{0.5}$ and $B_M = cte$ for the magnets, we have $C_{\lim} \propto l^{3.5}$ and the rotor mass is proportional to l^3, from which $C_{\lim} \propto m^{7/6}$.

In addition, for a maximum tolerated temperature elevation for the motor materials and for a given cooling mode of the motor, we have $P_J \propto l^2$ and therefore $P_J \propto m^{2/3}$.

The motor's effectiveness coefficient is therefore given by $K_m \propto m^{5/6}$ and the quality coefficient of these motors is therefore independent of the motor mass under consideration. In this way, the quality coefficient can be considered as a given technological characteristic of machines with magnets.

Figure 3.88 shows the profile of the quality coefficient for different types of motor seen above, in relation to their torque.

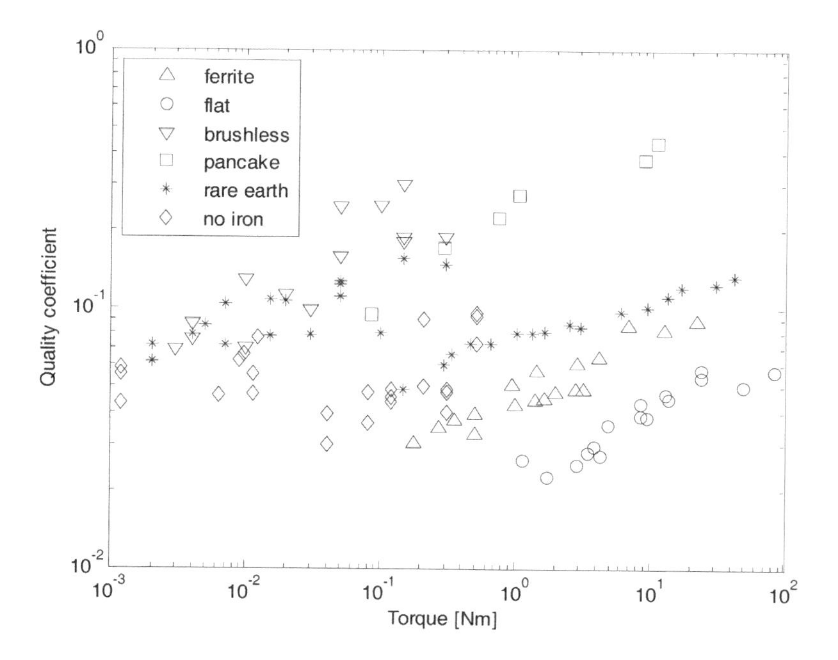

Figure 3.88. *Quality coefficient for different motor technologies*

The highest quality coefficients are obtained for the synchronous and pancake motor technologies. The DC motors with rare earth magnets and motors with no iron at the rotor have slightly lower quality coefficients. DC motors with ferrite magnets and flat motors have quality coefficients which are ten times smaller than synchronous motors, and they should therefore not be used in robotics except in a few rare cases. The comparisons and performances mentioned here are only relative however, as the mechanical load of the robot body parts can modify the values and conclusions mentioned above.

3.4.5. *Performances of transmission-actuator associations*

Only the complete analysis of the motor in relation to the robot body parts can give the best comparative criteria for the different motor technologies and the transmissions that can be associated to them. This analysis is complex as the number of combinations is very high and some of them are not always possible depending on the desired performances of the robot articulations. Only the most pertinent choice criteria are given here, as well as the transmission-actuators used the most. We have chosen the same motor types as in section 3.3.4, which are associated with a epicyclic reducer and a load which is represented by the bipedal robots' body parts. Figure 3.89 is a diagram of this association.

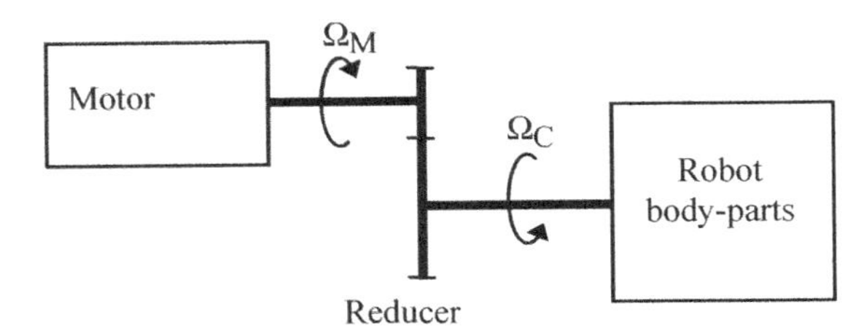

Figure 3.89. *Diagram of motor-reducer association*

Let us suppose that the robot's working function (in a walking cycle for example) requires the simultaneous application of a maximum torque C_{max} and a velocity Ω_{max} for the robot bodies.

This extreme working function necessitates, for direct drive, the choice of motor 1 and must satisfy the following inequalities:

$$C_{max} < C_{\lim 1} \quad [3.24]$$

$$\Omega_{max} < \Omega_{\lim 1} \quad [3.25]$$

$$C_{max}\Omega_{max} < P_{lim1} \quad [3.26]$$

The use of a reducer enables us to shift the motor's working range in (C_M, Ω_M) as defined by the reducer's behavioral equations:

$$C_M = C_{max}/\eta N \quad [3.27]$$

$$\Omega_M = N\,\Omega_{max} \quad [3.28]$$

where η is the reducer efficiency and $1/N$ its gear ratio. More precise reducer behavioral equations could be used, but what we are trying to show here are the major advantages of the reducer-motor association.

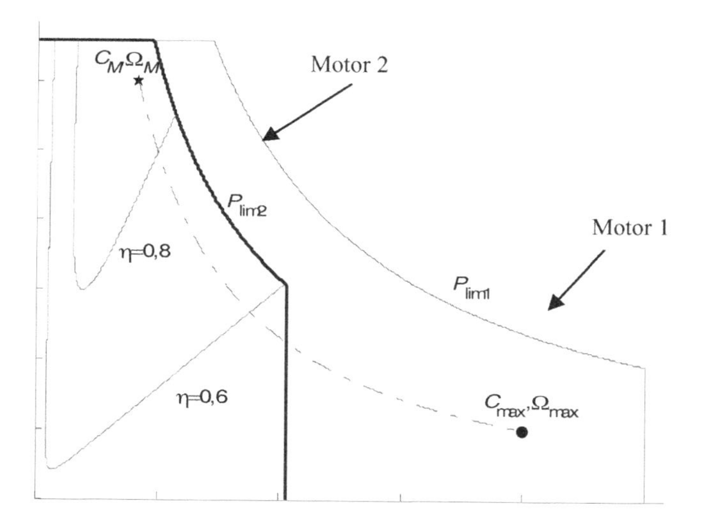

Figure 3.90. *Working range for a motor-reducer association*

Despite the reducer efficiency, we can see in Figure 3.90 that the new working range enables us to choose motor 2 with a lower power limit.

In the majority of design cases, the sum of the mass of motor 2 and of the reducer is lower than the mass of motor 1. The robot therefore has a significant mass loss.

In addition, the new working range (C_M, Ω_M) is situated in the zone where the motor efficiency is significantly higher. The product efficiencies of motor 2 and the reducer are therefore largely greater than the efficiency of motor 1 in its working range (C_{max}, Ω_{max}), a second advantage of the motor-reducer association.

The association's efficiency is given by the following formula:

$$\eta_T = \frac{C_C \Omega_C}{C_C \Omega_C + P_{Joule} + P_{meca}} \quad [3.29]$$

where (C_C, Ω_C) is the working range imposed by the robot gait, P_{Joule} the motor's Joule losses and P_{meca} the global mechanic friction losses.

The frictional losses, which are mainly concentrated in reducer, are proportional to the transmitted power. We can therefore incorporate the following:

$$P_{meca} = \alpha C_C \Omega_C \quad [3.30]$$

where α is a coefficient which depends on the reducer efficiency. The Joule[2] losses are expressed in the following way:

$$P_{Joule} = RI^2 = R\left(\frac{C_C(1+\alpha)}{k_e N}\right)^2 = \frac{C_C^2(1+\alpha)^2}{K_m^2 N^2} \quad [3.31]$$

Through the analogy of the motor's effectiveness coefficient given in equation [3.21], we can define the effectiveness coefficient of the motor-reducer association by:

$$K_A = \frac{C_{max}}{\sqrt{P_{meca} + P_{Joule}}} \quad [3.32]$$

By using equations [3.30] and [3.31] in equation [3.32], after simplification we obtain:

---

2. With the definition of the torque coefficient used previousy, the expression of the Joule losses given here is valid for both DC and synchronous motors.

$$K_A = \frac{1}{\sqrt{\frac{(1+\alpha)^2}{K_m^2 N^2} + \alpha \frac{\Omega_C}{C_C}}} = \frac{K_m N}{\sqrt{(1+\alpha)^2 + K_m^2 N^2 \alpha \frac{\Omega_C}{C_C}}} \quad [3.33]$$

Effectiveness coefficient K_A obviously depends on the motor and reducer characteristics but also on the working range.

By using equations [3.27] and [3.28], we can replace values (C_C, Ω_C) according to the motor's working range (C_M, Ω_M). The effectiveness coefficient becomes:

$$K_A == \frac{K_m N}{\sqrt{(1+\alpha)^2 + K_m^2 \alpha (1+\alpha) \frac{\Omega_M}{C_M}}} \quad [3.34]$$

The latter now only depends on the motor and reducer characteristics. We can therefore envisage the possibility of a comparison of motor-reducer association performances for each motor at the nominal working range given by the manufacturer. We therefore have a comparative criterion which is similar to that of the quality coefficient used for just the motor. The effectiveness coefficient, however, is not independent of the motor-reducer association mass and does not enable us to compare the technologies used for the association only. Figure 3.91 gives the comparative results for the case of a reducer with a efficiency of $\eta = 0.75$. The articulation velocity is fixed to $\Omega_C = 50$ rpm. The gear ratio is chosen in order to bring the motor's working range close to nominal point (C_N, Ω_N).

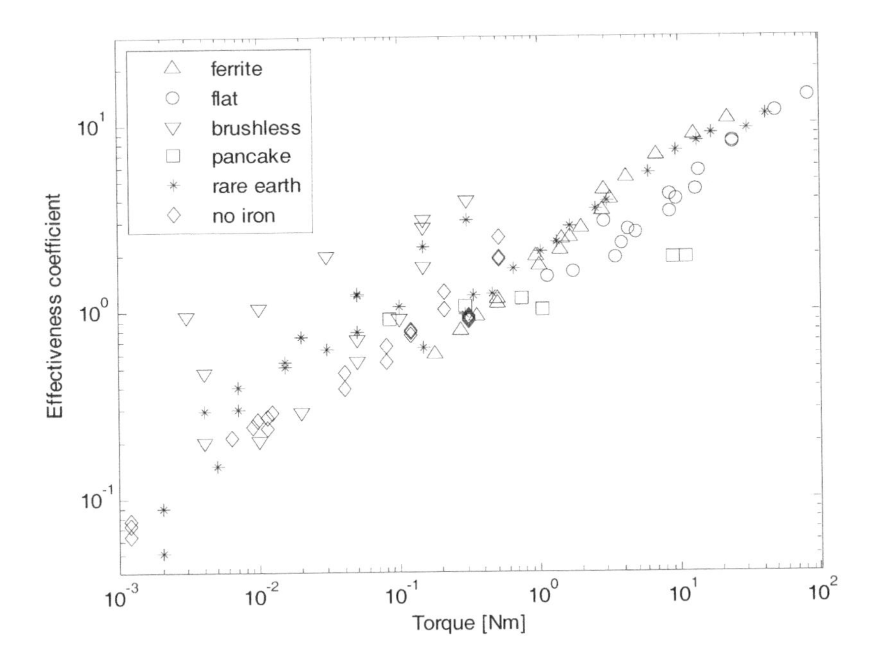

Figure 3.91. *Effectiveness coefficient for the motor-reducer association for different motor technologies, relative to the nominal torque*

Using the analogy once more with the definition which corresponds to the motor only, we define the quality coefficient by:

$$Q_A = \frac{K_A}{\left(m_m + m_R\right)^{5/6}} \qquad [3.35]$$

where m_m is the motor mass and m_R the associated reducer mass.

Figure 3.92 gives the line graph of the quality coefficient of the motor-reducer association in the specific aforementioned example. For pancake motors, there is no associated reducer. In this case, the comparison between the different technologies is now different. The synchronous motors and the DC motors with rare earth magnets have the best quality coefficients and their use should favored when building bipedal robots. In the case of articulations which need a low torque, motors with no iron at the rotor are also appropriate.

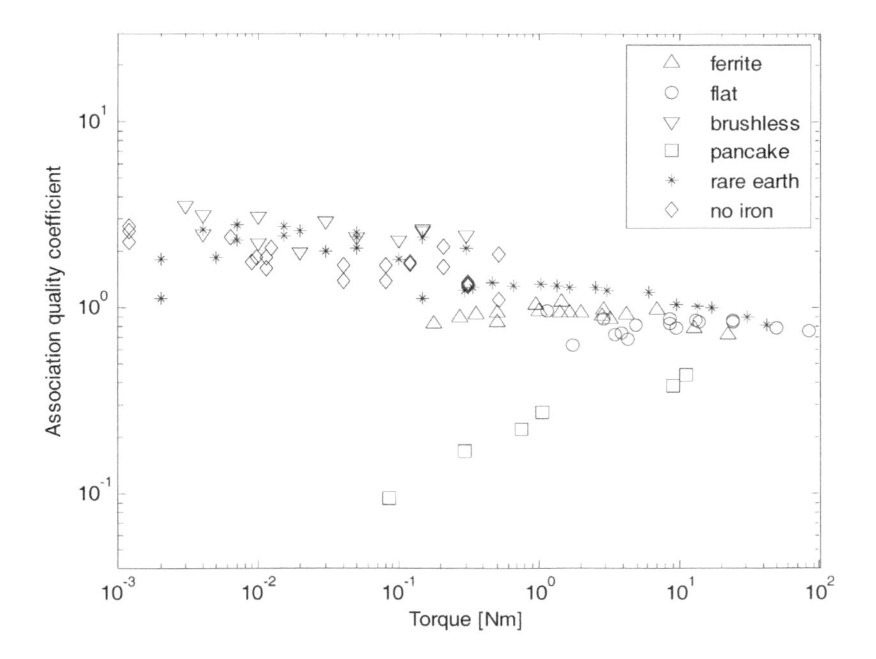

Figure 3.92. *Quality coefficient of the motor-reducer association for different motor technologies*

3.5. Sensors

3.5.1. *Measuring*

In general, a sensor is a device which when submitted to the action of a physical phenomenon delivers an electric signal. The physical sizes that need to be measured in the legged robotic domain are numerous and at times difficult to attain. There are many categories of sensors depending on the hierarchical position in their control, their accessibility, their exteroceptive or proprioceptive measurement properties and their degrees of complexity or cost.

Sensors give out signals linked to the sizes that have been measured. These signals can be analogical, numerical, Boolean or digital. Certain sensors directly measure the primary sizes, and others measure secondary sizes linked to the value to be measured by using a well-known and preferably linear physical law concerning measurements. Table 3.6 gives a few typical aspects of measuring.

Sensors	Exteroceptives	Proprioceptives	Accessibility
Location in space using a camera	Yes		Difficult
Location in space using a GPS	Yes		Average
Gyroscope	Yes		Average
Inclination		Yes	Average
Forces, moments	Yes	Yes	Average
Pressure	Yes	Yes	Easy
Acceleration		Yes	Average
Angular position		Yes	Easy
Angular velocity		Yes	Easy
Torque		Yes	Average
Contact	Yes	Yes	Easy
Proximity	Yes		Easy
End switches		Yes	Easy
Deformation		Yes	Average

Table 3.6. *Categorization of sensors used in robotics*

3.5.2. *Frequently used sensors*

The measurements which are taken from a bipedal robot mainly depend on its control modes and applications.

When incorporating the sensors within the control structure, many measurement characteristics intervene e.g. accessibility, permanent or temporary availability (the measurement can be read directly or it requires time to treat or transmit) and precision.

Numerous complete works [ASC 06, BUS 98, CHA 98] give overviews of sensors used in robotics. A thorough presentation of force sensors and the working constraints can be found in [GOR 97].

In this section we will consider two types of sensor: pressure or force sensors and location sensors using gyrometers and gyroscopes.

3.5.3. *Characteristics and integration*

The characteristics of the main sensors described below enable us to identify the main choices made when building a robot.

3.5.3.1. *Force sensors*

A force sensor is a mechanical arrangement that we insert into the robot's body in order to measure the effort exerted on it. For example, it is generally placed in the walking robot's leg, as close as possible to its extremity, to measure ground reaction.

The arrangement is designed in such a way that it has elastic deformations, which are measured with the assistance of constraint gauges according to the preferred directions and force components that we wish to determine.

Figure 3.93 shows a sensor which was fixed onto the legs of a biped at the University of Moscow. It enabled effort control of the biped during double support [GOR 97].

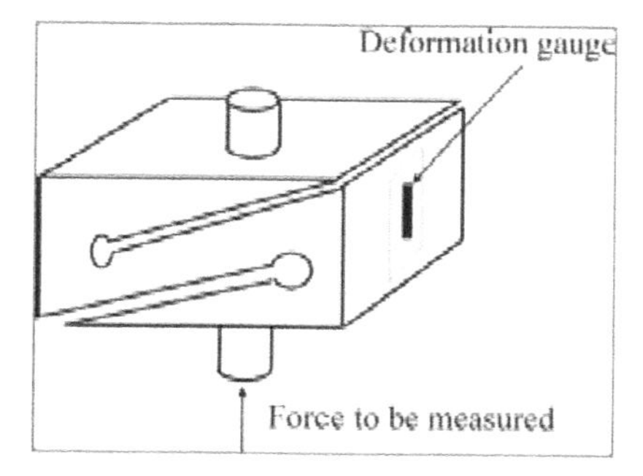

Figure 3.93. *Diagram of a force sensor which measures a single effort component and which is perpendicular to the base of the sensor [GOR 97]*

The construction of a sensor which is capable of measuring three effort components is fragile, especially with respect to impacts. When we consider the available existing technologies and the intense impact shocks during a robot's dynamically stable walk, being able to measure two single components is a good compromise.

A low range mono-dimensional sensor can also be useful for determining the impact instants. The advantages of this sensor are a good passing bandwidth and being robust against shocks. As we can see for the SemiQuad [AOU 06] in Figure 3.94, flat wheels were manufactured on the stabilization branch of the robot in the frontal plane. Constraint gauges were stuck to them in order to measure the impact of the legs with the ground on contact.

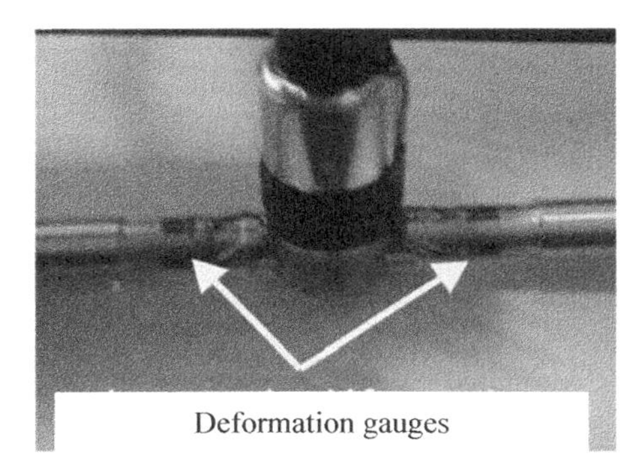

Figure 3.94. *Assembly of deformation gauges on the leg extremity of the SemiQuad robot to detect impact*

3.5.4. *Sensors of inertial localization*

There is a wide range of gyroscopes and accelerometers to capture inertial location. Today, some of these instruments have miniature components.

3.5.4.1. *Accelerometers*

These sensors measure forces that are applied to a mobile mass. Their relative precision is of one-ten-thousandth to one-hundred-thousandth of the mass weight. Although this precision seems impressive, their technological performances are not as great as those of gyroscopes, which are far more difficult to design and make.

3.5.4.2. *Gyroscopes*

A gyroscope is a revolutionary solid that turns around its axis at fast speeds and is perfectly suspended to a support by its mass center C. The first working gyroscope was created by Léon Foucault in the 1850s. The principle of the gyroscope is based on the application of the kinetic moment in C within an absolute Galilean frame. When the link of the gyroscope with its support is taken to be perfect and the sum of the exterior forces moment is zero, supposing the symmetry to be perfect, we have:

$$\frac{d\vec{\sigma}_C}{dt} = \frac{d(I\vec{\omega})}{dt} = 0 \qquad [3.36]$$

where ω is the gyroscope rotational velocity, $\vec{\sigma}_C$ is the kinetic moment expressed in C and I is the gyroscope's inertia. The vectorial sizes $\vec{\sigma}_C$ and $\vec{\omega}$ are orientated in the same direction. The gyroscope axis remains orientated towards the same point in space. The gyroscope on its support exerts a gyroscopic torque so that is regains its initial rotational axis when we submit it to perturbations. This therefore enables us to measure the mobile body's rotational angle to which it is fixed. The gyroscopes associated with acceleration and angular velocity sensors enable us to create an inertia central station or central inertia which is capable of giving, in real time, the profile of velocity and position vectors as well as the attitude (roll, pitch and yaw) of a walker robot [HIR 98].

Concerning precision, the typical derivative of gyroscopes are 100 deg h^{-1}. The derivative is the error in maintaining a fixed direction, which the gyroscope is supposed to give in relation to the stars.

There are gyroscopes with significantly smaller derivatives, but their high cost means that they cannot be used in robotics. Many different types of technology can correspond to different classes of precision, as well as to the operational constraints:

– floating gyroscopes (the debuts of the big inertias of the 1950s);

– spinning top gyroscopes with associated dynamic suspension;

– vibrating component gyroscopes (diapason, resonating hemispheric gyroscope);

– optic fiber gyroscope;

– electric (electrostatic) suspension gyroscopes.

Faced with such a variety of performances and technologies, it is difficult to know which to adopt in order to provide the best solution for the problems of piloting or navigation. This problem is the performance/cost ratio.

3.5.4.3. *Gyrometers*

A gyrometer is a sensor which enables us to determine variations in attitude at any instant. Depending on the type of technology, the velocity measurement span varies from the continuous to 100 deg s^{-1}.

Many types of technology are used to create gyrometers:

– laser gyrometers detect variations in attitude via the visible path of light entering an optic resonator [RAD 99];

– in optic fiber gyrometers, the variation in attitude directly modulates the light which is propagated in the fiber [RAD 99];

– in resonating gyrometers, a sensitive component is excited by a magnetic field at its resonating frequency. We then see voltage (which is proportional to the angular velocity) at the component bounds;

– in the case of piezoelectric gyrometers, the Coriolis force exerts a constraint on the piezoelectric material. The charge varies proportionally to the constraint.

Integration is therefore necessary to capture the attitude of the walker robot. Consequently the evaluation of the attitude is deteriorating over time and a resetting of the measurement is then necessary.

3.5.4.4. *GPS receptors*

These give precise geographical positions (latitude, longitude) but are subject to eclipses in urban zones where their receptor satellites are blocked. This is an obvious choice for inertial navigation but there may be deviations.

Since the 1980s GPS receptors have become smaller and are no more expensive than small electronic modules. They have become plug-ins which can be inserted into inertia units.

The determining factor today, however, is the quality of the algorithms which enable us to combine a GPS and dead-reckoning navigation (where the inertial navigation is the summit) to obtain the best possible integrated navigation as well as its sub-products: robot orientation and velocity.

3.6. Conclusion

Based on the desired performance objectives, the design of a legged robot must be based on choices of motorization, mass distribution over the whole structure and sensor.

As we saw in the case of two simple planar bipedal robots, leg mass has a greater influence on energy performance than trunk mass. This is also true for spherical feet shapes which favor a reduction in consumed energy.

These observations should only be taken into account for defining motorized articulations. As a robot walker uses its environment to move around in, by definition, localization sensors and ground contact sensors are essential for helping it to move.

This chapter has shown the multiple and complex compromises which need to be made at the design stage, as well as the criteria which must be respected. In the following chapter, further useful tools are provided so that the robot designer can define reference trajectory via optimization.

3.7. Appendix

This appendix gives the geometric and dynamic models for a three-link robot (as seen in Figure 3.1). The numeric values of the parameters used for the simulations are given in Table 3.7.

L_1 [kg]	m_1 [kg]	m_2 [kg]	m_3 [kg]	s_1 [m]
0.8	3.2	6.8	16.5	0.527
s_2 [m]	s_3 [m]	I_{r1} [kg m^2]	I_{r2} [kg m^2]	I_{r3} [kg m^2]
0.163	0.2	0.05	0.07	1.56

Table 3.7. *Nominal values of three-link robot parameters*

3.7.1. *Geometric model*

The geometric model includes:

– hip center position: $(x,\ z)$;

– position of stance leg extremity: $(x + L_1\sin(q_1), z - L_1\cos(q_1))$;

– position of swing leg extremity: $(x + L_1\sin(q_2), z - L_1\cos(q_2))$.

3.7.2. *Dynamic model*

The robot's dynamic model is given by the following equation:

$$D\ddot{q} + C\dot{q} + G = B\,\Gamma + E\,F$$

where $q, \dot{q}, \ddot{q}$ are the generalized vector positions, velocities and accelerations, $q^T = \begin{bmatrix} q_1 & q_2 & q_3 & x & z \end{bmatrix}$, $\Gamma^T = \begin{bmatrix} \Gamma_1 & \Gamma_2 \end{bmatrix}$ is the motor torque vectors at articulation level, D is the inertia matrix, C is the Coriolis and centrifugal force matrix, B is the control matrix, E is the Lagrange multiplier matrix for the external actions and $F^T = \begin{bmatrix} F_x & F_z \end{bmatrix}$ is the external force vector applied to the stance foot. These matrices are expressed in the following way:

$$D = \begin{bmatrix} d_{11} & 0 & 0 & d_{14} & d_{15} \\ 0 & d_{22} & 0 & d_{24} & d_{25} \\ 0 & 0 & d_{33} & d_{34} & d_{35} \\ d_{14} & d_{24} & d_{34} & d_{44} & 0 \\ d_{15} & d_{25} & d_{35} & 0 & d_{55} \end{bmatrix} \qquad C = \begin{bmatrix} 0 & 0 & 0 & 0 & 0 \\ 0 & 0 & 0 & 0 & 0 \\ 0 & 0 & 0 & 0 & 0 \\ c_{41} & c_{42} & c_{43} & 0 & 0 \\ c_{51} & c_{52} & c_{53} & 0 & 0 \end{bmatrix}$$

$$G^T = \begin{bmatrix} g_1 & g_2 & g_3 & 0 & g_5 \end{bmatrix} \qquad B^T = \begin{bmatrix} -1 & 0 & 1 & 0 & 0 \\ 0 & -1 & 1 & 0 & 0 \end{bmatrix}$$

where

$d_{11} = m_1 s_1^2 + m_2 s_2^2 + I_{r1} + I_{r2}$,

$d_{22} = m_1 s_1^2 + m_2 s_2^2 + I_{r1} + I_{r2}$,

$d_{33} = m_3 s_3^2 + I_{r3}$,

$d_{44} = m_1 + m_2 + m_3$,

$d_{55} = m_1 + m_2 + m_3$,

$d_{14} = (m_1 s_1 + m_2 s_2) \cos(q_1)$,

$d_{24} = (m_1 s_1 + m_2 s_2) \cos(q_2)$,

$d_{34} = -m_3 s_3 \cos(q_3)$,

$d_{15} = (m_1 s_1 + m_2 s_2) \sin(q_1)$,

$d_{25} = (m_1 s_1 + m_2 s_2) \sin(q_2)$,

$d_{35} = -m_3 s_3 \sin(q_3)$,

$c_{41} = (m_1 s_1 + m_2 s_2)\ \dot{q}_1 \sin(q_1)$,

$c_{42} = -(m_1 s_1 + m_2 s_2)\ \dot{q}_2 \sin(q_2)$,

$c_{43} = m_3 s_3\ \dot{q}_3 \sin(q_3)$,

$c_{51} = (m_1 s_1 + m_2 s_2)\ \dot{q}_1 \cos(q_1)$,

$c_{52} = (m_1 s_1 + m_2 s_2)\ \dot{q}_2 \sin(q_2)$,

$c_{53} = -m_3 s_3 q_3 \cos(q_3)$,

$g_1 = (m_1 s_1 + m_2 s_2)\ g \sin(q_1)$,

$g_2 = (m_1 s_1 + m_2 s_2)\ g \sin(q_2)$,

$g_3 = -\ m_3 s_3\ g \sin(q_3)$,

$g_5 = (m_1 + m_2 + m_3)\ g$.

3.8. Bibliography

[AOI 05] AOI S., TSUCHIYA K., "Locomotion control of a biped robot using nonlinear oscillators", *Autonomous Robots*, vol. 19, no. 3, p. 219–232, 2005.

[AOU 03] AOUSTIN Y., FORMAL'SKY A., "Control design for a biped: reference trajectory based on driven angles as functions of the undriven angle", *Journal of Computer and System Sciences International*, vol. 42, no. 4, p. 159–176, 2003.

[AOU 06] AOUSTIN Y., CHEVALLEREAU C., FORMAL'SKY A., "Numerical and experimental study of the virtual quadrupedal walking robot-semiquad", *Multibody System Dynamics*, vol. 16, p. 1–20, 2006.

[ASC 06] ASCH G., *Les capteurs en instrumentation industrielle,* 6th edition, Dunod, Paris, 2006.

[BES 04] BESSONNET G., CHESSÉ S., SARDAIN P., "Optimal gait synthesis of a sevenlink planar biped", *International Journal of Robotics Research*, vol. 23, no. 10–11, p. 1059–1073, 2004.

[BUS 98] BUSCH-VISHNIAC I.J., *Electromechanical Sensors and Actuators*, Springer Mechanical Engineering Series, New York, 1998.

[CHA 98] CHAVAND F., *Perception de l'environnement en robotique*, Hermes, Paris, 1998.

[CHE 03] CHEOL KI A., MIN CHEOL L., SEOK JO G., "Development of a biped robot with toes to improve gait pattern", *Proceeding of the 2003 IEEE/ASME International Conference on Advanced Intelligent Mechatronics, AIM 2003*, vol. 2, p. 729–734, 2003.

[CHEV 00] CHEVALLEREAU C., SARDAIN P., "Design and actuation optimization of a 4-axes biped robot for walking and running", *Proceedings of IEEE International Conference on Robotics and Automation, ICRA'00*, vol. 4, p. 3365–3370, 2000.

[CHEV 01] CHEVALLEREAU C., AOUSTIN Y., "Optimal reference trajectories for walking and running of a biped robot", *Robotica*, vol. 19, p. 557–569, 2001.

[CHEV 03] CHEVALLEREAU C., ABBA G., AOUSTIN Y., PLESTAN F., WESTERVELT E., CANUDAS DE WIT C., GRIZZLE J.W., "Rabbit: a testbed for advanced control theory", *IEEE Control Systems Magazine*, vol. 23, no. 5, p. 57–79, 2003.

[GOR 97] GORINNEVSKY D.M., FORMALSKY A.M., SCHNEIDER A.Y., *Force Control of Robotics Systems*, CRC Press, Boca Raton, 1997.

[GOS 96] GOSWAMI A., ESPIAU B., KERAMANE A., "Limit cycles and their stability in a passive bipedal gait", *IEEE International Conference on Robotics and Automation, ICRA'96*, Minneapolis, USA, p. 246–251, 1996.

[GOS 97] GOSWAMI A., ESPIAU B., KERAMANE A., "Limit cycles in a passive compass gait biped and passivity-mimicking control laws", *Autonomous Robots*, vol. 4, no. 3, p. 273–286, 1997.

[GRI 01] GRIZZLE J.W., ABBA G., PLESTAN F., "Asymptotically stable walking for biped robots: analysis via systems with impulse effects", *IEEE Transactions of Automatic Control*, vol. 46, no. 1, p. 51–64, 2001.

[HIR 98] HIRAI K., HIROSE M., HAIKAWA Y., TAKENAKA T., "The development of Honda humanoid robot", *ICRA'98*, Leuven, Belgium, p. 1321–1326, 1998.

[LAC 99] LACROUX G., *Les actionneurs électriques pour la robotique et les asservissements*, 2nd edition, Lavoisier, Paris, 1999.

[LEB 92] LE BORZEC R., *Réducteurs de vitesse à engrenages*, B5640, Sciences et Techniques, Paris, 1992.

[MCG 90] MCGEER T., "Passive dynamic walking", *International Journal of Robotic Research*, vol. 9, no. 2, p. 62–82, 1990.

[MOR 05] MORIS B., GRIZZLE J.W., "A restricted Poincare map for determining exponentially stable periodic orbits in systems with impulse effects: application to bipedal robots", *Re-print CDC'05*, Seville, Spain, 2005.

[MOR 00] MORISAWA M., YAKOH T., MURAKAMI T., OHNISHI K., "A comparison study between parallel and serial linked structures in biped robot system", *26th Annual Conference of the IEEE Industrial Electronics, IECON 2000*, vol. 4, p. 2614–2619, 2000.

[ONO 02] ONO K., LIU R., "Optimal biped walking locomotion solved by trajectory planning method", *J. Dynamic Systems, Measurement and Control*, vol. 124, no. 4, p. 554–565, 2002.

[PLA 05] PLATTENBURG D.H., "Pneumatic actuators: a comparison of energy-to-mass ratio's", *9th International Conference on Rehabilitation Robotics, ICORR*, p. 545–549, 2005.

[PLE 03] PLESTAN F., GRIZZLE J.W., WESTERVELT E., ABBA G., "Stable Walking of a 7-DoF Biped Robot", *IEEE Transactions on Robotics and Automation*, vol. 19, no. 4, p. 653–668, 2003.

[RAD 99] RADIX S., *Gyroscopes et gyromètres mécaniques avec élément rotatif*, R1940, Edition Sciences et Techniques, Paris, 1999.

[RIG 06] RIGHETTI L., IJSPEERT A.J., "Programmable central pattern generators: an application to biped locomotion control", *Proceedings of IEEE International Conference on Robotics and Automation, ICRA'06*, p. 1585–1590, 2006.

[SAG 02] SAGA N., NAKAMURA T., UEHARA J., IWADE T., "Development of artificial muscle actuator reinforced by Kevlar fiber", *IEEE International Conference on Industrial Technology, ICIT'02*, vol. 2, p. 950–954, 2002.

[SAR 99] SARDAIN P., ROSTAMI M., THOMAS E., BESSONNET G., "Biped robots: Correlation between technological design and dynamic behavior", *Control Engineering Practice*, vol. 7, p. 401–411, 1999.

[SCA 04] SCARFOGLIERO U., FOLGHERAITER M., GINI G., "Advanced steps in biped robotics: innovative design and intuitive control through spring-damper actuator", *4th IEEE/RAS International Conference on Humanoid Robots*, vol. 1, p. 196–214, 2004.

[SUG 04] SUGAHARA Y., KAWASE M., MIKURIYA Y., HOSOBATA T., SUNAZUKA H., HASHIMOTO K., HUN-OK L., TAKANISHI A., "Support torque reduction mechanism for biped locomotion with parallel mechanism", *Proceedings of IEEE/RSJ International Conference on Intelligent Robots and Systems, IROS'04*, vol. 4, p. 3213–3218, 2004.

[TAK 04] TAKUMA T., NAKAJIMA S., HOSODA K., ASADA M., "Design of self-contained biped walker with pneumatic actuators", *Annual Conference SICE*, vol. 3, p. 2520–2524, 2004.

[TON 00] TONDU, B., LOPEZ P., "Modeling and control of McKibben artificial muscle robot actuators", *IEEE Control Systems Magazine*, vol. 20, no. 2, p. 15–38, 2000.

[TZA 96] TZAFESTA S., RAIBERT M., TZAFESTA C., "Robust sliding-mode control applied to a 5-link biped robot", *J. of Intelligent and Robotic Systems*, vol. 15, no. 1, p. 67–133, 1996.

[VAN 05] VANDERBORGHT B., VERRELST B., VAN HAM R., VERMEULEN J., LEFEBER D., "Dynamic control of a bipedal walking robot actuated with pneumatic artificial muscles", *Proceedings of IEEE International Conference on Robotics and Automation, ICRA'05*, vol. 1, p. 1–6, 2005.

[WAL 95] WALLACE R.S., SELIG J.M., "Scaling direct drive robots", *Proceedings of IEEE International Conference on Robotics and Automation, ICRA'95*, p. 2947–2954, 1995.

[ZON 02] ZONFRILLI F., ORIOLO G., NARDI D., "A biped locomotion strategy for the quadruped robot Sony ERS-210", *IEEE International Conference on Robotics and Automation, ICRA'02*, p. 2768–2774, 200.

Chapter 4

Walking Pattern Generators

4.1. Introduction

Walking is an intrinsically unstable movement. This is why the very constraining conditions for dynamic balance must be compatible with an appropriate propulsive effect. The restrictive conditions of foot-ground interaction must also be respected (such as the non-sliding and unilaterality of the contact forces). In addition, the kinematic redundancies of the locomotion system can lead to unnecessary gesticulations, undesirable hyperextension and internal or external collisions. As all the internal degrees of freedom (DoF) are actuated, there is a possibility of antagonistic forces in the kinematic loop (which makes up the locomotion system during double support) with a correlative risk of loss of ground contact.

There is therefore a need to generate perfectly coordinated movements with guaranteed balance, which respect all their kinematic and dynamic constraints. Determining and putting to work the organizational principles of walking enables us to master the kinematic complexity and dynamic subtleties of this complex problem.

In increasing order of complexity, we can identify the following three different approaches which have already been developed for the creation of walking gaits:

– *passive-dynamics walking* which results from the biped's capacity of acting as an inverted jointed-pendulum during the single support;

– *static balance walking:* the principle here is to maintain, at all instants, the biped's ground projection center of mass on the inside of the support polygon;

– *minimum effort walking or minimum spent energy walking*. This approach gives us the possibility of treating the problem of the dynamics synthesis of walking globally. This approach calls for the theory of optimization.

The first two approaches are based on simple walking layouts and consider particular gait aspects. The third approach enables us to take into consideration all the kinematic and dynamic characteristics of bipedal locomotion. We can then design numerical walking generators which are capable of synthesizing all the conditions which define the dynamic laws and limits which the desired movement must satisfy. The development perspectives are vast and are limited only by the numerical robustness of the available computation codes which are used for solving the set optimization problems.

4.2. Passive and quasi-passive dynamic walking

During human walking, myographic analysis shows that muscular activity is very much reduced during the single support phase which constitutes about 80% of the walking cycle [MCM 84]. Muscular relaxation during moderate gaits therefore shows that the single support movement resembles passive pendulous movement, the result of an effective combination of an initial impulse and the coupled effects of gravity and biped inertia. The dynamic behavior of a walker is therefore close to that of an inverted jointed pendulum. This aspect of legged-locomotion has been the subject of numerous studies and has given rise to the creation of passive bipeds that exhibit a purely pendulum-like behavior.

4.2.1. *Passive walking*

It has theoretically been established (and verified experimentally) that compass-like bipeds (even those with two-link articulated legs) can ambulate passively along plane ground on a slight slope. The only motor for this movement is therefore gravity.

The notion of passive walking has been particularly developed by McGeer in [MCG 90a] where the author studies the spontaneous ambulation of a sagittal biped on an inclined plane. This biped has rigid legs and its only articulation is at the hip. Its feet are a continuation of the legs and have incurved contact surfaces. The originality of such a system is that of combining a pendulum movement with that of a wheel. McGeer formalizes this idea of a composite pendulum-wheel system by presenting the concept of a *synthetic wheel*. This "wheel" is reduced to two spokes which are freely articulated at the hub, and which have separate arched rims

(Figure 4.1). A difference in kinematics and dynamic behavior must be noted however: the *synthetic wheel* results in an alternate relative movement, whereas the movement of the wheel in itself is purely rotational.

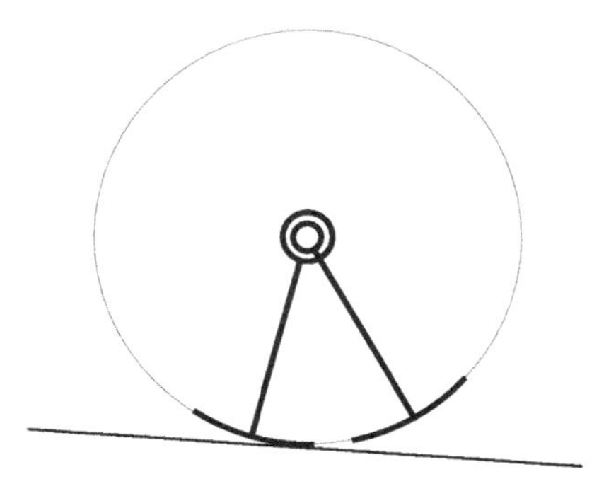

Figure 4.1. *Illustration of McGeer's concept of a synthetic wheel*

McGeer's idealized compass-gait biped requires preliminary safety guards. The biped is equipped with a retraction device for when the foot crosses the stance leg, in order to avoid an untimely collision with the ground. The study of the biped dynamic behavior is based on using dynamic equations that are linearized in the vicinity of the legs' average vertical position. An impact equation at touch-down completes the model. Energy dissipation during the successive non-controlled impacts is a moderating and regulating element for the cadenced movement which is sought after. This model shows that a cyclic and stable state can be established after a few steps.

In [ESP 94, GOS 96], the study of the gait of a compass-like biped is developed by using the concepts and methods of Poincaré's dynamic system analysis. This research shows that the dynamic behavior of this type of biped is mainly determined by three non-dimensional parameters: a length ratio, a mass ratio and the steepness of the support plane on which the movement occurs. The analysis of limit cycles reveals the existence of periodical movements which are stable or chaotic. A similar study of the periodicity and stability of possible movements can be found in [GAR 98a].

McGeer goes further in [MCG 90b] by showing that a planar robot with two-link legs having a knee-type articulation can walk passively on a slight slope. Contact soles with the appropriate geometry (as well as a judicious distribution of mass) can bring about a walking step which combines an inverted pendulum forward movement on the biped's stance leg, and a swing with a spontaneous flexion of the other leg.

It should be noted that a passive knee-locking device is necessary in order for the stance leg to avoid hyperextension. Another point of interest is the swing leg's spontaneous capacity to flex forwards, which avoids collisions with the ground when it moves past the stance leg. The reader can refer to [ALE 95] for a detailed analysis of McGeer's study results.

McGeer's study model was taken up and developed by Garcia *et al.* in [GAR 98b] who made a passive sagittal biped with knees. Such a system's aptitude at passive walking was established theoretically and proved through experimentation [COL 01a]. The same team went further in [COL 01b] by creating a biped with articulated legs and with arms which acted as a counterbalance. It was capable of three-dimensional passive walking [COL 01a].

The theoretical results obtained and the biped created show that the coupled effects of gravity and inertia can lead to a natural pendulum-like organization of complex movements such as walking, be it sagittal or three-dimensional. The cyclic nature of this movement is therefore made possible by the successive impacts at each touch-down which act as a movement regulator.

As is suggested by the authors of [COL 01b], these results could help in designing active bipeds which are capable of walking on horizontal ground (and not only on sloping grounds, as is required for purely passive bipeds) with minimal actuation and control of their movements as the only requirements.

With no inclination at ground level, keeping the pendulum movement going requires propulsion at each touch-down to compensate for the dissipated energy at foot-strike. This general idea is the basis for the notion of quasi-passive dynamic walking presented in section 4.4.2 below.

4.2.2. *Quasi-passive dynamic walking*

In [CHE 98, FOR 94], the created walking step has an instantaneous double support which reinitializes a passive single support. This is done through impulsional control torques. This pendulous movement can be determined as the solution for the following boundary value problem. In the absence of joint torques, we wish to determine the velocities which are associated with an initial given biped configuration and which allow a prescribed final configuration to be reached.

This problem can be solved numerically by using an iterative method [AOU 03]. For example, let us review the example of the five-link robot (Figure 4.2) with point ground-foot contacts.

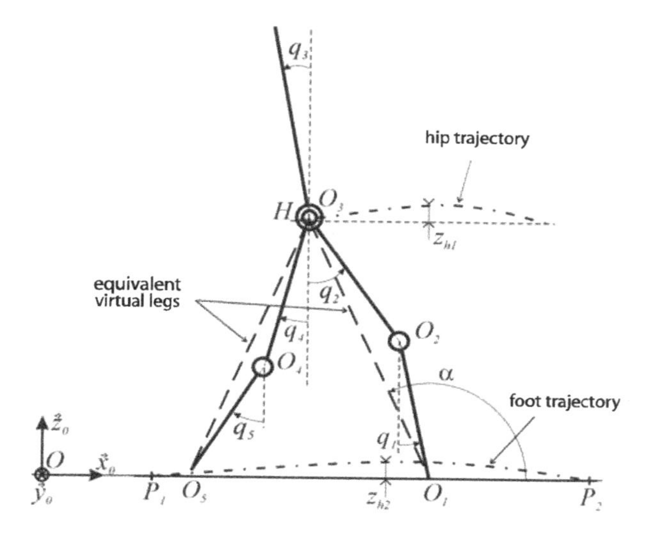

Figure 4.2. *Five-link biped with point ground-foot contacts*

A passive single support phase with duration of $T = 0.4$ s was defined. In Figure 4.3 we can see the monotonous behavior of the virtual leg's orientation, which is equivalent to the stance leg. The monotonous rotation of the virtual stance leg is an unchanging characteristic of bipedal walking. In Figure 4.4, the movement of the trunk is symmetric in relation to instant $T/2$ where, for $0 \leq t \leq T/2$, we have $q_3\left(T/2-t\right) = q_3\left(T/2+t\right)$.

This type of walking is very similar to passive gait: there is an instantaneous transfer of contact from one foot to the other with impact at foot-strike, followed by a passive single support. However, there is a major difference for a biped which needs to walk on horizontal ground. In passive walking, the energy which is dissipated by the impact at foot-strike is progressively restored by gravity during single support. For quasi-passive walking on horizontal ground, however, the mechanical energy of the biped is maintained during the single support. The restoration of its kinetic energy, which is necessary for the reinitializing of the single support, must occur at foot-strike. The energy recuperation is progressive for the first case and instantaneous for the second.

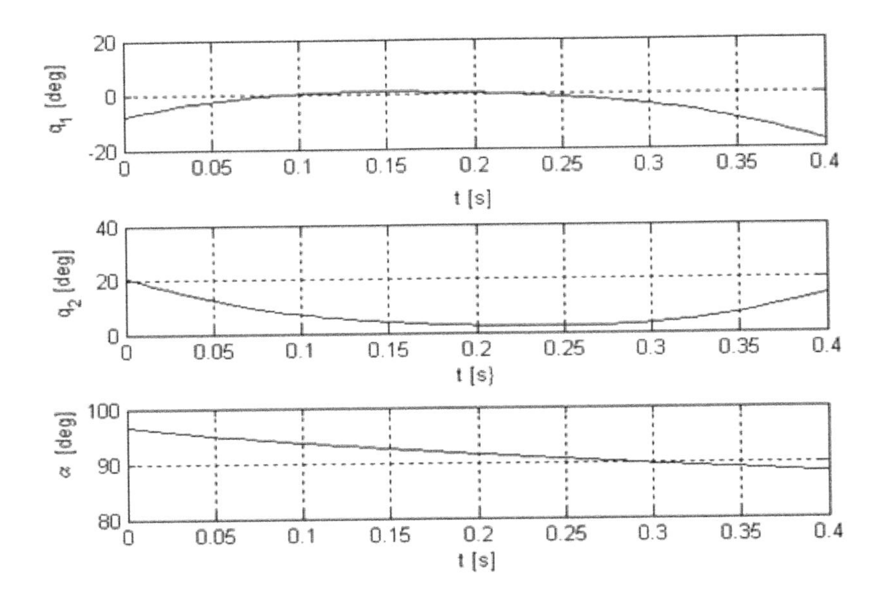

Figure 4.3. *Time charts of rotations of stance leg body-segments and equivalent virtual leg*

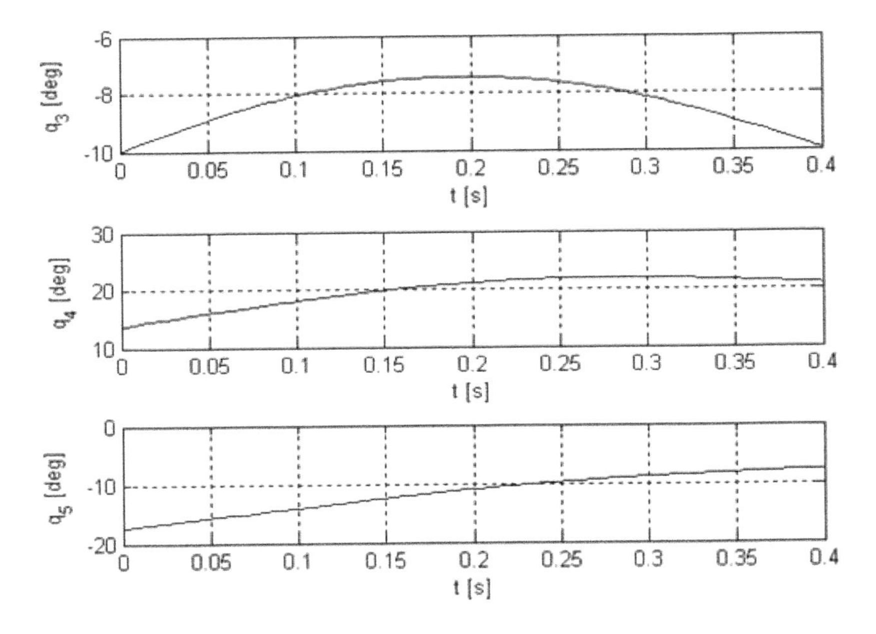

Figure 4.4. *Time charts of rotations of the trunk and swing leg body segments*

The instantaneous double support reinitializes (through impulsional joint torques) a passive and cyclic single support. This can be broken down into three main events, as indicated in Figure 4.5.

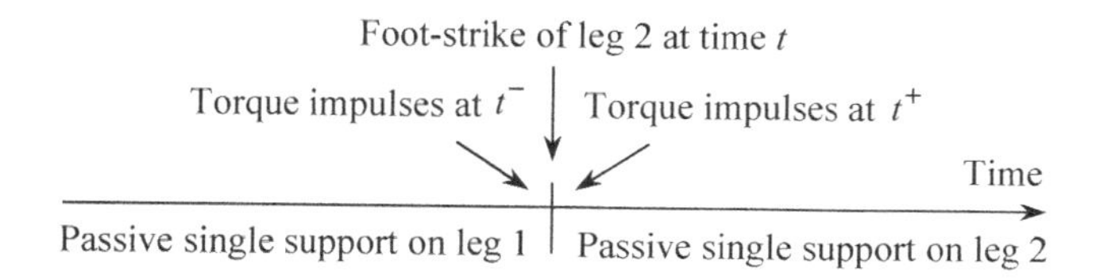

Figure 4.5. *Application of the driving torque impulses during the instantaneous double support*

The first event is the application of impulsional control torques which reduce the final velocities of the single support to those of impact. The second event is impact itself. The third event is the application of a complementary impulsional control which determines the initial velocities of the following single support. This problem generally has more unknown variables (control torques, speeds which define passive impact and ground reactions) than algebraic equations with which to define the instantaneous double support. Solving an optimization problem based on the minimization of an energetic criterion can resolve this indeterminacy.

Generating movements through dynamic impulses is put forward in [CHE 98, FOR 94] to motivate the authors' approach based on energy saving. Impulsional control of a dynamic system induces, in effect, the minimization of spent energy. It is well known that an optimizing criterion, which is purely energetic, generates "bang-off-bang" controls which tend toward impulsional controls when their bounds are released. Nevertheless, in the control model of the biped which completes the study presented in [CHE 98], double support is considered during a brief instant during which the delivered actuating torques are bounded. This is a more realistic approach which enables the contact constraints to fulfill the non-sliding condition and authorizes the saturation of the physical controls (joint torques at acceptable values). Starting a cyclic, pendulum-like movement using brief and finite actuating controls required full actuation for the mechanical system. This is the case for the planar biped considered in [CHE 98].

A study of the pendulum effect on horizontal plane walking was developed in [ROS 01] and focused on the aspects of movement dynamics and mass distribution. A sagittal model was studied (as seen in Figure 4.6) and the following hypotheses were included in the study:

– the stance foot remains flat during the single support;

– the stance leg is locked in extension;

– the ankle of the swing leg is locked. This simplification is close to that of human walking during single support.

The fact that there is also a trunk means that an active torque must be added at the hip.

The biped's seven body segments are therefore cut down to four rigid and articulated bodies. These are labeled S_{23}, S_4, S_5 and S_{67} in Figure 4.6. The chosen dimensions for the numerical simulations are the same as those used for the BIP biped in [SAR 98].

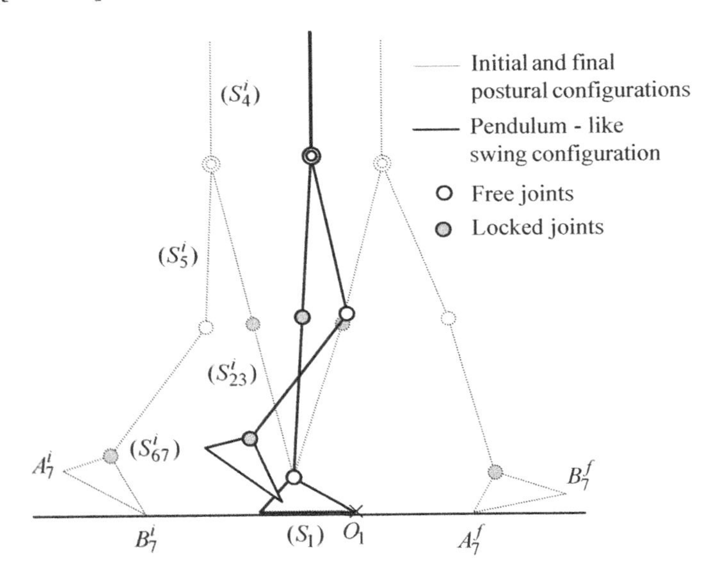

Figure 4.6. *Study of pendulous swing phase with an extended stance leg (knee locked)*

For this first approach the study of pendulous transfer only consists of determining a step length together with initial conditions for postural configuration and joint velocities, which could ensure a passive transfer of the swing leg to a forward position which maintains the step. The method used is based on the formulation of an optimization problem which aims to minimize the impact velocity and any shifts when repositioning. This is done by adjusting the initial conditions and the step length, as well as the angle of the rear foot ankle when it is at its initial position. The correlations between the final and initial conditions are made by integrating a complete dynamic model of the mechanical system.

In the second approach, the aim is to find design aids for conceiving a biped taking advantage of pendulous effects. The approach is based on the fact that pendulous behavior is closely linked to the mass distribution of the body segments. This aspect is focused on by the authors [COL 01b, GAR 98b] who highlight the importance of mass distribution when designing passive bipeds. With this in mind, [ROS 01] aims to make active walking bipeds benefit from the sufficiently marked coupled effects of gravity and biped inertia in order to create a natural organization of swing transfer with a reduced, or even zero, spent energy. The applied method utilizes the previously mentioned approach which introduced design parameters in terms of complementary optimization variables. These parameters consist of the local coordinates of centers of mass and the ratio of the body-segment masses in relation to the biped's total mass. The parameters are set within feasible bounds, predefined by design constraints, which specify a certain type of limited mass distribution and bounded geometric dimensions.

4.3. Static balance walking

McGheer and Franck [MCGH 68] studied a robot walker's locomotion along a horizontal plane. It had a rigid trunk and massless legs which could exert impulsional contact forces. They proposed the following static stability definition for this "ideal" robot walker. An ideal robot walker is statically stable at instant t if the legs maintain their contact when the biped immobilizes its movement at that same instant. McGheer and Franck also show that this definition of static equilibrium is equivalent to the condition that the center of mass should be located within the polygon of support. On this basis, they defined the margin of static stability at instant t as being the shortest distance between the vertical projection of the biped's center of gravity and the boundary of the support polygon. In the case of sagittal walking and during single support phases, the projection of the biped's center of gravity must be between the two extremities of the stance foot. During double support phases, the projection of the biped's center of gravity must be between the tip of the forward foot and the end of the rear foot. It must be noted that for a biped in single support on a flat foot, when the velocity and acceleration of the generalized coordinates of biped are zero, the projection of the inertia center and the contact pressure center between the sole and the ground are merged.

This concept of static stability was used to create walking trajectories. For the BIP robot [LYD 02], the authors obtained optimal evolutions of physical variables which were representative of the configuration of a biped in space (the center of mass of the trunk and pelvis, ankle positions, abduction and adduction movements and foot orientation) by solving a constrained minimization problem, the criterion to be minimized consisting of quadratic joint torques. Along with the constraints of joint rotation bounds, maximum torque values and the conditions of sole/ground contact, the authors added a constraint on the biped's static stability margin during single and double support.

4.4. Dynamic synthesis of walking

By dynamic synthesis we mean an optimization operation which consists of extracting a solution from the dynamic equations which can minimize a performance criterion on a set of feasible state variables and controls.

In practice, this computing process is aimed at generating reference gaits which incorporate various walking steps: start steps, stop steps, cyclic steps, climbing up or down steps and turning steps which respect all the characteristics of movement, especially that of the biped's dynamics.

Various constraints and movement equations concerning walking were formulated in Chapter 2. A minimizing criterion still needs to be formulated so that we can proceed to the development of optimization techniques which can be used to generate feasible optimal movements.

4.4.1. *Performance criteria for walking synthesis*

The aim is to define a dynamic criterion to be minimized which can determine optimal locomotion which is compatible with bipedal basic gaits. The criteria which are most often used in dynamic synthesis are aimed at minimizing driving torques or spent energy.

4.4.1.1. *Energetic criterion*

A desire to improve a biped's displacement autonomy is a strong motivation to minimize the energy cost. This cost is naturally evaluated on the basis of joint driving powers $P_i(t)$ which can be formulated as:

$$t \in [t^i, t^f],\ P_i(t) = \tau_{i^*+i}(t)\dot{q}_{i^*+i}(t),\ i = 1,\ldots, n_\tau$$

using the notation defined in Chapter 2. The subscript i^* depends on the chosen set of generalized coordinates and represents the number of generalized coordinates with no associated actuated torques: $i^* = 1$ if the biped is considered as rooted (the actuating torques are therefore subscripted from 2 to 13) and $i^* = 6$ if the biped is considered as a free mechanical system (the actuating torques are subscripted from 7 to 18; see sections 2.2.4.1 and 2.2.4.2). In both cases, $n_\tau = 12$.

Assuming that there is no recuperation of the mechanical energy when the actuating torques are "break-like", the energy used to generate and control the movement is the integral of the absolute joint powers on the time cycle:

$$J_0(\tau) = \int_{t^i}^{t^f} \sum_{i=1}^{n_\tau} |P_i(t)|\, dt$$

If we wish to focus on energetic autonomy, a more judicious choice consists of minimizing the spent energy per unit of distance traveled. This criterion is deduced from the above by setting:

$$J_1(\tau) = \frac{1}{L_{step}} J_0(\tau) \qquad [4.1]$$

where L_{step} is the step length.

The minimization of this criterion raises an objection as well as a difficulty. As is demonstrated in the theory of optimal control, a purely energetic criterion such as equation [4.1] generates control variables. In this case, joint actuating torques which are bang-off-bang (see [LEW 95]) i.e. which saturate and cancel successively during finite time intervals for the entire movement duration, are generated. The objection comes from the fact that a solution of this type can lead to jerky movements which are not very compatible with walking constraints, which are those of non-sliding unilateral contacts and tightly defined balance conditions. In addition, repeated jerks at joints could make these movements difficult to control. The difficulty arises when the optimization problem is dealt with using an exact or quasi-exact solving technique. The control variable discontinuities which appear in the differential constraint, i.e. the mechanical system's state equation, make problem-solving extremely difficult.

Nevertheless, this difficulty can be avoided by using parametric optimization techniques based on the representation of state variables by smooth approximation functions (section 4.5). This results in sub-optimal solutions with continuous controls. These are distanced approximations of the bang-off-bang exact controls. This approach is interesting because of its capacity to generate smooth movements with reduced energetic cost. Chevallereau and Aoustin [CHE 01] and Muraro *et al.*

[MUR 03] successfully used this technique to generate walking and running gaits for bipeds with simple kinematics.

4.4.1.2. *Sthenic criteria*

A sthenic criterion relates to forces which are generally driving forces or torques. The integral of quadratic actuating torques is most commonly used. As in equation [4.1], we will consider its value per unit of distance covered:

$$J_2(\tau) = \frac{1}{L_{step}} \int_{t^i}^{t^f} \tau(t)^T D_\tau \tau(t)\, dt \qquad [4.2]$$

where D_τ is a diagonal weight matrix.

This criterion was especially used in [CHE 01, MUR 03]. The above criterion is very advantageous to walking. Firstly, the resulting optimal control is continuous and cancels the risks of a jerky functioning (which can be the case for the energetic criterion). This smoothness property also guarantees a better numerical efficiency for the algorithms used for the optimization problem-solving.

In addition, as a biped is submitted to gravity, the minimization of the actuating torques will favor upright walking gaits which require a minimum amount of effort to counterbalance the effect of gravity. Such gaits therefore intrinsically obey one of the main characteristic rules of bipedia. This also means that fewer constraints for a feasible solution will be active during the numeric solving process. This makes problem solving easier.

In addition, we can re-consider the fact that a walking step includes a closed kinematic movement phase. In the rooted-system configuration with 13 generalized coordinates (see section 2.2.4.1), the opening of the kinematic loop directly injects contact forces under the forward foot into the dynamic model. These can therefore be interpreted as complementary actuating efforts which must be exerted to maintain the extremity of the free kinematic chain in the position assigned to it by the closure constraints.

We can therefore try to minimize such components of forces in the same way as the actuation torques by introducing the augmented criterion

$$J_3(\tau,\lambda) = J_2(\tau) + \frac{1}{L_{step}} \left(\int_{t_2^*}^{t_3^*} \sum_{i=6}^{10} \xi_i (\lambda_i^3(t))^2\, dt + \int_{t_3^*}^{t^f} \sum_{j=6}^{11} \varsigma_j (\lambda_j^4(t))^2\, dt \right) \qquad [4.3]$$

where the contact forces are represented by the Lagrange multipliers λ_i^3 and λ_j^4 associated with the closure constraints formulated at the front foot, and during double support (see section 2.3.1.2). The coefficients ξ_i and ζ_j are dimensionless weight factors.

The dynamic model obtained from a rooted-system configuration with 13 generalized coordinates is perfectly adapted to the formalization of this problem. Let us note that a similar criterion to that of equation [4.3] was used in [SEG 05] to generate a sagittal walking step with three sub-phases.

This can be extended to the set of ground interaction forces during the whole walking cycle, that is to say:

$$J_4(\tau,\lambda) = J_2(\tau) + \frac{1}{L_{step}} \sum_{\alpha=1}^{4} \int_{t_{\alpha-1}^*}^{t_\alpha^*} \lambda^\alpha(t)^T D_\alpha \lambda^\alpha(t)\, dt \qquad [4.4]$$

where the variable λ of J_4 is defined as the vector: $\lambda = ((\lambda^1)^T, ..., (\lambda^4)^T)^T$. A free-system configuration setting (see section 2.2.4.2) is therefore perfectly adapted to a formulation of this second approach.

Introducing the contact forces in criteria equations [4.3] and [4.4] will minimize the antagonistic efforts in the locomotion system when it is kinematically closed, and therefore over-actuated. The correlative minimization of the horizontal contact components will contribute to reducing the risk of sliding at foot level.

Such criteria (in particular, the second criterion during the single support), will also further a better distribution of normal contact efforts. The result should be a better dynamic balance for the biped.

4.4.1.3. *Mixed criteria*

Let us also note that we can use mixed effort-energy criteria by combining J_1 with any of the sthenic criteria seen above [BET 99, BOB 06]. These criteria are finally contained in the following general formulation:

$$J_{k,\mu}(\tau,\lambda) = \frac{1}{L_{step}} \int_{t_0}^{t^f} L_{k,\mu}(\tau(t),\lambda(t))\, dt \qquad [4.5]$$

where

$$L_{k,\mu}(\tau,\lambda^{\alpha}) := (1-\mu)\sum_{i=1}^{n_\tau}|P_i| + \mu[\tau^T D_\tau \tau + (\lambda^{\alpha})^T D_{\alpha,k}\lambda^{\alpha}] \qquad [4.6]$$

where $\mu \in [0,1]$; $k = \{1,\ldots,4\}$; and the weight matrix $D_{\alpha,k}$ can have a zero diagonal depending on the values of α and k.

The energetic criterion J_1 is found for $\mu = 0$ and the sthenic criteria are found for $\mu = 1$, with appropriate values for $D_{\alpha,k}$ versus k $(k = 2,3,4)$.

4.4.2. *Formalizing the problem of dynamic optimization*

The aim of this section is to give an overview of the general conditions which define the dynamic optimization problem to be dealt with. The problem can lead to very distinct formulations depending on whether the optimization method is based on the *direct* or *inverse* dynamic approach for dealing with the movement equations. The former usually leads to the formulation of an optimal control problem (also termed *differential optimization problem*). The latter leads to the construction of an algebraic optimization problem. Problems halfway between these two very distinct cases can also be formulated [BET 99].

The dynamic synthesis of optimal movements for mechanical systems submitted to control laws falls naturally within the scope of the optimal control theory. The typical mathematical tool with which it is associated is the Pontryagin Maximum Principle (PMP) [BRY 95, PON 62]. Using the PMP is appealing for several reasons. Firstly, its theoretical usefulness is particularly manifested through the *maximum condition* which ensures that constraints set on forces (sthenic constraints) are processed accurately. Secondly, the PMP can also lead to the development of algorithms computationally efficient for solving the dynamic optimization problem [BRY 95].

Nevertheless, it can be difficult to put the PMP into practice for dealing with movement synthesis of complex systems (such as mechanical bipeds). Obtaining the required formulations, such as the necessary conditions for optimality, can be a complex process. These conditions, when developed, can start to take on considerable proportions and could be quite intricate. In addition, the formally exact solutions produced by the PMP have mostly sharp variations, or can be discontinuous. This can be the source of numerical instabilities which can render the solving algorithm inoperative.

We should remember that a step is a continuous sequence of phases; each phase is characterized by its own dynamic model and constraints. During movement, the problem contents vary and this may be a noticeable complicating factor. The parameterization of the problem and converting it into an algebraic optimization problem is an efficient alternative when faced with these difficulties. This approach will be adopted from this point onwards.

Criterion [4.5], which has to be minimized, is the basis of the problem which must be formulated. When the PMP is used, the multipliers λ^{α} contained in λ can be considered as complementary control variables, independent of the physical controls τ_i (as was done in [BES 04]). When using the parametric optimization approach based on the prior operation of inverse dynamics, the variables τ and λ^{α} (which appear in criterion [4.5]) will not be considered as explicitly independent. They will be treated as given functions of the kinematic variables. Distinctions will then have to be made between two cases depending on the way contact forces are dealt with through the inverse dynamic process (see section 2.3.1.3).

The formulation of the simplest case is the result of the direct determination of τ and λ^{α} in relation to the movement kinematics (see sections 2.3.1.3.2 and 2.3.1.3.3), which leads to functional dependences such as equation [2.77] in $(q, \dot{q}, \ddot{q})$. The $J_{k,\mu}$ criterion is therefore formulated as:

$$J_{k,\mu}(\tau,\lambda) = \frac{1}{L_{step}} \int_{t^i}^{t^f} L_{k,\mu}(\tau(q(t),\dot{q}(t),\ddot{q}(t)),\lambda(q(t),\dot{q}(t),\ddot{q}(t))\, dt \qquad [4.7]$$

When the contact forces are parameterized (section 2.3.1.3.1), $J_{k,\mu}$ becomes (for the case represented by equations [2.59] and [2.60]):

$$J_{k,\mu}(\tau,\lambda) = \frac{1}{L_{step}} \int_{t^i}^{t^f} L_{k,\mu}(\tau(q(t),\dot{q}(t),\ddot{q}(t),\lambda^*(t)),\lambda^*(t))\, dt \qquad [4.8]$$

The feasible values for the actuating torques and the contact forces are defined by the set of constraints introduced in Chapter 2. To theses constraints are added those which relate directly to the state of the system described by the phase variables $(q, \dot{q})$. They can all be summarized according to the following categorization.

4.4.2.1. *Instantaneous constraints*

Such conditions are imposed at set instants. They regroup transition constraints which must be taken into account between adjacent steps, as well as between phases and sub-phases of the step about to be made. These were defined in equations [2.27], [2.31] and [2.32] (section 2.2.5), and can be summarized as follows:

$$g_k(t_k^*) = 0 \, (\in \Re^{n(k)}), \; k \in \{0,1,2,3,4\} \qquad [4.9]$$

where instants t_k^* are those of equation [2.1] (section 2.2.2). The exponent $n(k)$ indicates that the dimension of the vector function g_k varies with k, which was specified in equations [2.27], [2.31] and [2.32].

4.4.2.2. *Permanent equality constraints*

Permanent equality constraints are represented by the closure equations formulated in equations [2.12], [2.14], [2.19] and [2.21] (section 2.2.4.3). They must be fulfilled successively during each sub-phase. An abbreviated formulation is:

$$t \in I_\alpha, \; \Phi^\alpha(q(t)) = 0 \; (\in \Re^{n(\alpha)}), \; \alpha \in \{1,2,3,4\} \qquad [4.10]$$

We also have localization constraints for the center of pressure on the heel's edge of the forward foot after touch-down, or on the front tip of the propulsive foot. They are taken from equations [2.136] and [2.137], and are grouped together in the form of:

$$t \in I_\alpha, \begin{cases} \alpha \in \{2,3,4\}, \, g_p^\alpha(\lambda_p^\alpha(t)) = 0 \\ \alpha = 3, \, g_{p+1}^3(\lambda_{p+1}^3(t)) = 0 \end{cases} \qquad [4.11]$$

We must also add movement equations that are independent to actuating torques τ_is, which are obtained in the form (Chapter 2):

$$t \in I_\alpha, \; \Psi_i^\alpha(q(t), \dot{q}(t), \ddot{q}(t)) = 0 \qquad [4.12]$$

with $i = 6$ and $\alpha = 2$ in equation [2.72], $i = 1$ and $\alpha = 2$ in equation [2.74], and $i \le 6$, $\alpha \le 4$ in equation [2.84].

4.4.2.3. *Permanent inequality constraints*

Permanent inequality constraints are the most numerous. It is necessary to group them together according to the time intervals which define them.

Let us first consider those which vary in number and content depending on the time intervals under consideration. The location constraints for the centers of pressure defined in equations [2.134]–[2.137] are adjoined to the unilateral contact conditions. They can be grouped together in the following way:

$$t \in I_\alpha, \begin{cases} \text{Stance foot} \begin{cases} \alpha = 1,\ 1 \le k \le 6,\ h^1_{pk}(\lambda^1_p(t)) < 0 \\ \alpha \in \{2,3,4\},\ 1 \le k \le 3,\ h^\alpha_{pk}(\lambda^\alpha_p(t)) < 0 \end{cases} \\ \text{Front foot} \begin{cases} \alpha = 3,\ 1 \le k \le 3,\ h^3_{p+1,k}(\lambda^3_{p+1}(t)) < 0 \\ \alpha = 4,\ 7 \le k \le 12,\ h^4_{p+1,k}(\lambda^4_{p+1}(t)) < 0 \end{cases} \end{cases} \qquad [4.13]$$

The non-sliding constraints for the foot-ground contact equations [2.138] and [2.139] are defined analogously as:

$$t \in I_\alpha, \begin{cases} \text{Stance foot}: \alpha \in \{1,2,3,4\},\ h^{nsl}_p(\lambda^\alpha_p(t)) < 0 \\ \text{Front foot}: \alpha \in \{3,4\},\ h^{nsl}_{p+1}(\lambda^\alpha_{p+1}(t)) < 0 \end{cases} \qquad [4.14]$$

For formulations [4.11], [4.13] and [4.14], the multipliers λ^α_p and λ^α_{p+1} must be reformulated by inverse dynamics as was indicated in section 2.3.1.3.

The internal and external non-collision constraints must then be defined only during the monopodal phase equations [2.146] and [2.149]. The result is the following re-grouping:

$$t \in I_{SSP}, \begin{cases} h^{nci}(q(t)) < 0 \\ h^{nce}(q(t)) < 0 \end{cases} \qquad [4.15]$$

There remain the conditions to be fulfilled for the complete step. These are the box constraints expressing the limitations applied to the actuating torques, as well as the joint rotations and velocities in equations [2.141], [2.143] and [2.145] (see section 2.5.1) respectively. To summarize,

$$t \in I_{step}, \begin{cases} h^{\tau}(q(t), \dot{q}(t), \ddot{q}(t)) < 0 \\ h^{q}(q(t)) < 0 \\ h^{\dot{q}}(\dot{q}(t)) < 0 \end{cases} \quad [4.16]$$

In the constrained minimization problem defined by conditions [4.7] to [4.16], it must be noted that after the representation of τ and λ by functions of the type $\tau(q, \dot{q}, \ddot{q})$ and $\lambda(q, \dot{q}, \ddot{q})$ (section 2.3.1.3.2), or of the type $\tau(q, \dot{q}, \ddot{q}, \lambda_{p+1}^{\alpha *})$ and $\lambda_{p}^{\alpha}(q, \dot{q}, \ddot{q}, \lambda_{p+1}^{\alpha *})$ (section 2.3.1.3.1) obtained by inverse dynamics (Chapter 2), the movement equations will have been used completely and definitively. Once the problem has been reformulated in this way, it can be treated on the basis of the parameterized representation of the kinematics of movement and all, or some, of the multipliers.

4.5. Walking synthesis via parametric optimization

Fundamentally, the parameterization of a dynamic optimization problem consists of representing the functions to be optimized (these are generally the configuration variables or sometimes the movement control variables) by a finite number of discrete parameters which become the new optimization variables. This operation can be based on two quite different approaches, depending on which of the two types of preceding variables the parameterization represents. State variable parameterization is most often used. This parameterization is also the easiest to use when converting a dynamic optimization problem into a finite dimensional optimization problem. This is particularly true for walking, which is characterized by successive changes in state, many distinct contact modes and as many kinematic and dynamic models.

This parameterization leads to an approximate representation of the initial optimization problem. The built solutions are therefore sub-optimal. Their relative quality is closely linked to the number of independent parameters which were introduced to represent the problem under consideration. The quality also depends on the nature of the approximation functions employed. The problem of differential optimization is therefore converted into an algebraic optimization problem in which the time variable is eliminated. This is why the term of static optimization is also used [BRY 99]. This final problem can be solved by using computation codes implementing Sequential Quadratic Programming (SQP) [FLE 87, GIL 81] algorithms which have proved to be numerically robust.

4.5.1. *Approximating the control variables*

First, the time cycle is partitioned into sub-intervals defined using nodes t_k which can be uniformly distributed along the motion time:

$$\left\{t_1(=t^i), t_2, ..., t_k, ..., t_{N+1}(=t^f)\right\}, \ \Delta t_k = t_{k+1} - t_k, \ k = 1,..., N \qquad [4.17]$$

The control variables τ_is can then be defined as piecewise constant functions on the sub-intervals Δt_k [ROU 97], or they can be reconstructed via linear interpolation between the nodes t_k [AND 01]. The decision variables, i.e. the optimization parameters of the nonlinear programming problem to be formulated, are therefore the values of τ_is at nodes t_k. This approach has mainly been developed in the field of aerospace sciences for trajectory optimization [BET 99]. Its usage is limited when applied to the optimization of movements of multibody systems as it requires repeated integrations of the dynamic model for each set of calculated values of control variable τ_is .

Where a walking synthesis is concerned, this technique is applied in [ROU 97] to generate a walking step for a planar biped with four active joints. The state variables, which are represented by the configuration coordinates and their derivatives, are calculated step by step in relation to the $\tau_i(t_k)$ by using finite-difference approximations to deal with the movement equations. A criterion is introduced to minimize the integral of joint actuating torques as well as the velocity deviations when the forward foot touches down and when the rear foot toes off. A simplified dynamic model only takes into account the diagonal terms of the biped's mass matrix.

The application which is discussed in [AND 01] is based on the introduction of a musculoskeletal model of the locomotion system. Its actuating principle therefore becomes quite complex. The control variables of the mechanical system are no longer the joint torques τ_is but muscular activation levels. By following this model, which incorporates physiological parameters, the criterion to be minimized is the metabolic muscular energy which needs to be spent. The problem of dynamic optimization is then recast into a nonlinear programming problem by parameterizing the τ_is by their own values at nodes t_k, and by using linear interpolations between nodes. The solving technique is described in [PAN 92]. An initial state being given, the dynamic equations are integrated in order to reduce the evolution of the system to a generalized final state which is finally optimized. The accuracy of this method closely depends on the initial state under consideration. In addition, the retained muscular model introduces uncertainties through the phenomenological law which establishes a link between the degree of muscular activation and the produced muscular forces. Such a model also requires the introduction of a great number of optimization parameters (810 variables). This greatly increases the computational burden and the calculation time. In short, this approach does not seem very appropriate to the synthesis of movement which requires initial and final optimized states, as is the case for a walking cycle.

The parameterizations of the configuration variables which will be developed in the following sections give us the possibility of dealing with this difficulty in a simple way. In general, this approach is easy to use and enables us to deal with the characteristic constraints of walking as a whole.

4.5.2. *Parameterizing the configuration variables*

This approach is based on the representation of the configuration parameters using approximation functions. Polynomials are most often used, as are polynomial functions with compact support, spline functions or trigonometric functions (as truncated Fourier series). In all these cases, the method consists of using the following type of representation:

$$i \leq n_q,\ t \in [t^i, t^f],\ q_i(t) \cong \varphi_i(X^i, t) \tag{4.18}$$

where the variable X^i is a vector which regroups the shaping parameters of the approximation function φ_i. The components of X^i will be the new parameters to be optimized once the φ_i have been defined in relation to X^i and t. Two types of construction for these functions will be presented in the following sections.

4.5.2.1. *Approximation using polynomials*

The construction of the polynomial functions on the interval $[t^i, t^f]$ in equation [4.18] is carried out using the end configurations of the robot and inner configurations. Let us consider the example of set points for the actuated joint coordinates for a robot walker [CHA 90, CHE 01]. In short, the aim is to define the initial and final configurations and velocities. There is also the possibility of defining an inner configuration at t^{int} time to avoid, for example, a collision between the ground and the swing leg. In this case the number of parameters or coefficients for each polynomial is equal to four and is deduced from the following five equalities:

$$
\begin{aligned}
& q_i\left(t^i\right)=\varphi_i\left(X^i,t^i\right),\ \dot{q}_i\left(t^i\right)=\dot{\varphi}_i\left(X^i,t^i\right),\\
& q_i\left(t^{int}\right)=\varphi_i\left(X^i,t^{int}\right),\\
& q_i\left(t^f\right)=\varphi_i\left(X^i,t^f\right),\ \dot{q}_i\left(t^f\right)=\dot{\varphi}_i\left(X^i,t^f\right),
\end{aligned}
\quad [4.19]
$$

in which the following simplified notation has been introduced:

$$
\dot{\varphi}_i(X^i,t) := \frac{\partial \varphi_i}{\partial t}(X^i,t)
$$

For each generalized coordinate q_i associated with a powered joint, the polynomial is defined only by the values $q_i\left(t^i\right)$, $\dot{q}_i\left(t^i\right)$, $q_i\left(t^{int}\right)$, $q_i\left(t^f\right)$, $\dot{q}_i\left(t^f\right)$ and T $(T = t^f - t^i)$.

A low polynomial degree can mean no solution. High degree polynomials can lead to difficulties, such as the risk of inappropriate oscillations for the approximated functions.

During an under-actuated robot's unbalanced phase, the manner in which a control follows a reference movement may lead to final configuration errors just before the impacts on the ground during its walking and running gaits. So that we can limit these errors, reference movements can be defined by two polynomials on the interval defined in equation [4.17] when experimenting with single support. The first polynomial depends explicitly on the running time. The second polynomial is simply a constant:

$$
q_i(t) = \begin{cases} \varphi(X^i,t), & t \in [t^i,t^*] \\ q_i(t^{*+}), & t \in [t^*,t^f] \end{cases} \quad [4.20]
$$

The objective is to enable the closed-loop control of the robot walker to attain the final chosen configuration before impact. Certain authors [MUR 03, PLE 03] have used Besier polynomials for their curve-fitting properties.

When considering a set of $n+1$ control points $p_0, p_1,..., p_n$, the general formulation of Bezier polynomials is:

$$
\varphi_i\left(X^i,\tau\right) = \sum_{j=0}^{n} B_j^n(\tau)\, p_j, \quad 0 \le \tau \le 1 \quad [4.21]
$$

where $B_j^n(\tau)$, $j = 0,...,n$ is defined:

$$B_j^n(\tau) = \frac{n!}{j!(n-j)!}\tau^j(1-\tau)^{n-j}$$

The derivative with respect to τ is given by:

$$\frac{d\varphi_i(X^i,\tau)}{d\tau} = n\sum_{j=0}^{n-1}\left[p_{j+1} - p_j\right]B_j^{n-1}(\tau) \qquad [4.22]$$

With five control points with which to define an initial configuration, an initial velocity, a middle configuration, a final configuration and a final velocity, relationship [4.21] becomes:

$$\varphi_i(X^i,\tau) = (1-\tau)^4 p_0 + 4\tau(1-\tau)^3 p_1 + 6\tau^2(1-\tau)^2 p_2 + 4\tau^3(1-\tau)p_3 + \tau^4 p_4$$

The definition of the control points when taking into account equations [4.21] and [4.22] and by considering the time variable τ as having no dimension $(\tau = (t-t^i)/(t^f - t^i))$ is:

$$\begin{aligned}
p_0 &= \varphi_i(X^i,t^i) \\
p_1 &= p_0 + \frac{(t^f - t^i)}{4}\dot{\varphi}_i(X^i,t^i) \\
p_3 &= p_4 - \frac{(t^f - t^i)}{4}\dot{\varphi}_i(X^i,t^f) \\
p_4 &= \varphi_i(X^i,t^f)
\end{aligned} \qquad [4.23]$$

The control point p_2 is chosen in order to enable the curve to pass through a midway position. In this way, the Bezier curve does not pass through control point p_2. This characteristic can have the advantage of reducing the oscillations which occur with usual polynomials (when many intermediary points are used, for example). The spline method, which is based on linking polynomials up with inner points, is better adapted to a case where many control points need to be specified. This type of approach will be developed further in section 4.5.2.2.

4.5.2.2. *Approximation using spline functions*

The construction of spline functions is based on the prior partitioning of their definition range, as in equation [4.17] for the time interval $[t^i, t^f]$. A spline function defined on $[t^i, t^f]$ is obtained as the concatenation of N polynomial functions which are successively defined on the sub-intervals Δt_k, and which are attached in pairs at control times t_ks. This type of construction can be created in different ways, depending on the degree of polynomials used and the required smoothness level of spline functions at connecting nodes t_ks.

The general properties of spline functions were developed in works such as [DEB 78, MAT 92]. The main interest of these functions is that of not having to use high degree polynomials (because of the risk of undesirable oscillations), while allowing for a sufficient number of control points to be taken into account. The splines to be constructed must have a smoothness order at nodes adapted to the problem under consideration. Splines of class C^2 are most often used (functions with continuous second derivative). Using 3-degree polynomials is sufficient to ensure the concatenation of the second derivatives at nodes. This low degree eliminates the risk of unwanted oscillations between the control points while providing continuous second derivatives. This last point is important for the representation of the configuration parameters in equation [4.18].

Another advantage of splines in our application is that they allow the coefficients of connected polynomials to be expressed as functions of the values $q_i(t_k)$ at nodes t_k of the approximated functions $q_i(t)$. This transformation is of great interest in dynamic optimization because when varying the $q_i(t_k)$ to search for a solution, the conditions for optimality will have a greater sensitivity to these variations than those of the polynomial coefficients which have reduced physical meaning.

The representation of the q_is by splines of class C^2 was carried out in [SAI 03] for the construction of optimal walking gaits for a seven-link planar biped. Polynomial coefficients can then be eliminated in a very efficient way by solving linear systems with tri-diagonal matrix. The smoothness level obtained at connecting points, however, means that the second derivatives, i.e. the accelerations $\ddot{q}_i$s, have sharp (or jerky) variations.

This type of local behavior can be amplified at certain nodes in the joint torques. There is therefore the risk of generating jerky movements. This can be solved by constructing splines of class C^3 in order to ensure the continuity of the third order derivatives (the jerks). A construction of this type was used in [BES 05, SEG 05] to create sagittal walking gaits for the biped BIP [SAR 98]. The following presentation is a generalization of this approach to walking steps comprising four sub-phases (see section 2.2.2).

The partition of the time cycle must take into account the successive step phases and their relative lengths (equation [2.1]). It must be noted, however, that in what follows, the final instant t^f will not be fixed but will depend on a specified walking velocity and on the step length under optimization.

In the same way, the relative and non-specified length of the two main phases can be considered as a variable which needs to be optimized. Consequently, instant t_2^* of the transition between single and double support is considered as unknown. To simplify the problem, the instants t_1^* and t_3^* between the sub-phases of single and double support will be fixed with respect to t_2^* and t^f.

We therefore define the following partitions with a fixed number of sub-intervals which are of constant length on each main phase:

– single support:

$$\left\{t_1(=t^i),\ldots,t_k,\ldots,t_{n_2}(=t_2^*)\right\},\ t_{k+1}-t_k=\delta_{ss},\ 1\leq k\leq n_2-1 \qquad [4.24]$$

– double support:

$$\left\{t_{n_2},\ldots,t_j,\ldots,t_N,t_{N+1}(=t^f)\right\},\ t_{j+1}-t_j=\delta_{ds},\ n_2\leq j\leq N \qquad [4.25]$$

Each of these partitions is assumed to be created so that there is a subscript n_1 for the single support and a subscript n_3 for the double support, such that:

$$t_{n_1}=t_1^*,\ t_{n_3}=t_3^*$$

The following notation is used in Figure 4.7:

$$q_{ik}:=q_i(t_k),\ k=1,\ldots,n_1,\ldots,n_2,\ldots,n_3,\ldots,N+1 \qquad [4.26]$$

These values of the q_is at nodes result in as many variables of the parametric optimization problem which also need to be formulated. Moreover, the search for optimal transition states requires adding the derivatives of the q_is as further variables. These are noted in the form of (notation between the parentheses in Figure 4.7):

$$\dot{q}_{ij} := \dot{q}_i(t_j), \; j = 1, n_1, n_2, n_3, N+1 \tag{4.27}$$

For each of the sub-intervals defined by equations [4.24] and [4.25], we define a 4-degree interpolation polynomial as follows:

$$1 \le k \le N, \; t \in [t_k, t_{k+1}], \\ \begin{cases} \tau = (t - t_k)/\delta, \begin{cases} \delta = \delta_{ss}, \text{si } k < n_2 \\ \delta = \delta_{ds}, \text{si } k \ge n_2 \end{cases}, \\ q_i(t) \approx P_{ik}(C_k^i, \tau) = c_{k0}^i + c_{k1}^i \tau + c_{k2}^i \tau^2 + c_{k3}^i \tau^3 + c_{k4}^i \tau^4 \end{cases} \tag{4.28}$$

where $C_k^i = (c_{k0}^i, c_{k1}^i, c_{k2}^i, c_{k3}^i, c_{k4}^i,)$.

We note that the polynomials P_{ik} are all defined on the normalized interval $[0,1]$.

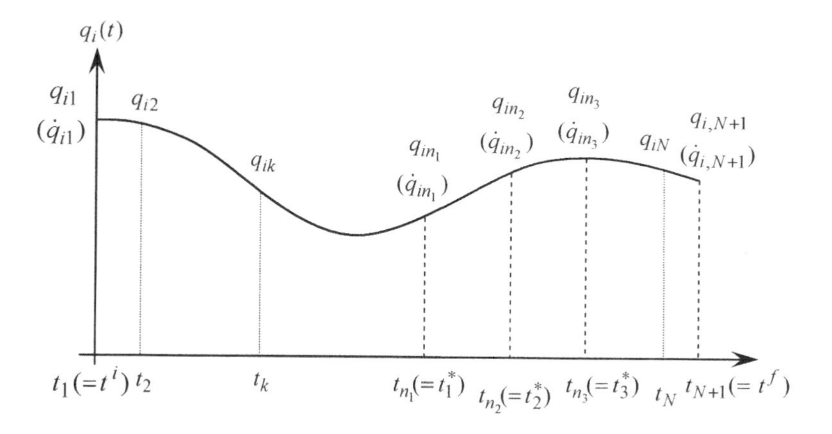

Figure 4.7. *Time cycle partitioning and values at nodes t_ks of a function $q_i(t)$*

The representation of the q_is by splines of class C^3 is to be carried out separately on each sub-phase, but not at the transition instant where there are admissible discontinuities in the accelerations and velocities when heel-touch takes place with impact. When there is no heel-strike, the connection conditions for the P_{ik} polynomials can be summarized in the following way:

$$i \leq n_q, \begin{cases} k \neq 1, n_1, n_2, n_3, N+1, \begin{cases} P_{i,k-1}(1) = P_{i,k}(0) \\ \dot{P}_{i,k-1}(1) = \dot{P}_{ik}(0) \\ \ddot{P}_{i,k-1}(1) = \ddot{P}_{ik}(0) \\ \dddot{P}_{i,k-1}(1) = \dddot{P}_{ik}(0) \end{cases} \\ k = n_1, n_2, n_3, \begin{cases} P_{i,k-1}(1) = P_{ik}(0) \\ \dot{P}_{i,k-1}(1) = \dot{P}_{ik}(0) \end{cases} \end{cases} \quad [4.29]$$

where, for the sake of simplification, the first variable C_k^i of the polynomials P_{ik} in equation [4.28] is omitted.

The first set of conditions in equation [4.29] expresses the connection of the P_{ik} up to the order 3 at the inner nodes for each sub-phase. In the second set, the connections are limited to the order 1 at the transition instants. This is nevertheless a high regularity condition and, by prescribing the continuity of the velocities, ensures the absence of impacts during the phase changes. It is possible to reinforce the smoothness of the q_is at the transition times by connecting the accelerations. Sagittal walking steps, which fulfill this condition, were generated in [SEG 05]. However, this even smoother global movement results in a significant increase in spent energy by about 50% per distance traveled, and this is only for a walking velocity of 0.7 m s^{-1}.

With conditions [4.29], construction [4.28] formally leads to the following set of spline functions ϕ_i^*s defined on the interval $[t^i, t^f]$:

$$i \leq n_q,\ k \leq N,\ t \in [t_k, t_{k+1}],\ \phi_i^*(C^i, t) = P_{ik}(C_k^i, t) \quad [4.30]$$

where $C^i = (C_1^i, \ldots, C_N^i)$. The ϕ_i^*s are therefore of class C^3 on the successive intervals I_1, I_2, I_3 and I_4 and of class C^1 on the global interval I_{step}. Equation [4.29] remains to be completed by giving P_{ik} the sought-after values of the q_is at nodes, represented in equation [4.26], as well as the initial and final values for the $\dot{q}_i$s and their values at the transition times between each phase, as is represented in equation [4.27]:

$$i \leq n_q, \begin{cases} 1 \leq k \leq N, \ P_{ik}(0) = q_{ik} \ ; \ P_{iN}(1) = q_{i,N+1} \\ j = 1, n_1, n_2, n_3, \ \dot{P}_{ij}(0) = \dot{q}_{ij} \ ; \ \dot{P}_{iN}(1) = \dot{q}_{i,N+1} \end{cases} \qquad [4.31]$$

The q_{ik} and $\dot{q}_{ij}$ of equation [4.31] represent values which are intended to be optimized and that we can regroup in the set of vectors X^is such that:

$$i \leq n_q, \ X^i = (x_1^i, ..., x_{N+6}^i)^T \equiv (q_{i1}, ..., q_{i,N+1}, \dot{q}_{i1}, \dot{q}_{in_1}, \dot{q}_{in_2}, \dot{q}_{in_3}, \dot{q}_{i,N+1})^T \qquad [4.32]$$

which leads to the number

$$N_q = n_q(N+6) \qquad [4.33]$$

of discrete variables to be optimized.

For every superscript *i*, the vector X^i is a linear function of the coefficients of the polynomial P_{ik} through the relationships [4.29] and [4.31]. A further interesting operation is to solve the obtained linear system with respect to the polynomial coefficients, in order to eliminate them from the final problem.

An overview of available equations and unknown variables which need to be determined is as follows: for every subscript *i*, equation [4.29] and [4.31] represent $5N-4$ equations, and there are N polynomials with five coefficients for each spline function. Four new equations must therefore be found, or four unknown variables must be eliminated. This condition is easy to fulfill: four 3-degree polynomials must substitute the same amount of polynomials, which were initially intended as 4-degree polynomials. As the connecting conditions have reduced order at the transition times in the chain of the N connected polynomials, we can choose to introduce a 3-degree polynomial on the final interval of each sub-phase, or:

$$k = n_1 - 1, \ n_2 - 1, \ n_3 - 1, \ N \ ; \ P_{ik}(C_k^i, \tau) = c_{k0}^i + c_{k1}^i \tau + c_{k2}^i \tau^2 + c_{k3}^i \tau^3$$

We therefore obtain a linear system with a square matrix of dimension $5N-4$. This diagonal sparse matrix, written *M*, is the same for every superscript *i*. The linear systems that need to be solved can therefore be written:

$$i \leq n_q, \ MC^i = AX^i \qquad [4.34]$$

where C^i is the column matrix of the coefficients to be determined and A is a distribution matrix of the vector X^i components, with dimension

$(5N-4)\times(N+6)$. Assuming that the matrix M is non-singular, we extract C^i from equation [4.34] as the result:

$$C^i(X^i) = M^{-1}AX^i$$

Each spline function of equation [4.30] can therefore be re-written as an X^i and time function:

$$t \in [t^i, t^f],\ \varphi_i(X^i, t) = \varphi_i^*(C^i(X^i), t)$$

This transformation completes the parameterization of the q_is in the form of equation [4.18]. This is the basis of the possible reformulation of the dynamic optimization problem summarized from equations [4.7]–[4.16], in the form of a parametric optimization problem. This study base can also be extended to the parameterization of multipliers. This operation will be described in section 4.5.3 which follows.

4.5.3. *Parameterizing the Lagrange multipliers*

The Lagrange multipliers represent the components of ground-foot interaction forces and moments at the freed contacts (see section 2.3.1.2). Their parameterization can be analogous to that of the q_is in the preceding section. Such a representation therefore enables us to directly determine the actuating torques without having to use the operation of pseudo-inversion described in section 2.3.1.3.2. However, it must be understood that this approach implies a significant increase in the number of optimization variables which will have to be processed in this new parametric optimization problem.

Concerning the representation of forces, the smoothness order for the approximation functions to be defined can be reduced to the simple continuity on each sub-phase. Therefore, splines of class C^0 are an appropriate choice. If, however, we wish to avoid the problem of having angular variations, we can raise the order of smoothness by one unit by introducing splines of class C^1.

For the sake of simplicity, and with an aim of limiting the number of new discrete variables to optimize, the construction is limited to the case described at the end of section 2.3.1.3.1. Let us note that the Lagrange, or better still the Newton-Euler, dynamic model is then formulated using 13 q_is, with an opening of the kinematic loop under the forward foot during the double support. An inverse dynamic model in τ_i (actuating torques) can therefore be obtained by the parameterization of the contact efforts under the forward foot. These are represented by the multipliers $\lambda_{p+1,i}^{\alpha}$ for $\alpha = 3$ and $i \in \{1,...,5\}$ for the first sub-phase, and $\alpha = 4$ and $i \in \{1,...,6\}$ for the second sub-phase. We notice that the two multipliers with the same subscript i less than or equal to 5 represent the same contact component during the whole double support. But all phase changes may give rise to effort discontinuities at the transition time. We must therefore approximate the $\lambda_{p+1,i}^{\alpha}$ separately during their respective sub-phases.

The principle for the construction of the approximation functions is the same as that used for the q_is in section 4.5.2.

We substitute equation [4.18] for a representation of the same type for the multipliers under consideration:

$$\begin{cases} t \in I_3,\ i \leq 5,\ \lambda_{p+1,i}^{3}(t) \cong \psi_i^3(Y^{3i}, t) \\ t \in I_4,\ i \leq 6,\ \lambda_{p+1,i}^{4}(t) \cong \psi_i^4(Y^{4i}, t) \end{cases} \quad [4.35]$$

The variables $Y^{\alpha i}$ $(\alpha = 3, 4)$ are the vectors of shaping parameters (to be optimized) of the approximation functions ψ_i^{α}s. The aim is to build the ψ_i^{α}s as piecewise functions on the time sub-intervals defined by the time-slicing equation [4.24]. This is detailed in the following form where the transition instant t_3^* between the sub-phases of the double support is explicitly shown:

$$\left\{t_{n_2}(=t_2^*),...,t_k,...,t_{n_3}(=t_3^*),...,t_j,...,t_N, t_{N+1}(=t^f)\right\} \quad [4.36]$$

As in equation [4.24], we have:

$$t_{k+1} - t_k = \delta_{ds},\ n_2 \leq k \leq N$$

To simplify, we formulate the values of $\lambda_{p+1,i}^{\alpha}$ at nodes in the following abbreviated form (in which the subscript $p+1$ can be omitted):

$$\Lambda_{ik}^{\alpha} := \lambda_{p+1,i}^{\alpha}(t_k),\ k = n_2,...,n_3,...,N+1 \quad [4.37]$$

These values will make up as many new discrete variables for optimization, which have to be added to the N_q variables introduced in equation [4.32].

In an analogous way to the re-grouping in equation [4.32], they can be put together in the two following sets of vectors $Y^{\alpha i}$ defined for $\alpha = 3, 4$:

$$\begin{cases} i \le 5, \alpha = 3, \ Y^{3i} = (y_1^{3i}, \dots, y_{n_3-n_2+1}^{3i})^T \equiv (\Lambda_{in_2}^3, \dots, \Lambda_{in_3}^3)^T \\ i \le 6, \alpha = 4, \ Y^{4i} = (y_1^{4i}, \dots, y_{N-n_3+2}^{4i})^T \equiv (\Lambda_{in_3}^4, \dots, \Lambda_{i,N+1}^4)^T \end{cases} \quad [4.38]$$

These vectors are then put together in turn in the unique vector Y as in:

$$Y = (y_1, \dots, y_l, \dots, y_{N_\lambda}) \equiv ((Y^{31})^T, \dots, (Y^{35})^T, (Y^{41})^T, \dots, (Y^{46})^T)^T \quad [4.39]$$

where:

$$N_\lambda = 5(n_3 - n_2 + 1) + 6(N - n_3 + 2) \quad [4.40]$$

The approximation of the $\lambda_{p+1,i}^\alpha$ by splines of class C^0 is done simply via an interpolation of their Λ_{ik}^α values at nodes, using 1-degree polynomials defined as follows:

$$\begin{aligned} & n_2 \le k \le N, \ t \in [t_k, t_{k+1}], \ \tau = (t - t_k)/\delta_{ds}, \\ & \begin{cases} i \le 5; \ k = n_2, \dots, n_3 - 1; \ \lambda_{p+1,i}^3(t) \cong Q_{ik}^3(\tau) = c_{k0}^{3i} + c_{k1}^{3i}\tau \\ i \le 6; \ k = n_3, \dots, N; \ \lambda_{p+1,i}^4(t) \cong Q_{ik}^4(\tau) = c_{k0}^{4i} + c_{k1}^{4i}\tau \end{cases} \end{aligned} \quad [4.41]$$

We wish to determine the polynomial coefficients of the Q_{ik}^α in relation to the Λ_{ik}^α. To do this we have the connecting conditions at the inner nodes on each interval I_3 and I_4:

$$i \le 5; \ k = n_2 + 1, \dots, n_3 - 1; \ Q_{i,k-1}^3(1) = Q_{ik}^3(0) \quad [4.42]$$

$$i \le 6; \ k = n_3 + 1, \dots, N; \ Q_{i,k-1}^4(1) = Q_{ik}^4(0) \quad [4.43]$$

as well as the allocation relationships at Q_{ik}^{α} of the sought-after values of the $\lambda_{p+1,i}^{\alpha}$ at nodes, or:

$$i \leq 5;\ k = n_2,\dots,n_3 - 1;\ Q_{i,k}^3(0) = \Lambda_{ik}^3,\ Q_{i,n_3-1}^3(1) = \Lambda_{in_3}^3 \qquad [4.44]$$

$$i \leq 6;\ k = n_3,\dots,N;\ Q_{i,k}^4(0) = \Lambda_{ik}^4,\ Q_{i,N}^4(1) = \Lambda_{i,N+1}^4 \qquad [4.45]$$

The number of conditions to be fulfilled and the number of coefficients to be determined amount jointly to:

– $N_3 = 10(n_3 - n_2)$ for $\alpha = 3$ (conditions [4.42] and [4.44]); and

– $N_4 = 12(N - n_3 + 1)$ for $\alpha = 4$ (conditions [4.43] and [4.45]).

The two corresponding linear systems can be solved explicitly. For $\alpha = 3$, conditions [4.42] and [4.44] can be recombined in the form:

$$k = 0,1,\dots,n_3 - n_2 - 1;\begin{cases} Q_{i,n_2+k}^3(0) = \Lambda_{i,n_2+k}^3 \\ Q_{i,n_2+k}^3(1) = \Lambda_{i,n_2+k+1}^3 \end{cases} \qquad [4.46]$$

When taking into account formulations [4.41], this system leads to the determination of the Q_{ik}^3 polynomial coefficients in relation to the newly defined variables in equation [4.38]:

$$i \leq 5; k = 0,1,\dots,n_3 - n_2 - 1;\begin{cases} c_{n_2+k,0}^{3i} = y_{k+1}^{3i} \\ c_{n_2+k,1}^{3i} = y_{k+2}^{3i} - y_{k+1}^{3i} \end{cases} \qquad [4.47]$$

For $\alpha = 4$, the corresponding system, represented by equations [4.43] and [4.45], gives the following solution which is formally identical to its precedent:

$$i \leq 6; k = 0,1,\dots,N - n_3;\begin{cases} c_{n_3+k,0}^{4i} = y_{k+1}^{4i} \\ c_{n_3+k,1}^{4i} = y_{k+2}^{4i} - y_{k+1}^{4i} \end{cases} \qquad [4.48]$$

By carrying these results over to equation [4.41], we obtain through equations [4.37] and [4.38] the parameterization of the multipliers in the form of the sought-after equation [4.35]. We can then arrive at a complete formulation of the parametric optimization problem.

4.5.4. *Formulation of the parametric optimization problem*

The transformation must lead to the elimination of time in all the conditions which define the initial optimization problem. This operation can be carried out in the following stages.

4.5.4.1. *Criterion transformation*

The objective here is to reformulate the integral criterion for minimization in the form of a function of parameters x_k^i of equation [4.32] and y_l of equation [4.39].

In fact, this set of parameters is insufficient for the global optimization of an SSP-DSP walking cycle. The cycle is associated with essential variables such as the duration of the cycle, the step length and the step's average velocity. The cycle is also made up of the relative duration of one of the two main phases: the ratio of the single support duration, for example, to that of the whole step. The first three quantities are:

$$T_{step}\ (T_{step} = t^f - t^i),\ L_{step} \text{ and } V_{step}$$

and the fourth is

$$x_{sp} = T_{ss} / T_{step}\ (T_{ss} = t_2^* - t^i)$$

These variables are not independent. By considering V_{pas} as basic data, the two correlations:

$$T_{step} = L_{step} / V_{step}\ ;\ T_{ss} = x_{sp} L_{step} / V_{step} \qquad [4.49]$$

show that L_{step} and x_{sp} can be considered as independent variables which lead to the determination of T_{step} and T_{ss} or, in an equivalent manner, of instants t^f and t_2^*.

It is therefore possible to make L_{step} and x_{sp} play the role of two supplementary optimization parameters for the determination of a globally optimized step. The general criterion equation [4.5] is therefore minimized in the form of:

$$\text{Min } J_{k,\mu}(\tau, \lambda, L_{step}, x_{sp}) \tag{4.50}$$

where the explicit dependency with respect to x_{sp} is effective using t_2^* which is considered as a middle integration bound for the evaluation of $J_{k,\mu}$ on the interval $[t^i, t^f]$.

By using the second expression of equation [4.49], this dependency is explicitly expressed by the relationship:

$$t_2^*(x_{sp}) = t^i + x_{sp}\, L_{step} / V_{step}$$

We group the completed optimization parameters together in the N^*-vector:

$$X = (x_1, \dots, x_{N^*}) \equiv ((X^1)^T, \dots, (X^{n_q})^T, L_{step}, x_{sp})^T$$

where $N^* = n_q(N+6)+2$ (equation [4.33]).

By using equation [4.18], we can therefore write:

$$t \in I_{step},\ q(t) \cong \varphi(X,t) = (\varphi_1(X^1,t), \dots, \varphi_{n_q}(X^{n_q},t))^T \tag{4.51}$$

then by referring to equations [4.38] and [4.35], we can define:

$$\lambda^*(t) \equiv \psi(Y,t) = \begin{cases} (\psi_1^3(Y,t), \dots, \psi_5^3(Y,t))^T, t \in I_3 \\ (\psi_1^4(Y,t), \dots, \psi_6^4(Y,t))^T, t \in I_4 \end{cases} \tag{4.52}$$

where λ^* is the 5-vector then 6-vector of the parameterized multipliers.

By carrying over the approximations $\varphi(X,t)$ of $q(t)$, and $\psi(Y,t)$ of $\lambda^*(t)$ to criterion [4.8] which is considered in its completed form, equation [4.50]; it is then approximated by the following function $F_{k,\mu}$ of variables X and Y:

$$J_{k,\mu}(\tau,\lambda,L_{step},x_{sp}) \cong F_{k,\mu}(X,Y) = \frac{1}{L_{step}}\sum_{\alpha=1}^{4}\int_{t_{\alpha-1}^{*}}^{t_{\alpha}^{*}} L_{k,\mu}(\tau^{*}(X,Y,t),\psi(Y,t))\,dt \qquad [4.53]$$

where:

$$\tau^{*}(X,Y,t) = \tau(\varphi(X,t),\frac{\partial}{\partial t}\varphi(X,t),\frac{\partial^{2}}{\partial t^{2}}\varphi(X,t),\psi(Y,t))$$

The problem under consideration therefore consists of minimizing the function $F_{k,\mu}$ as follows:

$$\underset{(X,Y)\in D_{ad}\subset\Re^{N^{*}\times N_{\lambda}}}{Min} \quad F_{k,\mu}(X,Y) \qquad [4.54]$$

on an feasible set D_{ad} included in $\Re^{N^{*}\times N_{\lambda}}$ and limited by constraints [4.9]–[4.16], after they are recast into functions of X and Y with the elimination of time. This transformation is developed further below.

It is worth noting that in the absence of the parameterization of the multipliers, i.e. when τ and λ are conjointly determined by the same inverse dynamics operation (see section 2.3.1.3.2 or 2.3.1.3.3), X is the only outstanding variable. Function $F_{k,\mu}$ is therefore minimized as follows:

$$\underset{X\in D_{ad}\subset\Re^{N^{*}}}{Min} \quad F_{k,\mu}(X)$$

In this case we have the benefit of reducing the number of variables to be optimized, but this means a greater computational complexity for the inverse dynamics.

4.5.4.2. *Treating the constraints*

First we must note that all the control points t_ks are defined in equations [4.24] and [4.25] in relation to t^f and t_2^*, then using equation [4.49] as functions of the optimization variables L_{step} and x_{sp}. They are therefore known instants as a function of X. As a first consequence, the point constraints [4.9] remain unchanged.

All the constraints which are formulated from equations [4.10]–[4.16] are functions of q, $\dot{q}$, $\ddot{q}$ and multipliers λ_p^α, λ_{p+1}^α and λ^α. By using approximations [4.51] and [4.52], we immediately arrive at a general representation of all the constraints as functions of (X, Y, t) which we can summarize in the following way:

$$t \in I_\alpha, \begin{cases} G^\alpha(q(t)) \cong G^{*\alpha}(X,Y,t) = 0 \\ H^\alpha(q(t),\dot{q}(t),\ddot{q}(t),\lambda^*(t)) \cong H^{*\alpha}(X,Y,t) < 0 \end{cases}, \alpha \in \{1,2,3,4\} \qquad [4.55]$$

where H^α is a vector function which groups all of the components of the constraint functions (from [4.13] to [4.16]) together, when we consider that constraints [4.15] are identically fulfilled on the time intervals I_3 and I_4 which are complementary to I_{SS}.

The problem now consists of treating constraints $G^{*\alpha}$ and $H^{*\alpha}$ of equation [4.55] so that we can neglect the time variable. The two following procedures are possible.

4.5.4.2.1. Fulfilling the constraints at nodes

It is possible to fulfill the preceding constraints only at the control points if the number of these points appears to be sufficient. Conditions [4.55] are then considered in the form:

$$\alpha = 1,2,3,4;\ n_{\alpha-1} + 1 \le k \le n_\alpha, \begin{cases} G^{*\alpha}(X,Y,t_k) = 0 \\ H^{*\alpha}(X,Y,t_k) < 0 \end{cases} \qquad [4.56]$$

where n_α represents subscripts n_1, n_2, n_3 and $N+1$ as they appear in equations [4.24], [4.26] and [4.27] (with $n_0 = 0$ and $n_4 = N+1$).

It should be noted that the number of these point constraints, denoted N_{pc}, can be high. This is represented by the sum:

$$N_{pc} = \sum_{\alpha=1}^{4}(N_{G^\alpha} + N_{H^\alpha})n_\alpha$$

where N_{G^α} and N_{H^α} are the number of constraints defined on each interval I_α by the vector functions G^α and H^α, respectively. If we suppose that all these constraints have been taken into account, and the time cycle is sliced into 10 time intervals ($N = 10$, 11 nodes) with five, two, one and two intervals for the four successive sub-phases, the value of N_{pc} would be close to 900, which is a considerable amount. In this evaluation, the box constraints [4.16] are by far the most numerous: on each line of equation [4.16] the 12 permanent constraints are doubled for the torques, the positions and the joint velocities successively. The result is 792 point constraints. A high number of bounds are not attained when the generated movement is not very fast. This remark enables us to considerably reduce the number of these constraints by limiting ourselves to those which are revealed to be indispensable.

This approach was chosen in [SEG 05] to generate sagittal walking steps for a 7-link robot. The computation code used (routine *fmincon* of Matlab®'s *Optimization Toolbox*) has the capacity to fulfill these point constraints with great accuracy. However, this excellent result is made relative to the fact that the original constraints [4.55] are free to vary between the control points where they are generally infringed. In the quoted example, the infringement amounts are acceptable and can easily be corrected by slightly shifting the authorized bounds.

This technique is efficient, but only within its own limits. Firstly, there is the risk of significant constraint infringements between nodes. A rise in the number of control points may reduce the transgression, but at a price. This is the difficulty of the correlative increase of the number of constraints to be treated. The result is an increase in the complexity of the problem which can have a negative outcome on the calculation time, as well as on the capacity of the solving process to converge. It is possible to avoid this problem by using the following alternative method.

4.5.4.2.2. Penalty method

This technique, which is inspired here by the well-known exterior penalty method used in mathematical programming [BON 97], consists of minimizing the constraint violations when they are infringed. We define the violations of inequality constraints over the running time as follows:

$$t \in I_{\alpha}, \begin{cases} H_j^{*\alpha+}(X,Y,t) = \text{Max}\,(H_j^{*\alpha}(X,Y,t),0),\ 1 \le j \le N_{H^{\alpha}} \\ H^{*\alpha+}(X,Y,t) = (H_1^{*\alpha+}(X,Y,t),\ldots,H_{N_{H^{\alpha}}}^{*\alpha+}(X,Y,t)) \end{cases}$$

We can therefore envisage the possibility of fulfilling the double set of both the equality and inequality constraints of [4.55] by minimizing the augmented criterion:

$$F^{*}_{k,\mu,r_1,r_2}(X,Y) = F_{k,\mu}(X,Y)$$
$$+\frac{1}{2}\sum_{\alpha=1}^{4}\int_{t_{\alpha-1}}^{t_{\alpha}}\left(r_1\left\|G^{*\alpha}(X,Y,t)\right\|^2 + r_2\left\|H^{*\alpha+}(X,Y,t)\right\|^2\right)dt$$

in which coefficients r_1 and r_2 are penalty factors. The cost function F^* must be minimized for these coefficients with fairly high given data, so that we obtain a minimizing effect of the quadratic integral norms of functions $G^{*\alpha}$ and $H^{*\alpha+}$, which is sufficient to bring about residual values close to zero.

A few precautions need to be taken when using this technique. First, the beginnings of a convergence generally require a guess solution which is compatible with low or zero penalty factor values. The problem solving can then go ahead iteratively for moderately increasing values of r_1 and r_2. As a general rule, the greater the imposed final coefficient values, the better the constraints are respected. It must be noted, however, that high values of the penalty factors are likely to deteriorate the numerical conditioning of the problem to be solved. A compromise must therefore be found between the desired solution accuracy and the algorithm's capacity of mastering a numerical conditioning which has a tendency to degenerate.

4.5.5. *A parametric optimization example*

Cyclic steps were created using parametric optimization [MIO 06] for a five-link planar walker without feet (Figure 4.8). The physical parameters which were introduced (sizes, masses and moments of inertia) were those of the *Rabbit* biped [CHE 03]. The movement generalized coordinates are the relative rotations at joints and the absolute rotation of the stance leg. Let us note that the time variation of the latter rotation is monotonous during the single support, as is the orientation of the dummy foot (see section 4.2.2). This approach facilitates the treatment of under-actuation, as is indicated in the following section.

The steps generated are made up of a single support phase, but also of a double support phase with a non-zero finite duration. The touch-down of the swing leg occurs with no impact velocity, which eliminates the propagation of velocity jumps in the articulated chain. The approximation functions which are introduced (polynomials noted δ_i) to describe the single support phase are original because they do not depend explicitly on running time, but on the rotation angle α of the stance leg: $\delta_i = \delta_i(\alpha)$ ($i = 1,...,4$). This choice, which is justified by the monotonous variation of the variable α, enables us to rid ourselves of the specification of an impact instant and to easily treat the problem of the biped's under-actuation during single support. During double support, the biped only has three DoF. The rotation α is re-considered as a time function. With a view to coinciding with the single support formulation, the two remaining parameters of the independent internal rotations can be defined in relation to α : $\delta_i = \delta_i(\alpha)$ $(i = 1,2)$. The degree of the polynomials which are introduced and the cyclic conditions lead to a number of optimization parameters equal to 18.

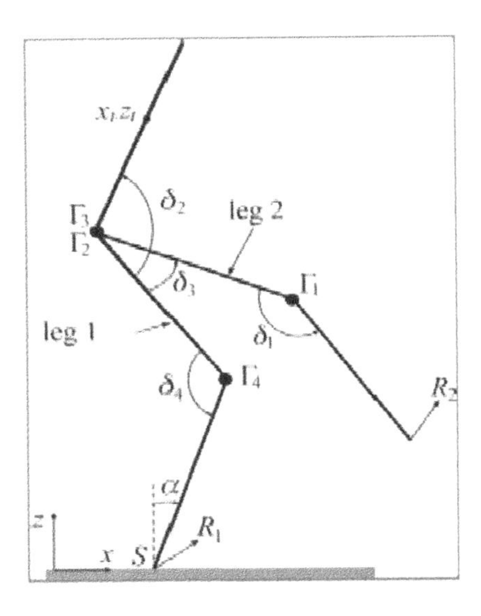

Figure 4.8. *Diagram of biped with indications of the distribution of the generalized coordinates, torques and ground reaction forces*

The criterion, which is adopted for the whole step, is the integral:

$$J = \frac{1}{L_{step}} \left(\int_{\alpha_{iss}}^{\alpha_{fss}} \frac{\Gamma^T \Gamma}{\dot{\alpha}} d\mu + \int_0^{T_{ds}} \Gamma^T \Gamma dt \right) \quad [4.57]$$

where α_{iss} and α_{fss} represent the values of α at the beginning and at the end of single support. T_{ds} is the duration of double support.

To improve the convergence of the optimization algorithm, the gradient of the criterion to be minimized was developed in relation to the 18 parameters for optimization. Another particularity of the movement to be optimized is the calculation of the joint torques during double support. During this phase, the biped is over-actuated: four torque controls are required to create a movement with only three DoF. Faced with the multiplicity of admissible solutions (by respecting the constraints), it is possible to extract one solution by solving the optimization problem in which one of the ground reaction forces is treated as an optimization variable. The horizontal component of one of these two forces was chosen as an optimization parameter. As seen in Chapter 2, for any given movement there is a unique solution for the vertical components R_{1y} and R_{2y} of the reaction forces. This is not the case for the horizontal components R_{1x} and R_{2x} (see Figure 4.9).

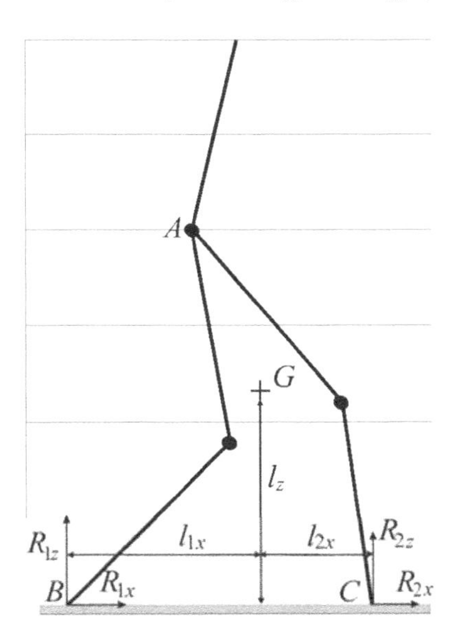

Figure 4.9. *Diagram of biped during double support with localization of the center of mass G*

By choosing R_{2x} as a variable, the biped's necessary torques during the double support phase are determined by minimizing the criterion:

$$\min_{R_{2x}} \Gamma^{*T}\Gamma^{*} \qquad [4.58]$$

under non-sliding and non-roll-off constraints for the feet.

Figure 4.10 shows the velocities of four angular actuated variables which are the result of the parametric optimization of a cyclic walk with single and double support phases. These velocities clearly remain on the inside of the limited domain of the velocity/torque template provided by *Rabbit's* motor. These types of curves can be used to select the biped's motors and, more generally, to give the biped its dimensions by taking the existing motors into account [CHE 00].

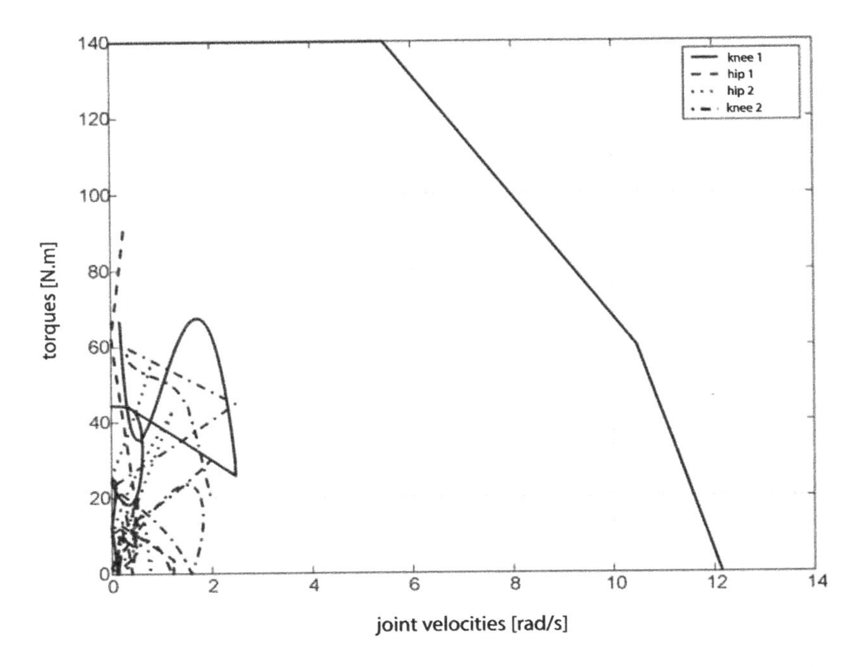

Figure 4.10. *Torque variations versus the joint velocities and a representation of the velocity/torque template of the Rabbit motors*

Figures 4.11 and 4.12 show the time variations of torques at the hips and knees and normal components of the ground reaction forces. The single and double support phases are visible. The transition which occurs at touch-down of the swing leg at instant $t = 0.93\,\text{s}$ is marked by the torque and contact force discontinuities. The final configuration of double support triggers the single support of the following step, which is a translation of the cyclic nature of the step which has been generated. Figure 4.13 shows the variations of the criterion residual value obtained after minimization, depending on the biped's forward velocity. We can observe a noticeable rise in the criterion when the required velocity is higher.

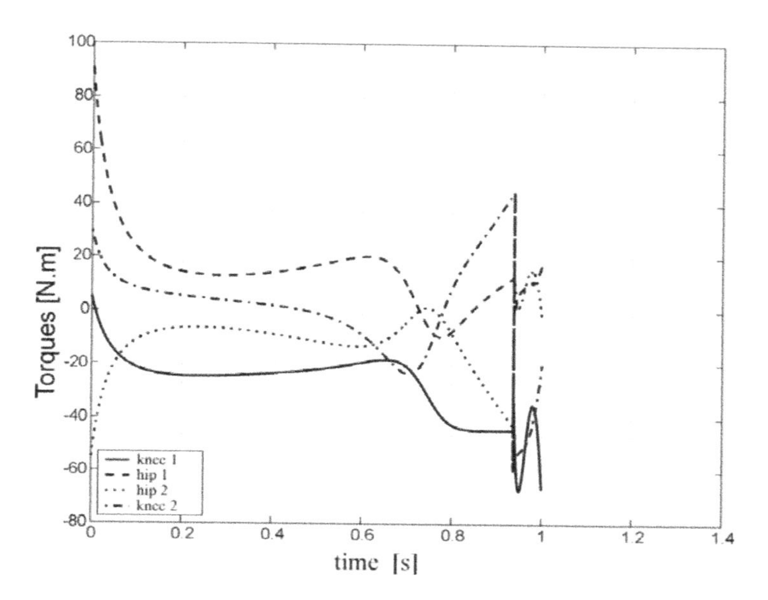

Figure 4.11. *Time charts of actuating torques during a step*

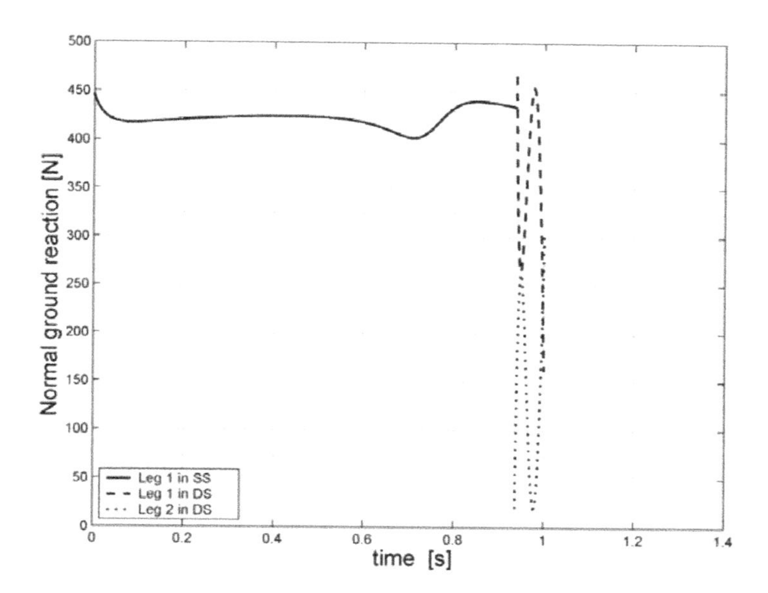

Figure 4.12. *Time chart of normal components of reaction forces during a step*

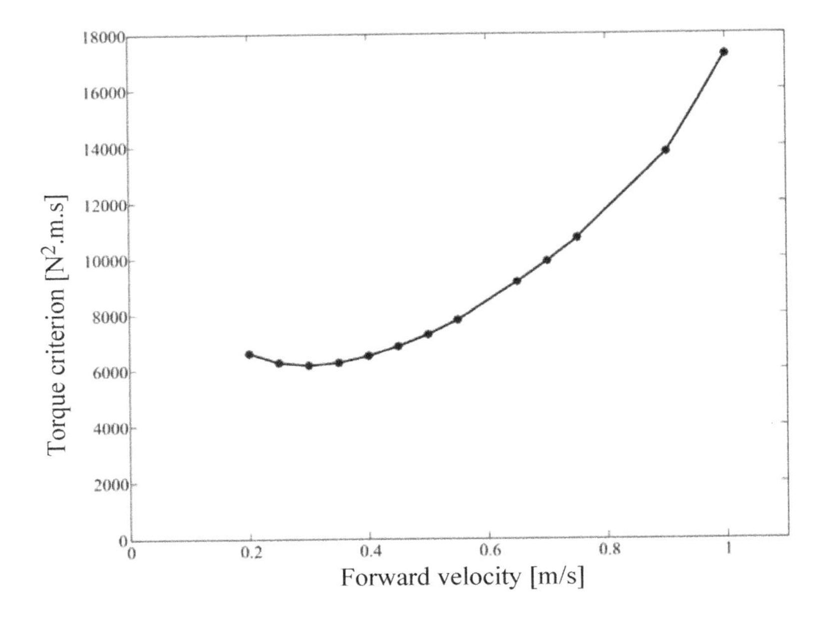

Figure 4.13. *Variation of the minimized criterion versus the biped's walking velocity*

This rapid rise in the effort of walking leads to the observation that running is more efficient in required effort when the biped is above a certain forward velocity level.

Indeed, in past studies [CHE 01], a comparison was made between cyclic walking made up of single support phases followed by impacts and a cyclic race. The optimization criterion was of the same type as that presented in equation [4.57].

Figure 4.14a shows that from 1 m s^{-1} it is more advantageous for the biped under consideration to run than walk.

In the same way, the higher the biped's velocity, the higher the ratio between the aerial phase and single support during a step (see Figure 4.14b).

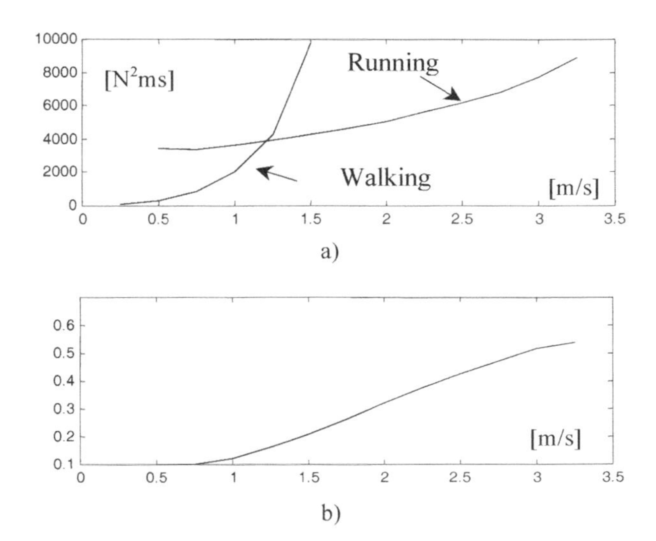

Figure 4.14. *(a) Value comparisons of the minimized criterion in relation to the biped's walking and running velocities and (b) variations in ratio between the duration of the aerial phase and the total duration of a step in relation to the forward velocity*

4.6. Conclusion

The creation of passive bipeds has shown that walking on a slight slope can be generated spontaneously via pendulous effects. This reaction is the result of the dynamic gravity-inertia coupling linked to an appropriate mass distribution for the biped. This property can help us to improve the dynamic behavior of active walking bipeds. However, in a quasi-reciprocal way, it is also possible (for a given mass distribution) to create walking steps which combine gravity and inertia with minimum effort cost or minimum spent energy. In this way, this general organizing principle replaces the passive approach, but without eliminating it all together.

This approach is made concrete when it is based on the biped's complete dynamic modeling, and calls for the use of the optimization theory which enables us to deal efficiently with the dynamic synthesis of movement. Basically, in order to create this movement according to the data, dynamic laws and limits must be fulfilled. To this is added a selection aspect, as the obtained movement is also the one which answers best to an execution criterion. This type of construction has the advantage of creating movements with well-organized kinematics and fully coordinated joint dynamics. This is also a guarantee of energetic efficiency, as well as of security in terms of good distribution of contact forces, and therefore of good control of the biped balance.

The optimization problems under consideration are generally solved by using parametric techniques. Formulating such problems and the general presentation of solving techniques were the corpus of this chapter. Optimal dynamic synthesis techniques must be developed and furthered to contribute to an in-depth knowledge of the dynamics of walking, for robots as well as humans. This is also a reliable means of organizing walking steps of a biped robot with a view to controlling its movements. This essential aspect of walking control *in situ* for bipedal robots will be the subject of the following chapter.

4.7. Bibliography

[ALE 95] ALEXANDER R.MC.N., "Simple models of human movement", *Applied Mechanic Reviews*, no. 48, p. 461–469, 1995.

[AND 01] ANDERSON F.C., PANDY M.G., "Dynamic optimization of human walking", *Journal of Biomechanical Engineering*, no. 123, p. 381–390, 2001.

[AOU 03] AOUSTIN Y., FORMAL'SKY A.M., SARVROVSKY E., "Ballistic run of an anthropomorphic biped", *Proceedings of the Conference on Climbing and Walking Robots, CLAWAR'03*, p. 399–408, 2003.

[BES 04] BESSONNET G., CHESSE S., SARDAIN P., "Optimal gait synthesis of a seven-link planar biped", *International Journal of Robotics Research*, no. 33, p. 1059–1073, 2004.

[BES 05] BESSONNET G., SEGUIN P., SARDAIN P., "A parametric optimization approach to walking pattern synthesis", *International Journal of Robotics Research*, no. 24, p. 523–536, 2005.

[BET 99] BETTS J.T., "Survey of numerical methods for trajectory optimization", *Journal of Guidance, Control and Dynamics*, vol. 21, no. 2, p. 193–207, 1999.

[BOB 06] BOBROW J.E., PARK F.C., SIDERIS A., "Recent advances on the algorithm optimization of robot motion", Lecture notes in *Control and Information Sciences*, Springer-Verlag, New York, p. 21–41, 2006.

[BON 97] BONNANS J.F., GILBERT J.C., LEMARECHAL C., SAGASTIZABAL C., *Optimization Numérique*, Springer-Verlag, Berlin, 1997.

[BRY 95] BRYSON A.E., HO Y.C., *Applied Optimal Control*, John Wiley & Sons, New York, 1995.

[BRY 99] BRYSON A.E., *Dynamic Optimization*, Addison-Wesley, Reading, Massachusetts, 1999.

[CHA 88] CHATELIN F., *Valeurs propres de matrices*, Masson, Paris, 1988.

[CHA 90] CHANNON P.H., HOPKINS S.H., PHAM D.T., "Simulation and optimization of gait for a bipedal robot", *Math and Computer Modelling*, no. 14, p. 463–467, 1990.

[CHE 98] CHEVALLEREAU C., FORMAL'SKY A., PERRIN B., "Low energy cost reference trajectories for a biped robot", *Proceedings of IEEE International Conference on Robotics and Automation*, p. 1398–1404, Leuven, Belgium, 1998.

[CHE 00] CHEVALLEREAU C., SARDIN P., "Design and actuation optimization of a 4 axes biped robot for walking and running", *Proceedings of IEEE International Conference on Robotics and Automation*, p. 3365–3370, San Francisco, USA, 2000.

[CHE 01] CHEVALLEREAU C., AOUSTIN Y., "Optimal reference trajectories for walking and running of a biped robot", *Robotica*, no. 19, p. 557–569, 2001.

[CHE 03] CHEVALLEREAU C., ABBA G.Y., AOUSTIN Y., PLESTAN F., WESTERVELT E.R., CANUDAS DE WIT C., GRIZZLE J.W., "Rabbit: a testbed for advanced control theory", *IEEE Control Systems Magazine*, vol. 23, no. 5, p. 57–78, 2003.

[COL 01a] www.ijrr.org./contents/20_07/abstract/607.html.

[COL 01b] COLLINS S.H., WISE M., RUINA A., "A three-dimensional passive-dynamic walking robot with two legs and knees", *International Journal of Robotics Research*, no. 20, p. 607–615, 2001.

[DEB 78] DE BOOR C., *A Practical Guide to Splines*, Springer-Verlag, New York, 1978.

[ESP 94] ESPIAU B., GOSWAMI A., "Compass gait revisited", *4th IFAC Symposium On Robot Control*, p. 839–846, 1994.

[FLE 87] FLETCHER R., *Practical Methods of Optimization*, 2nd edition, John Wiley and Sons, 1987.

[FOR 94] FORMAL'SKY A., "Impulsive control for anthropomorphic biped", *Proceedings of Ro-Man-Sy 10, Theory and Practice of Robots and Manipulators*, p. 387–393, Udine, Italy, 1994.

[GAL 01] GALLIER J., *Geometric Methods and Applications*, Springer-Verlag, New York, 2001.

[GAR 98a] GARCIA M., CHATTERJEE A., RUINA A., COLEMAN M., "The simplest walking model: stability, complexity, and scaling", *ASME Journal of Biomechanical Engineering*, no. 120, p. 281–286, 1998.

[GAR 98b] GARCIA M., CHATTERJEE A., RUINA A., "Speed, efficiency, and stability of small-slope 2-D passive dynamic bipedal walking", *Proceedings of IEEE International Conference on Robotics And Automation*, p. 2351–2356, 1990.

[GIL 81] GILL P. E., MURRAY W., WRIGHT M. H., *Practical Optimization*, Academic Press, London and New York, 1981.

[GOS 96] GOSWAMI A., TUILOT B., ESPIAU B., "Compass-like biped robot. Part I: stability and bifurcation of passive gaits", *INRIA Rhône-Alpes, Research report no. 2996*, 1996.

[HIR 03] HIRUKAWA H., KANEHIRO F., KAJITA S., FUJIWARA K., YOKOI K., KANEKO K., HARADA K., "Experimental evaluation of the dynamics simulation of biped walking of humanoid robots", *Proceedings of IEEE International Conference on Robotics and Automation*, Taipei, China, p. 14–19, 2003.

[KAN 04] KANEHIRO F., HIRUKAWA H., KAJITA S., "OpenHRP: open architecture humanoid robotics platform", *International Journal of Robotic Research*, no. 23, p. 155–165, 2004.

[KHA 86] KHALIL W., KLEINFINGER J.F., "A new geometric notation for open and closed loop robots", *Proceedings of IEEE International Conference on Robotics and Automation*, San Francisco, USA, p. 1174–1179, 1986.

[LEB 03] LEBOEUF F., BESSONNET G., LACOUTURE P., "Travail des efforts internes en analyse du geste sportif: étude comparative de divers bilans", *Actes du 16e Congrès Français de Mécanique*, Nice, 2003.

[LEW 95] LEWIS F.L., SYRMOS V.L., *Optimal Control*, John Wiley and Sons, New York, 1995.

[LYD 02] LYDOIRE F., AZEVEDO C., ESPIAU B., POIGNET P., "3D parameterized gaits for biped walking", *Proceedings of 5th International Conference on Climbing and Walking Robots*, Paris, p. 749–757, 2002.

[MAT 92] MATHEWS J.H., *Numerical Methods for Mathematics, Science and Engineering*, Prentice Hall International Editions, India, 1992.

[MCG 90a] MCGEER T., "Passive dynamic walking", *International Journal of Robotic Research*, no. 9, p. 62–82, 1990.

[MCG 90b] MCGEER T., "Passive walking with knees", *Proceedings of IEEE International Conference on Robotics and Automation*, p. 1640–1645, Cincinnati, USA, 1990.

[MCGH 68] MCGHEE R.B., FRANK R.B., "On the stability properties of quadruped creeping gaits", *Journal of Mathematical Biosciences*, no. 3, p. 331–351, 1968.

[MCM 84] MCMAHON T.A., "Mechanics of locomotion", *International Journal of Robotic Research*, no. 3, p. 4–28, 1984.

[MIO 06] MIOSSEC S., AOUSTIN Y, "Dynamical synthesis of a walking cyclic gait for a biped with point feet", in S. MIOSSEC (ed.), *Control and Information Sciences*, Springer Verlag, vol. 2, no. 12, p. 233–253, 2006.

[MUR 03] MURARO A., CHEVALLEREAU C., AOUSTIN Y., "Optimal trajectories of a quadruped robot with trot, amble and curvet gaits for two energetic criteria", *Multibody System Dynamics*, no. 9, p. 39–62, 2003.

[OUE 03] OUEZDOU F.B., KONNO A., SELLAOUTI R., GRAVEZ F., MOHAMED B., BRUNEAU O., "ROBIAN biped project – a tool for the analysis of the human-being locomotion system", *Proceedings of 6th International Conference on Climbing and Walking Robots*, Catania, Italy, p. 375–382, 2003.

[PAN 92] PANDY M.G., ANDERSON F.C., and HULL D.G., "A parameter optimization approach for the optimal control of large-scale musculoskeletal systems", *Journal of Biomechanical Engineering*, vol. 114, 450–460, 1992.

[PFE 03] PFEIFFER F., LÖFFLER K., GIENGER M., "Humanoid robots", *Proceedings of 6th International Conference on Climbing and Walking Robots*, p. 505–516, Catania, Italy, 2003.

[PLE 03] PLESTAN F., GRIZZLE J.W., WESTERVELT W., ABBA G., "Stable walking of a 7-DOF biped robot", *IEEE Transaction on Robotics and Automation*, vol. 19, no. 4, p. 653–668, 2003.

[PON 62] PONTRYAGIN L., BOLTIANSKY V., GAMKRELITZE A., MISHCHENKO E., *The Mathematical Theory of Optimal Processes*, Wiley Intersciences, New York, 1962.

[ROS 01] ROSTAMI M., BESSONNET G., "Sagittal gait of a biped robot during the single support phase. Part 1: passive motion", *Robotica*, no. 19, p. 163–176, 2001.

[ROU 97] ROUSSEL L., CANUDAS DE WIT C., GOSWAMI A., "Generation of energy optimal complete gait cycles for biped", *Proceedings of IEEE International Conference on Robotics and Automation*, p. 2036–2042, 1997.

[SAI 03] SAIDOUNI T., BESSONNET G., "Generating globally optimized sagittal gait cycles of a biped robot", *Robotica*, no. 21, p. 199–210, 2003.

[SAR 98] SARDAIN P., ROSTAMI M., BESSONNET G., "An anthropomorphic biped robot: dynamic concepts and technological design", *IEEE Transactions on Systems Man and Cybernetics*, no. 28a, p. 823–838, 1998.

[SEG 05] SEGUIN P., BESSONNET G., "Generating optimal walking cycles using spline-based state-parameterization", *International Journal of Humanoid Robotics*, no. 2, p. 47–80, 2005.

Chapter 5

Control

5.1. Introduction

There are two main robot walker categories: robots with statically stable gaits and those with dynamically stable gaits. For the first category, the dynamic effects can be neglected because their locomotion is sufficiently slow. In this case, the robot configuration is said to be stable if the center of mass is projected within the convex polygon defined by the ground footprints. These gaits often concern robots which have more than three stance feet [SON 88]; they can also concern bipedal robots [ELA 99].

In this presentation we will limit ourselves to the case of gaits which are not statically stable. The dynamic effects are not negligible. Apart from the specifications due to the model of robot walkers, the task to be accomplished is in itself particular. We could consider that the aim is to make a marked robot frame (e.g. its head) obey a request, but this approach is not generally adopted as it is not easy to manage the problems linked to equilibrium. Let us instead start from the perception that human walking is "putting one foot in front of the other and then starting over again". The locomotive task corresponds to the different articulations converging towards a cyclic movement, and results in a forward movement along a path. As our study concerns a hybrid bipedal robot model (a dynamic model which is structurally different depending on the support phases), the task will be the execution of a cyclic movement. This is why the study of the evolution of the system in the phase plane and the stability analyses (with the help of the Poincaré map) will be particularly useful. These tools will be presented in section 5.2 and their usage will be illustrated by different applications in sections 5.4–5.6 and 5.8.

To make the robot execute the locomotive task, a joint reference motion is usually defined and a subsequent control ensures that the movement is properly tracked. This approach will be developed in the first part of this chapter. We can distinguish the control of a robot walker with that of a robot manipulator from the following features:

– the different walking phases are described by different models;

– the impact is described by algebraic equations with discontinuous robot velocities;

– the contact robot/ground is unilateral;

– during multiple support phases, the robot can be over-actuated. An identical movement of the legs therefore corresponds to different motor torques and contact forces.

These features are linked to the robot modeling, as seen in Chapter 2. The contact intermittence is made possible because the contact constraint between the robot and the ground is naturally unilateral (a stance leg's foot can lose contact with the ground because of the robot's movement). Certain difficulties arise as a result of this important characteristic. Indeed, a large majority of current day controls are based on a predefined succession of the different contacts: flat-footed grounding, two-footed grounding, impact and rotation along the heel. It is therefore necessary to ensure that the real contact corresponds to the desired contact. In particular, it is difficult to ensure that there is an effective flat-footed grounding.

The conditions described in section 2.4.1, which use the zero moment point (ZMP) were therefore defined with this aim in view. Many of today's humanoid prototypes use control methods based on the notion of the ZMP. Examples of their application will be presented in sections 5.3 and 5.4. In section 5.3, an initial approach will be suggested. This is where the robot shifts from its reference movement to fulfill the contact constraint. The disadvantage of this approach is that to fulfill the mechanical constraints, the gap between the reference and the tracked movements can widen and result in high torques before going back to the reference, when this becomes possible. Section 5.4 shows how to limit this effect by modifying the reference online, while maintaining it in a given set of reference. In the case of a temporal modification of the reference, a complete stability study is carried out for a planar robot whose walk is defined as a succession of single support and impact phases. The introduction of a point support phase will be considered in section 5.5.

The introduction of a non-instantaneous double support phase will be covered in section 5.6.

There are more global approaches where the generation of references and controls are closely linked. This is the case e.g. for predictive control, intuitive and neuronal approaches or approaches which stem from passive movements on a slope. The first approach (predictive control) is attractive because of its adaptation possibilities but it has a high calculation cost and its efficiency is yet to be proved [AZE 02]. The second and third methods were validated through experimentation on different robots [AOU 06, PRA 98, SAB 04]. These methods will be discussed in section 5.7. A robot (with no trunk) which can perform purely passive movements on a slope can be control on flat horizontal terrain if it is completely actuated. This approach is of interest because the generation of reference and the walking stability are managed simultaneously [SPO 03]. Stable walking for an under-actuated and very simple compass robot was also obtained via a control of the energy levels of the mechanical system [GOS 97]. However, it is difficult to use this method with more complex structures and generally semi-passive approaches are applied instead. In this case, actuators are used for certain joints, simple control laws are defined and the study of the convergence towards cyclic movements is carried out. These aspects will be developed in section 5.8.

5.2. Hybrid systems and stability study

A walking gait can be broken down into a succession of single support phases, impact and double support phases. Depending on the type of gaits under consideration, rotational phases along the extremity of the stance foot can be taken into account. Two examples of walking are represented in Figure 5.1.

The system's state is described by x, which contains configuration and velocity variables. For walking which is made up of single support and impact, the complete system model is hybrid:

$$\begin{aligned} &\dot{x} = f(x) + g(x)u \quad \text{if } x \notin S \\ &x(t^+) = \Delta\left(x(t^-)\right) \quad \text{if } x \in S \end{aligned} \qquad [5.1]$$

where u is the control vector and f, g and Δ are nonlinear continuous functions which are deduced from equations [2.39] and [2.115]. The impact set S defines the commutation surface. This, in turn, contains all the robot states which will lead to an impact with the ground. The second equation is an expression of an instantaneous conditional jump (discontinuity) in the biped's velocity which occurs when the state trajectory (which is described by the first equation) runs into the commutation area defined by $S: = \{ x \mid \phi(x) = 0 \}$ where ϕ is a continuous application. We will use $\phi(x) = y_2(x)^2 + \max(0, \dot{y}_2(x))$ for example, where y_2 is the height of the swing foot [CHE 05a]. This function is canceled when the swing foot reaches the ground with negative vertical velocity. Just after impact, the swing foot's vertical velocity is positive and this function is not zero.

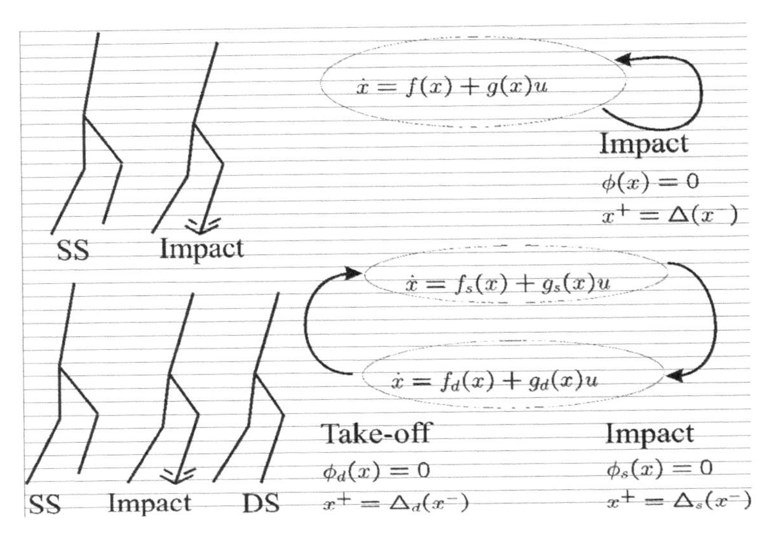

Figure 5.1. *During single support, the robot model is continuous. When the height of the free foot is equal to the height of the ground ($\phi(x) = 0$), impact is detected. After the variation in the robot's velocity there is a single or double support phase on the other foot. The double support phase is terminated by a control decision, by producing a vertical acceleration of the foot so that it leaves the ground*

The aim of a walking robot control is generally to obtain a cyclic movement. These cycles are made up of many phases (single, double support, etc.) and impacts.

To reach this objective, it is not an *a priori* prerequisite that the control be stable during each independent phase. What is required is the convergence towards a limit cycle.

Within this context, the description of the nonlinear dynamic system in the phase plane is a useful analytical graphic tool [SLO 91]. Starting with the initial given conditions, the robot's movement is traced in the phase plane and the features of the obtained curves are then studied. As robot walkers have more than two states, we will represent each articulation's movement in its phase plane. Its joint velocity is represented according to its joint position. The robot's movement corresponds to the succession of the different phases (Figure 5.2). A cyclic movement gives a closed curve at each phase plane.

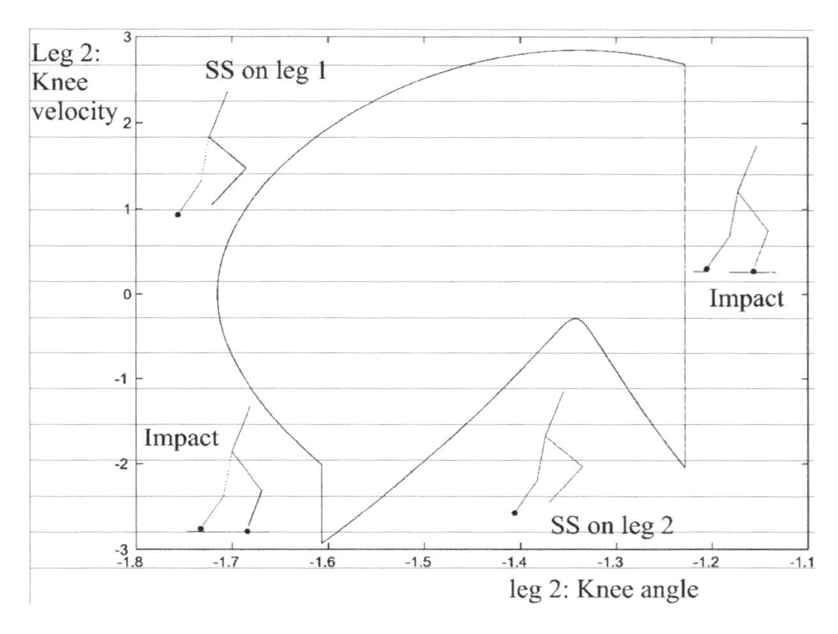

Figure 5.2. *A robot's cyclic movement corresponds to a closed cycle for each movement projection in the phase plane of a joint. The impacts are represented by vertical lines which correspond to a variation in joint velocity with no variation in position*

If this closed cycle is isolated, it is a limit cycle which can be stable, unstable or semi-stable. Movements which start in the limit cycle's proximity may or may not converge towards it. Henri Poincaré [POI 04] developed a technique for analyzing the stability of the dynamic systems. He designed a hyper-surface with a dimension of $n - 1$ which is transversal to the limit cycle. In this way, the flow's intersection with this hyper-surface can be observed. This leads to the creation of a discrete-time system called a Poincaré return map [GUC 85]. In the case of robot walkers, the chosen Poincaré section is generally defined when the free leg makes contact with the ground [GOS 97, HUR 94]. Between two contact indexes k and $k + 1$, the states X that correspond to the flow intersection with the Poincaré section are linked by:

$$X_{k+1} = P(X_k) \qquad [5.2]$$

A cyclic movement corresponds to a fixed point X^* for the Poincaré application: $X^* = P(X^*)$. When the intersection of the flow and the Poincaré section can be described by a scalar p, a graphic tool allows the visualization of the convergence of the robot's movement towards the limit cycle. The discrete function $p(k)$ is defined, and takes the value of the p scalar at the kth flow intersection with the Poincaré section.

We trace $p(k + 1)$ as a function of $p(k)$. The fixed point corresponds to the intersection of this graph with the bisector. The position of the graph in relation to this bisecting line indicates the convergence towards the fixed point (Figure 5.3).

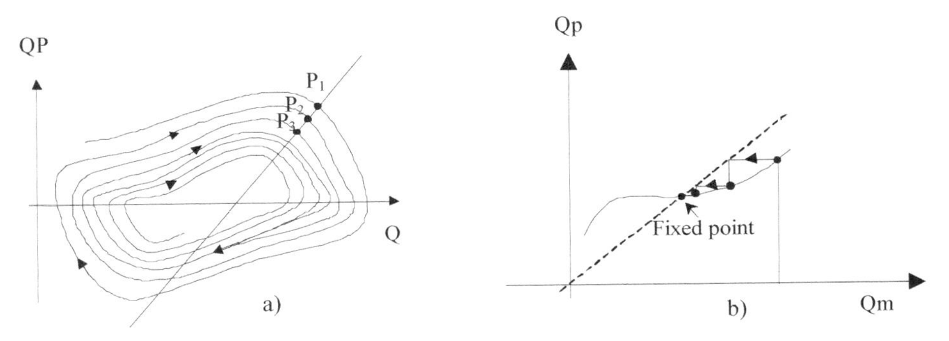

Figure 5.3. *For a system described by 2 state variables, the successive intersections of the flow with the Poincaré section (dimension 1) are described by a succession of P_k points or p(k) scalar points. We represent p(k+1) as a function of p(k). A fixed point corresponds to the intersection with the bisecting line. A cycle is stable if the local slope, at a fixed point, is below 45°. The arrow illustrates the convergence towards the fixed point*

The convergence towards the limit cycle is linked to the convergence of function P which can also be studied via the linearization of equation [5.2] around X^*. We have:

$$X_{k+1} - X^* = J_P(X^*)(X_k - X^*) \qquad [5.3]$$

where $J_P(X^*)$ is the Jacobian in X^* of application $P(X)$. If the eigenvalues of $J_P(X^*)$ are inside of the unit circle, the limit cycle is stable in the Lyapunov sense. In [GRI 01] it was shown that these results can be applied for continuous autonomous systems, but non-continuous in the Lipchitz sense and in the presence of impacts. In the case of robot walkers, the state dimension is often high. The analytic calculation of $J_P(X^*)$ is generally impossible and the stability test is carried out numerically. The disadvantage of a numeric method is that the obtained information is poor. The walking cycle may be stable or non-stable, but this does not give us any information about how this result was obtained. This is why it is useful to be able to reduce the spatial dimension in which the Poincaré return map is studied.

5.3. Taking into account the unilateralism of the contact constraint

During a biped's flat-footed contact with the ground, the robot can be considered as a mechanical tree structure which is fully actuated. The traditional control techniques of a rigid robot manipulator can therefore be applied. These consist of independent PD or PID control at each joint, a dynamic control (with a complete or approached model) or a robust control via sliding mode [PLE 03, TZA 03].

The aim of this chapter is not to give an overview of all these different control approaches. We will focus on computed torque or input/output linearization. We will highlight the limitations of these controls due to the unilateralism of the ground contact constraint when faced with perturbations. We will also look at the different strategies which can be used to improve the results.

5.3.1. *Computed torque control*

This type of control is based on a dynamic model. During the flat-footed stance phase, and when using minimal parameterization of the system (section 2.2.4.1 with a fixed value of q_1 because the foot is flat), the equation corresponding to q_1 is unused and the dynamic model can be written as follows:

$$M(q)\ddot{q}+C(q,\dot{q})+G(q)=D(q,\dot{q})+\tau \qquad [5.4]$$

To follow the reference $q^d(t)$, the robot's acceleration must be:

$$\ddot{q}=w^d=\ddot{q}^d+K_v(\dot{q}^d-\dot{q})+K_p(q^d-q) \qquad [5.5]$$

and the dynamic control is written:

$$\tau=M(q)\left(\ddot{q}^d+K_v(\dot{q}^d-\dot{q})+K_p(q^d-q)\right)+C(q,\dot{q})+G(q)-D(q,\dot{q}) \qquad [5.6]$$

If there is no modeling error, the robot's behavior in closed loop is:

$$\ddot{e}+K_v\dot{e}+K_pe=0 \quad \text{with} \quad e=q-q^d \qquad [5.7]$$

The gains K_v and K_p must be chosen to ensure the convergence of the error towards zero.

Dynamic model [5.4], which is used to write the control law, is based on certain hypotheses which describe contact (here the stance foot is flat and the swing leg is not in contact with the ground). If the torque, which is calculated by equation [5.6], is such that the desired contact is not achieved (for example, if the foot rotates along one of its extremities) then the behavior which is preset by equation [5.7] will not be obtained. The robot may fall.

Chapter 2, section 2.4 gives details of the dynamic constraints. These constraints are non-take-off, non-slide and non-rotation constraints. The last constraint is generally the hardest to fulfill. In the case of a robot with a flat-footed contact, the constraint is written in the form of equation [2.134].

This condition expresses the fact that the ZMP must be situated inside the convex hull of the foot support region (the ZMP's position is calculated by equation [2.128]). Let us note that this is not a stability condition, but a constraint which ensures that the foot does not start to rotate along one of its edges.

The principle of the first method presented here is to begin by ensuring that the dynamic constraints are respected. The convergence towards the reference movement can be sacrificed if it is necessary.

The control which is calculated by equation [5.6] will only be applied if the corresponding position of the ZMP is correct.

The ZMP's position, denoted C_p, is calculated from the measured values of q and $\dot{q}$ and from the desired value for the acceleration w^d equation [5.5]:

$$\overrightarrow{OC}_p = \frac{\vec{z}_0 \wedge \left(\vec{\delta}_G(q,\dot{q},w^d) + \overrightarrow{OG} \wedge \left(m(\vec{\gamma}_G(q,\dot{q},w^d) - \vec{g})\right)\right)}{\vec{z}_0 \cdot \left(m(\vec{\gamma}_G(q,\dot{q},w^d) - \vec{g})\right)} \quad [5.8]$$

where $\vec{\gamma}_G(q,\dot{q},w^d)$, $\vec{\delta}_G(q,\dot{q},w^d)$ are the desired acceleration for the gravity center and the desired rate of moment of momentum computed around the gravity center, respectively. If C_p, which is defined by equation [5.8], is not inside the convex hull of the foot support region, then the torque is not applied. We determine the acceleration w which is the closest to w^d so that the ZMP position is on the inside of the convex hull of the foot support region ($C_p \in \text{Conv}(S)$), and we determine the torque to obtain this acceleration:

$$\tau = M(q)w + C(q,\dot{q}) + G(q) - D(q,\dot{q}),$$

where w minimizes $\left(\left\|w - w^d\right\|\right)$ under the constraint $C_p \in \text{Conv}(S)$

$$\text{with } \overrightarrow{OC}_p = \frac{\vec{z}_0 \wedge \left(\vec{\delta}_G(q,\dot{q},w) + \overrightarrow{OG} \wedge \left(m(\vec{\gamma}_G(q,\dot{q},w) - \vec{g})\right)\right)}{\vec{z}_0 \cdot \left(m(\vec{\gamma}_G(q,\dot{q},w) - \vec{g})\right)}. \qquad [5.9]$$

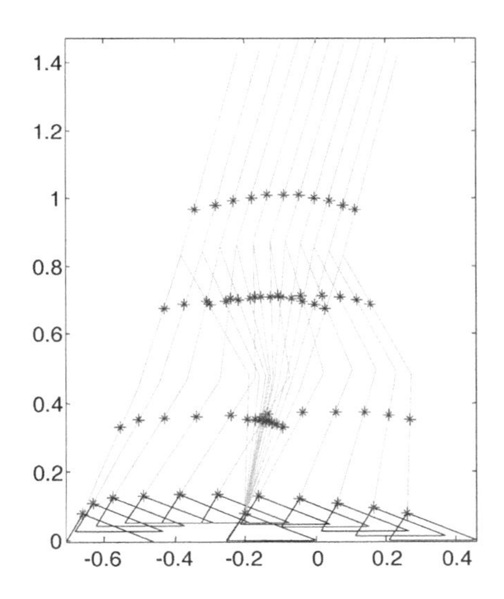

Figure 5.4. *Captured images which correspond to the reference movement*

	Trunk	Thigh	Leg	Foot
Mass (kg)	17	6.8	3.23	1
Length (m)	0.6	0.4	0.472	0.25
Inertia (kg m^2)	2.22	0.25	0.1	0.012
Motor inertia (kg m^2)	0.83	0.83	0.83	0

Table 5.1. *Robot parameters used for the simulation [DJO 07]*

This approach, which was used with reference q^{d} that fulfills the pre-set contact conditions and enables us to allow for limited perturbations, is illustrated in the following case. We considered a planar robot with a gait made up of flat-footed single support phases, which were separated by impacts. The parameters of the simulated robot are given in Table 5.1, and the reference corresponds to the robot's profile described by the stick diagram in Figure 5.4.

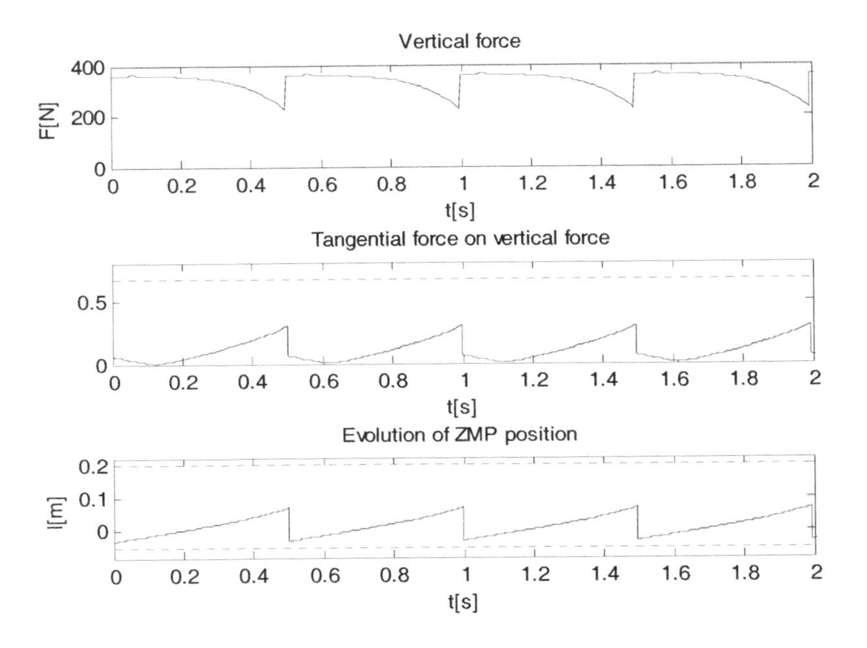

Figure 5.5. *The desired movement is such that when there are no perturbations, no model errors and no initial errors, the constraints on the ground reaction forces are fulfilled. The foot does not take off, as the vertical reaction forces are positive. The foot does not slide as the ground reaction force is within the friction cone defined by the friction coefficient (2/3). There is no foot rotation because the ZMP is on the inside of the convex hull of the foot support region. The foot measures 0.25 m*

When the references are followed perfectly, the ground reaction forces are compatible with flat-footed contact. As is seen in Figure 5.5, which shows the constraint profiles for just over four cyclic walking steps, the non-take-off, non-slide and non-rotation constraints are fulfilled.

The robot is submitted to a perturbation. This consists of a horizontal force applied to the robot's center of mass. This perturbation is applied for 0.1 s at t = 0.2 s, which is nearly halfway through the first step.

An amplitude of force 70 N can be applied without the robot falling by using control law [5.5] with gains $K_p = 400$ and $K_v = 2\sqrt{K_p}$.

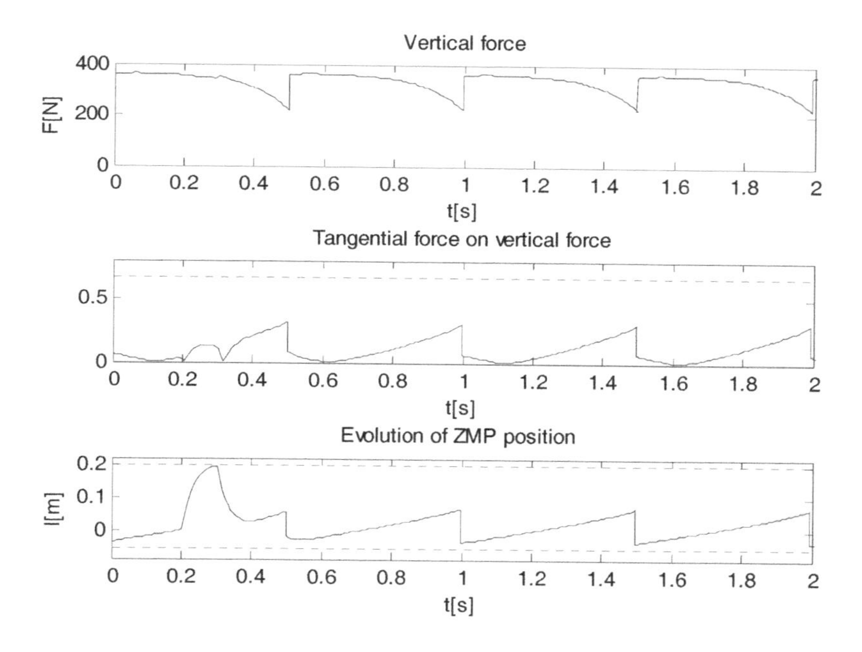

Figure 5.6. *Constraints on the ground reaction forces after a perturbation of 70 N using normal control equation [5.5]. The ZMP position is significantly affected by the presence of this perturbation*

It is quite difficult to choose gains. We may wish for high gains to ensure a rapid convergence towards the desired movement (with respect to the duration of a step). However, an error in an end of step position means a different robot configuration for the impact model and this generally leads to a significant velocity error. Variations in the error at the start of a step, with high gains, will consequently lead to significant torque variations and result in contact force variations. It therefore becomes difficult to ensure that the contact constraints are fulfilled. The behavior which is obtained at the contact constraints for perturbations of 70 N and 120 N are shown in Figures 5.6 and 5.7, respectively.

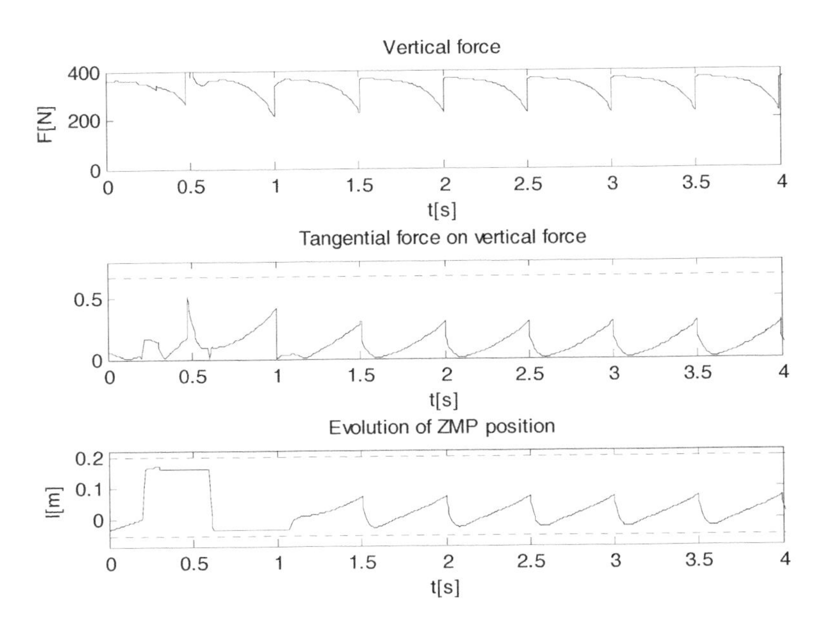

Figure 5.7. *Constraints on the ground reaction forces after a perturbation of 120 N, with control [5.9]. The control enables us to maintain the ZMP position on the inside of the supporting sole, in order to avoid a pedal rotation*

We can manage perturbations of up to 130 N by using the strategy described in equation [5.9]. Tracking errors are permitted for the fulfillment of the contact constraints. The foot size constraints are defined within a certain margin. Here, the ZMP position is constrained to evolve between -0.03 m and 0.16 m. However, the real foot limits are -0.05–0.2 m as when the perturbation is applied the real ZMP position is not the position predicted by the dynamic model (which does not take the perturbation into account). After the perturbation there is a tracking error. The ground collision does not occur at the pre-set instant and there are therefore other more significant errors at the start of the next step. The result is a cyclic movement where the errors do not cancel each other out, but remain cyclical. The chosen gains do not enable us to have a perfect convergence towards the reference. The free leg's ground impact does not occur at the pre-set instant.

Tracking errors therefore appear at the start of a step, just after impact. Nevertheless, they tend to cancel themselves out at the end of a step, just before impact (Figure 5.8).

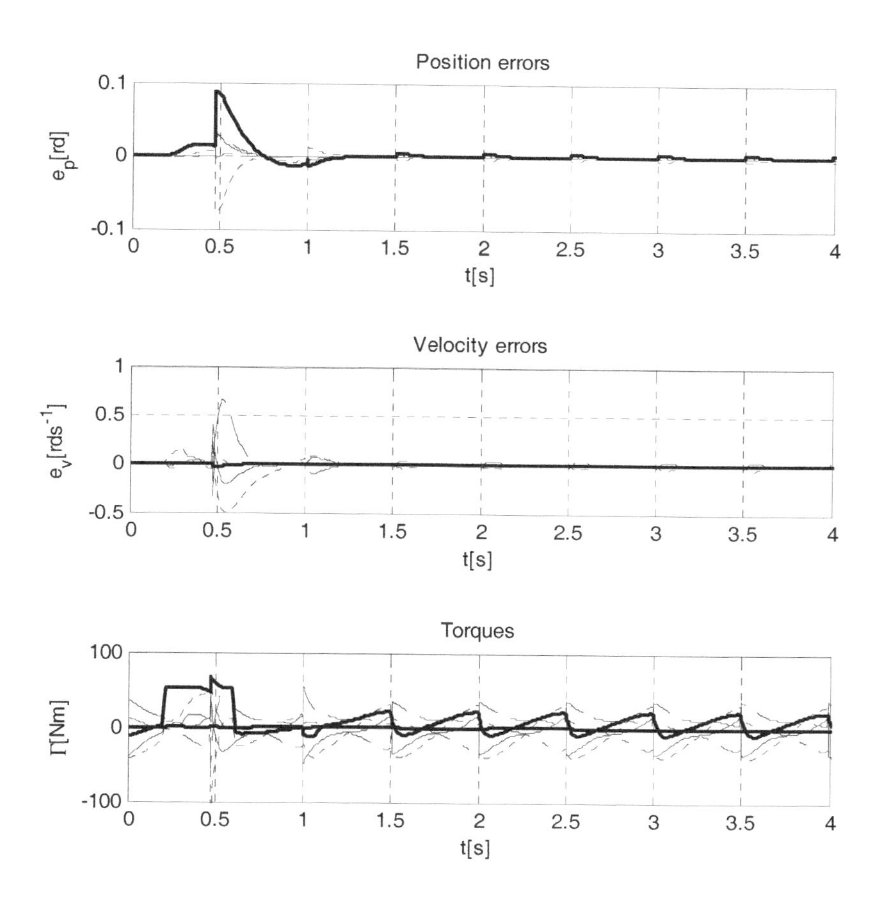

Figure 5.8. *After the perturbation, the robot's movement finds stability in the cyclic movement, which is not the exact desired movement. There are cyclic errors and the duration of the step is not exactly the preset duration*

The configuration error during impact can be limited by focusing on certain favoring direction errors, rather than by minimizing the error norm.

As the impact occurs on a geometric and non-temporal condition, it seems pertinent to ensure that the joint trajectory reference is correctly followed by allowing for a variation in the step duration. The geometric synchronization of the joint trajectories is more certain if we choose a control such that:

$$\tau = M(q)w + C(q,\dot{q}) + G(q) - D(q,\dot{q})$$

where $w = \alpha w^d$ and α minimizes $(\|\alpha - 1\|)$ under the constraint

$$C_P \in \text{Conv}(S) \qquad [5.10]$$

$$\text{with } \overrightarrow{OC}_P = \frac{\vec{z}_0 \wedge \left(\vec{\delta}_G(q,\dot{q},\alpha w^d) + \overrightarrow{OG} \wedge \left(m(\vec{\gamma}_G(q,\dot{q},\alpha w^d) - \vec{g})\right)\right)}{\vec{z}_0 \cdot \left(m(\vec{\gamma}_G(q,\dot{q},\alpha w^d) - \vec{g})\right)}.$$

With this kind of correction, there are more acceptable perturbations and we can attain 170 N. The way the robot reacts to this perturbation is seen in Figures 5.9 and 5.10. As the errors during impact are reduced, the robot movement converges towards the reference movement even when using relatively small gains.

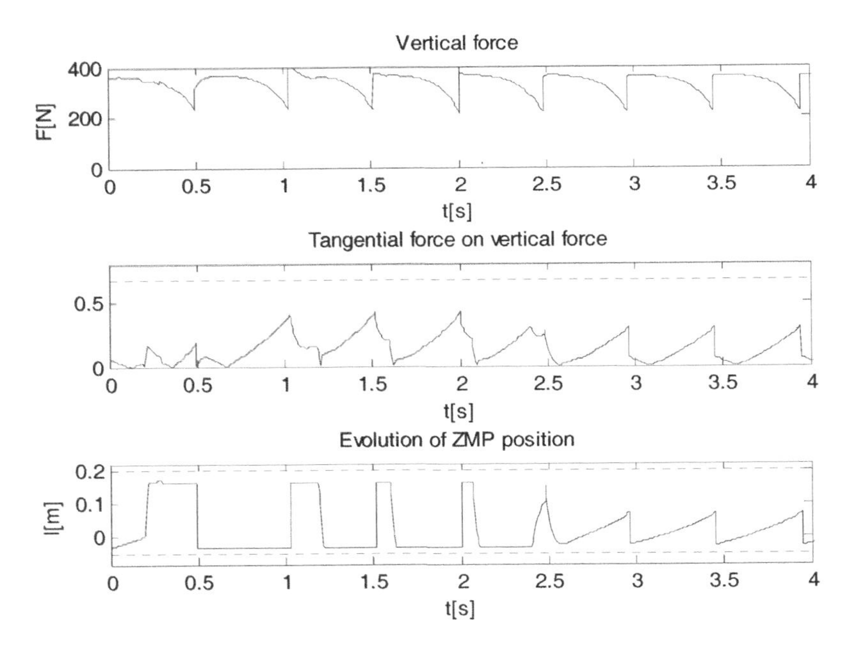

Figure 5.9. *Constraints on the ground reaction forces after a perturbation of 120 N. Equation [5.10] control strategy is used*

With these approaches, the initial reference does not take the perturbation and the subsequent contact constraints into account. The errors increase.

High torques (caused by high amounts of gains and errors) favor the activation of the non-rotation constraints.

In Figure 5.6 we notice that these constraints remain active for more than 3 steps after the perturbation has been suppressed. The robot must synchronize itself to the temporal reference, although it is not important for the robot to align itself with a predefined constraint.

To limit the errors in the following steps, it is preferable to modify the joint references online, so that the foot's non-take-off, non-slide and non-rotation constraints are fulfilled.

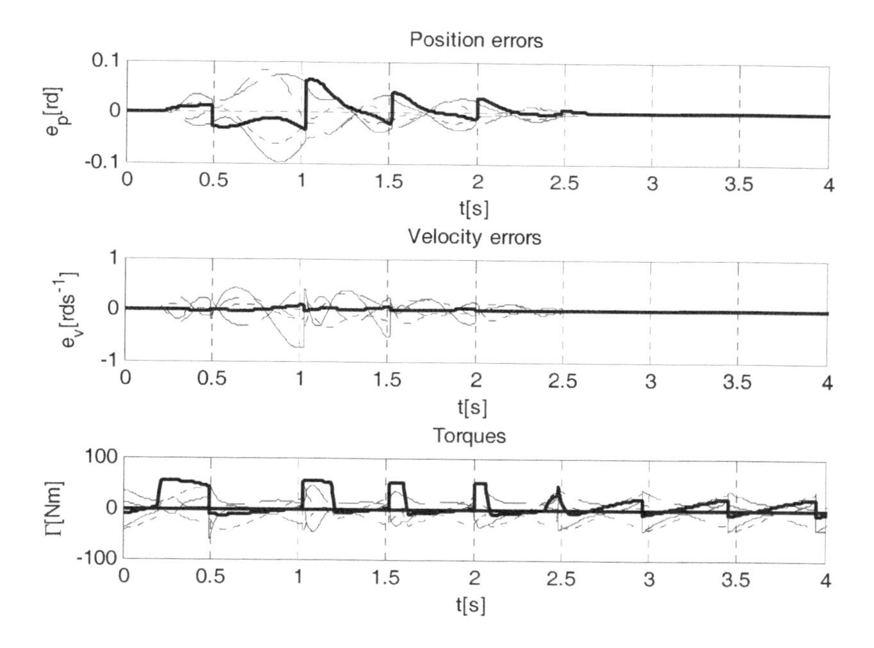

Figure 5.10. *After significant tracking errors, the desired reference is regained. Faced with such errors the torques can be high, in particular at the start of a step*

5.4. Online modification of references

As was shown in the preceding illustrations, even when we use an appropriate reference movement, the ZMP position is sensitive to the presence of a perturbation (such as the robot being pushed) and this can lead to difficulties arising in the robot control.

This problem can be dealt with by modifying online the robot's references to fulfill the constraints. This principle was used for the Honda robots, as will be seen in the first example. Like most robot walker controls, the control strategy is based on the use of two levels. Low-level control guarantees that the temporal references are followed for each joint. High-level control modifies the reference movement to ensure that the ZMP remains in the support zone.

We will now study a planar robot and a temporal reference modification. We will study the stability of the obtained walk when the ZMP evolution is prescribed. We will also study the case where this ZMP profile is simply bounded.

5.4.1. *General principle*

The principle of online modifications will be illustrated by the example of the Honda robot [HIR 98]. The real pressure center is measured via the intermediary of several effort sensors which are placed on the foot.

Temporal joint references are defined according to the desired walk which corresponds to the desired foot's situations (positions/orientations). The trunk and arms are considered globally (Figure 5.11).

A horizontal acceleration of G (the center of mass) is sought. The desired position of the ZMP corresponds to a robot's zero rate of moment of momentum at G (Figure 5.11).

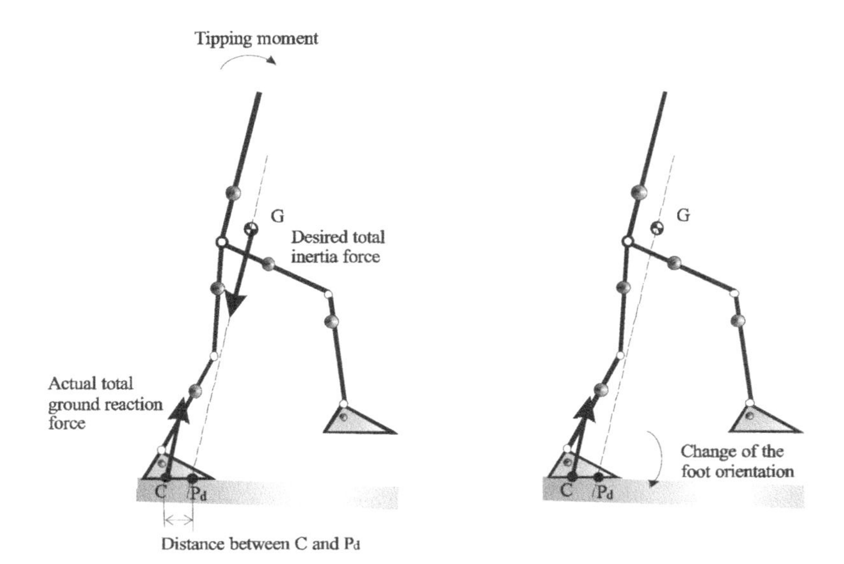

Figure 5.11. *The chosen references correspond to a zero rate of moment of momentum at G. The desired ZMP position is such that the vector which links P^d and G is directed according to the ground reaction direction. The effort sensors which measure the contact forces acting on the foot enable us to define the real position of pressure center C. If P^d and C are not merged, the foot orientation is modified in order to adjust the ground-foot contact and shift C to P^d (the contact is not completely rigid) [HIR 98]*

If the conditions are ideal, then C the center of pressure (CoP) coincides with P^d (the desired ZMP).

In reality, the terrain is often irregular and C can differ from P^d. There is therefore a tipping moment, which is evaluated by the following equation:

$$M_b = (P_x^d - C_x)R_z^d \qquad [5.11]$$

where R_z^d is the normal component for the calculated ground reaction force. This equation will be the basis for two actions: the correction of position C to reduce the tipping moment and the modification of position P^d to create a tipping moment, but with no fall.

Figure 5.12 shows how these actions are integrated into the robot's control law.

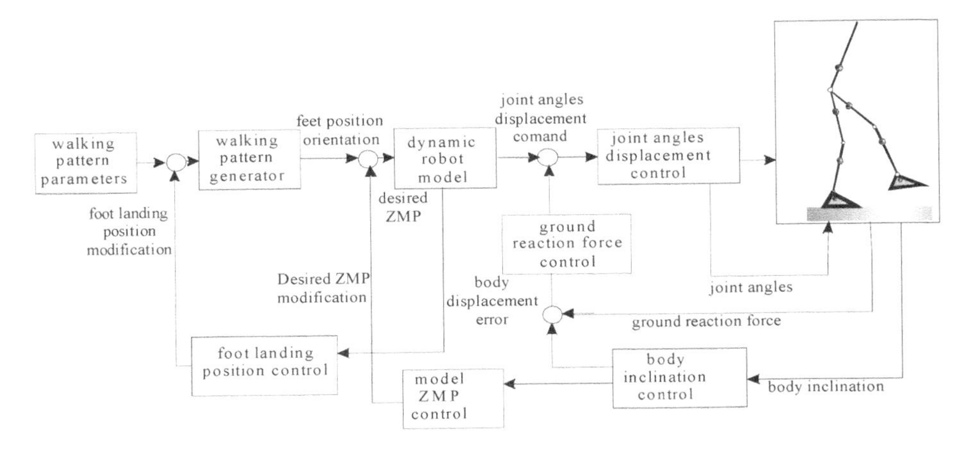

Figure 5.12. *Organization of Honda robot's control. The control algorithm uses the references which are generated from a desired foot profile: from the ZMP and from a joint servo-control. A first correction is made to ensure appropriate ground-foot contact. If the inclination of the trunk attains certain limits, the desired ZMP profile is modified and the free foot's landing can also be modified [HIR 98]*

If a gap is detected between C and P^d, the foot position is adjusted to shift C to P^d and avoid the creation of an impetuous tipping movement. During single support, for example, the robot body tips over; the robot lowers the front of its foot in order to increase the efforts exerted on the ground by the toes. This brings the pressure center forward and cancels the tipping torque. During double support, a foot rotation around P^d is required (see Figure 5.11).

If the robot's trunk is inclined too far towards the front, a fall is possible even in the absence of a forward tipping torque. A backward tipping torque must therefore be created so that the trunk's inclination corrects itself. The P^d position will be shifted towards the back, and the first action will maintain C on the initial P^d value. The inclination of the trunk will progressively return to its correct values. As the relative joints are controlled, the free foot's position is controlled with respect to the trunk. In the case of a forward inclination, the robot will touch the ground with a non-desired configuration. The foot's trajectory relative to the trunk during transfer is therefore modified in accordance with the online modification of P^d.

These modifications are using a heuristic approach. They have been validated through experimentation on prototypes, but a global stability analysis was not conducted.

5.4.2. *The ZMP's imposed evolution*

We will now suggest a control approach and study the conditions which lead to stable walking for a planar robot.

The control principle is as follows. The reference, in which online adaptation is possible, corresponds to the same trajectory in joint space (see Figure 5.13). The choice of a temporal evolution along this joint reference enables us to obtain the desired ZMP evolutionary profile during the single support phase.

The control principle is as follows. A dynamic control is used to make the robot follow reference $q^d(s(t))$. The evolution of $s(t)$ is defined online through the intermediary of its second derivative to obtain a desired evolution for the current ZMP position: $l^d(s)$.

When there is a perturbation, the duration of a step will not be constant but the desired joint path will be followed. The ZMP will follow the desired evolution, will therefore remain in the contact surface area and the stance foot will remain flat.

The ZMP position is linked to the acceleration which must be produced by the control, according to relation [5.8]. When the references are considered as being functions of the new variable s, the desired acceleration becomes (starting from equation [5.5]):

$$w^d = \frac{\partial q^d(s)}{\partial s}\ddot{s} + \frac{\partial^2 q^d(s)}{\partial s^2}\dot{s}^2 + K_v\left(\frac{\partial q^d(s)}{\partial s}\dot{s} - \dot{q}\right) + K_p\left(q^d(s) - q\right) \qquad [5.12]$$

The robot's configuration and joint velocity are measured. The values of s and of its first derivative are obtained via integration. The value of $\ddot{s}$ is calculated so that the ZMP position, which is defined by equation [5.8], has the desired value of $l^d(s)$. The calculation is possible if the kinetic momentum calculated around the ZMP is non-zero [DJO 07]. By using this strategy, we can allow for perturbations of up to 350 N. The behaviors obtained for a perturbation of 300 N are shown in Figures 5.14–5.16.

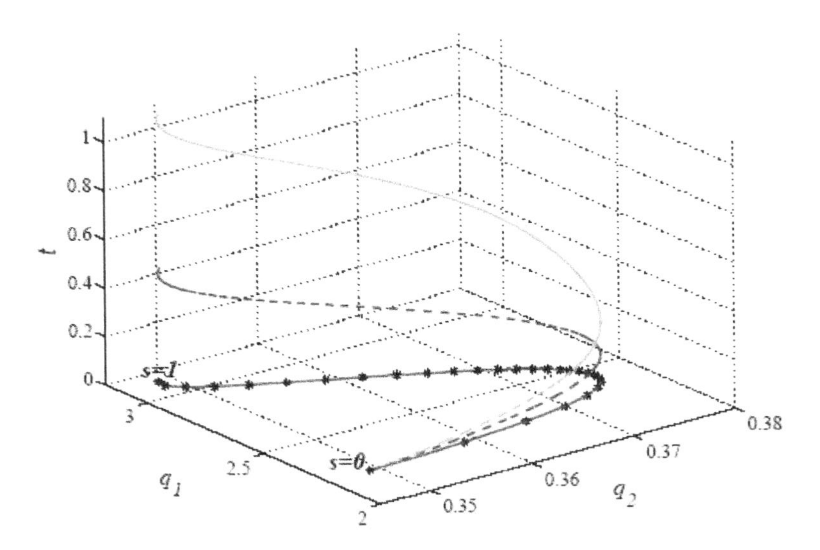

Figure 5.13. *The dotted curves are two movements represented by profiles $q_1(t)$ and $q_2(t)$, which correspond to the same trajectory (represented by the continuous line in plane (q_1, q_2)). A trajectory is a curve in joint space; this curve can be graduated as a function of the new variable s. The trajectory is given by ($q_1(s)$, and $q_2(s)$). This s function is defined such that the initial configuration corresponds to $s = 0$, the final configuration corresponds to $s = 1$ and the graduation of s between 0 and 1 is monotonous. All monotonous functions $s(t)$ define a movement which corresponds to a trajectory $q(s)$. For example, $s = t/T$ defines a movement where the duration is T. If a joint variable (for example q_2) has a monotonous profile, the trajectory can also be defined by $q_1(q_2)$*

After a perturbation, the robot's movement can move away from the initial cyclic movement. The control will enable the robot to quickly regain the reference movement defined by $q^d(s)$ and $l^d(s)$ because the control ensures that the most constraining contact conditions are fulfilled (no foot rotation). However, the convergence towards a cyclic movement will only be ensured if $s(t)$ converges towards a cyclic movement.

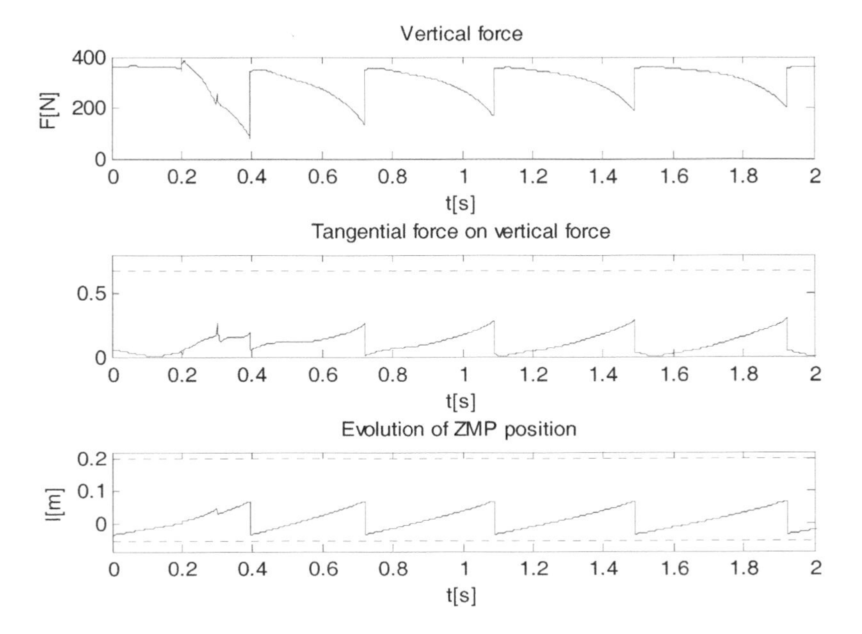

Figure 5.14. *Constraints on the ground reaction forces after a perturbation of 300 N. Control strategy [5.12] is used. During the application of the perturbation, the ZMP position is not exactly the desired position because the perturbation is not known by the control. When there is no perturbation, the ZMP profile is as prescribed. The duration of the steps is reduced just after the perturbation. The reaction force is perturbed just after the perturbation and then goes back to the profile which corresponds to the cyclic movement*

The references $q^d(s)$ are chosen so that they correspond to a cyclic path in joint space and by taking into account the impact phase. The equality $s = 1$ corresponds to the impact configuration. For this configuration, the transferring foot touches the ground whereas for all the configurations where $0 < s < 1$, the transferring foot is above the ground.

The fact that the impact condition is geometric guarantees that whatever the profile $s(t)$ after perturbation, the robot is in its double support phase for $s = 0$ and $s = 1$ only.

At the start of each new step we will therefore have $s = 0$. The only unknown variable is therefore the initial velocity for step k, denoted $\dot{s}_k(0)$.

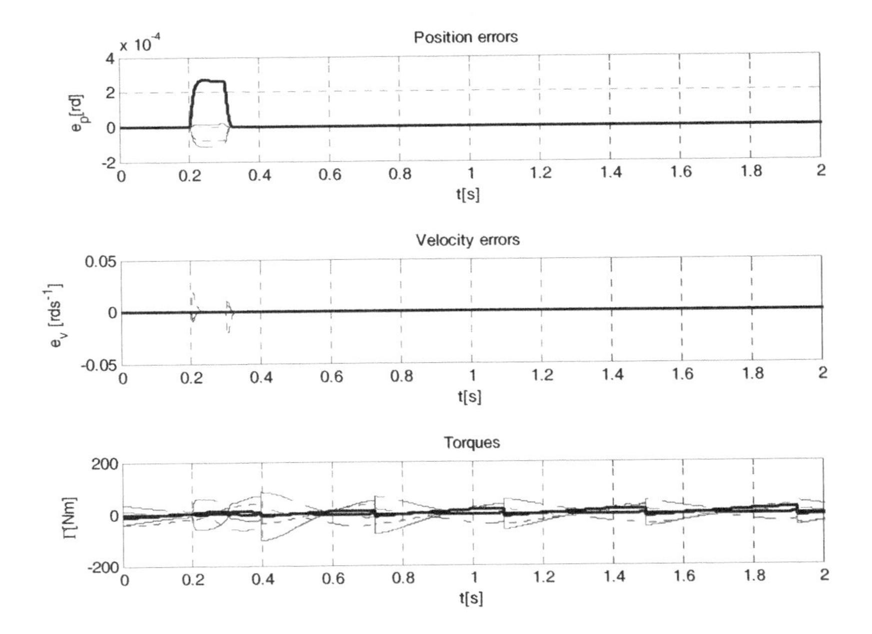

Figure 5.15. *There are tracking errors only during the application of the perturbation. At other times the reference is modified. The torques used to produce the desired movements are a little different to the torques which correspond to the optimal movement, but the control enables us to have quasi-zero tracking errors*

To study the robot's movement stability (with the help of the Poincaré method), we only have to study the successive values of $\dot{s}_k(0)$ for the whole of k. This method is similar to that described in section 5.2.

The control enables us to reduce the state variable dimensions to just two variables: s and $\dot{s}$. The Poincaré section defined at the beginning of a step enables us to reduce the space dimension to 1 (as s = 0), and to focus on the study of a succession of scalars.

The section can be defined for s = 1 or for any other value of s. A numerical study could be carried out but, for this case, an analytical study is also possible. The study can be based on an analysis of the physical properties of the robot walker and its control.

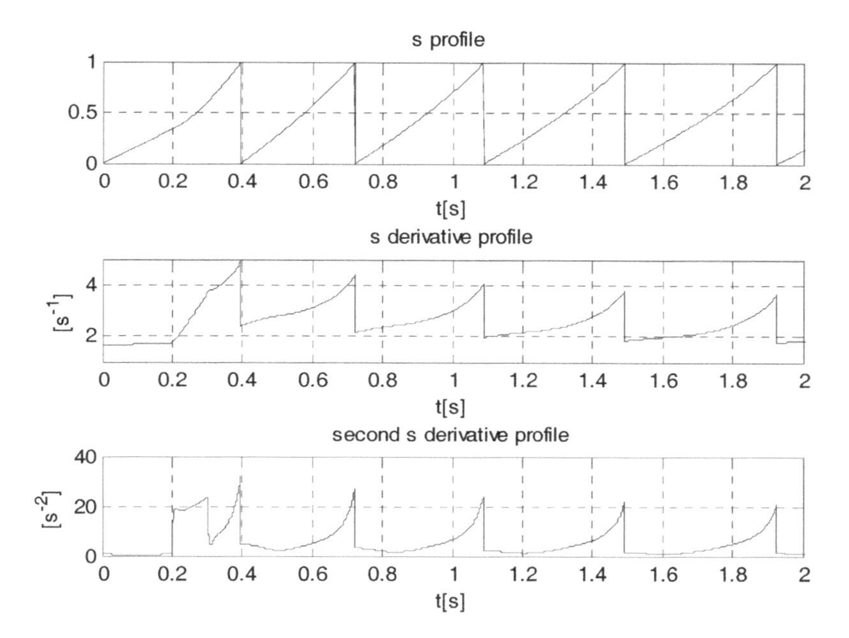

Figure 5.16. *The profile of variable s and its derivatives in relation to time is illustrated. For the successive steps, s varies from 0 to 1. During the perturbation, the acceleration* $\ddot{s}$ *is formally modified and then* $\dot{s}$ *naturally converges towards a cyclic movement*

During the single support phase, the robot walker is submitted to two external forces: gravity and ground reaction. The robot's equilibrium during rotation can be formulated at any given point.

Formulating this rotational equilibrium around the ZMP (point where the planar robot's contact wrench reduced to a force) implies that we do not have to use the intervention of the force which is exerted by the ground, as it does not generate a moment around this point. Using the theory of rate of moment of momentum, the rotational equilibrium is formulated:

$$\vec{\delta}_C = \overrightarrow{CG} \wedge (m\vec{g}) \quad [5.13]$$

As the robot is planar, the rate of moment of momentum only has one component and we can therefore use the scalar equation:

$$\delta_C = -mg(x_G - x_C) \quad [5.14]$$

where x_C and x_G are the abscissa of the ZMP and the center of mass. As the ZMP is not a fixed point, and by using the expression of the rate of moment of momentum in relation to the angular momentum at C, we obtain:

$$\dot{\sigma}_C + m\dot{z}_G\dot{x}_C = -mg(x_G - x_C) \qquad [5.15]$$

The proposed control law ensures that the references $q^d(s)$ and $l^d(s)$ are followed, and this equation becomes:

$$\dot{\sigma}_C + m\frac{\partial z_G}{\partial q}\frac{\partial q^d}{\partial s}\frac{\partial l^d}{\partial s}\dot{s}^2 = -mg(x_G(s) - l(s)) \qquad [5.16]$$

By definition, the angular momentum is linear with respect to the joint velocities, which are linear with respect to $\dot{s}$ when using the control law. We therefore also have:

$$\sigma_C = I(s)\dot{s} \quad \text{or} \quad \dot{s} = \frac{\sigma_C}{I(s)} \qquad [5.17]$$

These equations can be combined to formulate the derivative of the angular momentum. This can be done without taking time into consideration. If ζ_C is non-zero, the angular momentum derivative as a function of s can be expressed:

$$\frac{d\sigma_c}{ds} = -mg\frac{I(s)}{\sigma_C}(x_G(s) - l(s)) - m\frac{\partial z_G}{\partial q}\frac{dq^d}{ds}\frac{dl^d}{ds}\frac{\sigma_C}{I(s)} \qquad [5.18]$$

By introducing a new variable $\zeta = \frac{\sigma_C^2}{2}$ this equation becomes:

$$\frac{d\zeta}{ds} = -mgI(s)(x_G(q^d(s)) - l(s)) - 2\kappa(s)\zeta$$
$$\text{with} \quad \kappa(s) = \frac{m}{I(s)}\frac{\partial z_G(q^d(s))}{\partial q}\frac{dq^d(s)}{ds}\frac{dl^d(s)}{ds}. \qquad [5.19]$$

This first-order differential equation with variable coefficients with respect to s has an explicit solution if $\dot{s} \neq 0$ for the single support phase.

For a current single support phase, $0 \leq s \leq 1$, we have:

$$\zeta(s) = \delta(s)^2 \zeta(0) - \Phi(s),$$

where $\delta(s) = \exp\left(-\int_0^s \kappa(\tau)d\tau\right)$ [5.20]

$$\Phi(s) = mg\int_0^s \exp\left(-\int_\tau^s 2\kappa(\tau_1)d\tau_1\right)I(\tau)(x_G(q^d(\tau)) - l(\tau))d\tau.$$

The hypothesis $\dot{s} \neq 0$ for the single support phase is translated by a constraint on $\zeta(0)$, we should therefore have:

$$\zeta(0) > \max_{0 \leq s \leq 1}\left(\frac{\Phi(s)}{\delta(s)^2}\right) \quad [5.21]$$

Equation [5.20] enables us to determine, for the k step, the value of $\dot{s}_k(s)$ during the step in relation to value $\dot{s}_k(0)$ at the start of a step:

$$\dot{s}_k(s) = \sqrt{\frac{\delta(s)^2 I(0)^2 \dot{s}_k(0)^2 - 2\Phi(s)}{I(1)^2}} \quad [5.22]$$

In particular, the value of $\dot{s}_k(1)$ can be deduced at the end of the step.

The impact phase was described in section 2.3.3. The joint velocity after impact is proportional to the joint velocity before impact and can be written:

$$\dot{q}_{k+1}(0) = \Delta(q_k(1))\dot{q}_k(1) \quad [5.23]$$

where the index k corresponds to the step number and for the velocities expressed as a function of s.

The reference is chosen in order to fulfill impact. The velocity at the beginning of a step fulfills:

$$\dot{s}_{k+1}(0) = \delta_S \dot{s}_k(1) \quad \text{where} \quad \delta_S = \left(\frac{dq^d(0)}{ds}\right)^+ \Delta(q_k(1))\frac{dq^d(1)}{ds}. \quad [5.24]$$

The combination of equation [5.22] for $s = 1$ and equation [5.24] enables us to deduce the relation between $\dot{s}_{k+1}(0)$ and $\dot{s}_k(0)$:

$$\dot{s}_{k+1}(0) = \sqrt{\frac{\delta_S^2 \delta(1)^2 I(0)^2 \dot{s}_k(0)^2 - 2\delta_S^2 \Phi(1)}{I(1)^2}} \qquad [5.25]$$

We can conclude from this equation that there is a cyclic movement of the robot defined by:

$$\dot{s}(0) = \sqrt{\frac{-2\delta_S^2 \Phi(1)}{I(1)^2 - \delta_S^2 \delta(1)^2 I(0)^2}} \qquad [5.26]$$

which fulfills equation [5.21], only if the references $q^d(s)$, $l^d(s)$ are such that:

$$-\frac{\delta_S^2 I(0)^2 \Phi(1)}{I(1)^2 - \delta_S^2 \delta(1)^2 I(0)^2} > \max_{0 \le s \le 1} \left(\frac{\Phi(s)}{\delta(s)^2} \right) \qquad [5.27]$$

The robot's movement will therefore converge towards this cyclic movement if and only if:

$$\frac{\delta_S^2 \delta(1)^2 I(0)^2}{I(1)^2} < 1 \qquad [5.28]$$

In the cases illustrated by Figures 5.14–5.16, the two conditions [5.27] and [5.28] are fulfilled. If we choose another ZMP profile (for example, a linear profile of 3 cm at the start of a step and 15 cm at the end of a step), the robot movement does not converge towards a periodic movement even if the ZMP position remains on the inside of the footprint. The ZMP evolutionary profile and the $\dot{s}$ position during 5 steps illustrate the robot's behavior in Figure 5.17. The robot movement slows down and then stops.

5.4.3. *Bounded evolution of the ZMP*

In reality, to avoid a stance foot rotation, the evolution of the ZMP during walking must stay within the contact surface area. However, no condition imposes that the effective position of the ZMP must perfectly follow a predefined profile $l^d(s)$. In the case of the planar biped in Figure 5.4 (and to have a certain safety margin), the limited profile position of the ZMP must belong to the domain $[l_{min}, l_{max}]$ with $l_{min} > -0.03$ m and $l_{max} < 0.16$ m.

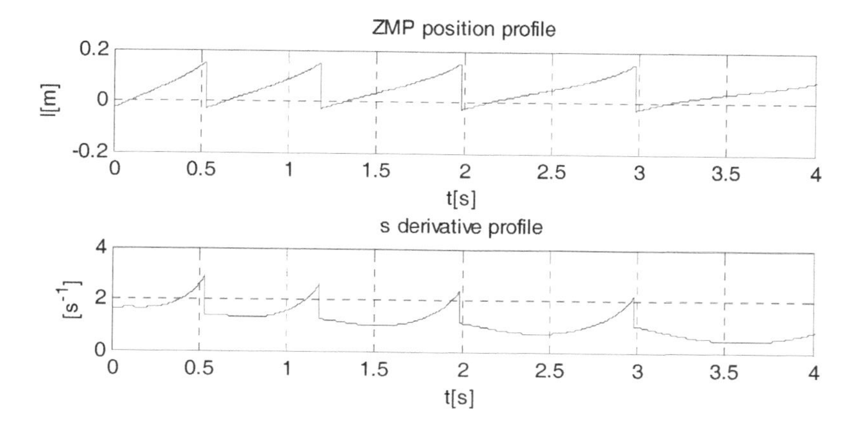

Figure 5.17. *Robot movement behavior with control [5.12] for a reference which does not fulfill conditions [5.27] and [5.28]. The movement does not converge towards a periodic movement even if the ZMP profile remains within the footprint*

If there is a cyclic movement which corresponds to $q^d(s)$, $l^d(s)$ (condition [5.27]), it is preferable that the biped's movement quickly converges towards this cyclic movement. We define the cyclic movement from the cyclic profile of velocity $\dot{s}$ written $\dot{s}_c$. To return to the cyclic movement, we define the error between the biped's current velocity $\dot{s}$ and the cyclic velocity $\dot{s}_c(s)$ with the following function:

$$e_v = \frac{\dot{s} - \dot{s}_c(s)}{\dot{s}_c(s)} \qquad [5.29]$$

When error [5.29] is cancelled out, the biped's movement returns to the cyclic movement. With the objective of canceling this error out, we choose the biped's desired acceleration $\ddot{s}$ so that:

$$\dot{e}_v + K_{vs} e_v = 0 \qquad [5.30]$$

where K_{vs} is a gain which defines the convergence velocity towards this cyclic movement. In this way, the desired acceleration is such that:

$$\ddot{s}^d = \frac{d\dot{s}_c(s)}{ds} \frac{\dot{s}^2}{\dot{s}_c(s)} + K_{vs}\left(\dot{s}_c(s) - \dot{s}\right) \qquad [5.31]$$

Acceleration $\ddot{s}$ and the ZMP position are linked by equation [5.8]. Even if the constraint on the specific ZMP profile is relaxed, the constraint on its limited profile zone still remains valid in order to avoid a stance foot rotation. The ZMP position must be between l_{min} and l_{max}. The control law therefore becomes:

$$\tau = M(q)(\frac{\partial q^d(s)}{\partial s}w + \frac{\partial^2 q^d(s)}{\partial s^2}\dot{s}^2 + K_v(\frac{\partial q^d(s)}{\partial s}\dot{s} - \dot{q}) + K_p(q^d(s) - q)) + C(q,\dot{q}) + G(q) - D(q,\dot{q})$$

where w minimizes $\left\| w - \frac{d\dot{s}_c(s)}{ds}\frac{\dot{s}^2}{\dot{s}_c(s)} + K_{vs}\left(\dot{s}_c(s) - \dot{s}\right) \right\|$ under the constraint [5.32]

$$l_{min} \leq \frac{\vec{x}_0 . \left(\vec{\delta}_G(q,\dot{q},w) + \overrightarrow{OG} \wedge \left(m(\vec{\gamma}_G(q,\dot{q},w) - \vec{g})\right)\right)}{\vec{z}_0 \cdot \left(m(\vec{\gamma}_G(q,\dot{q},w) - \vec{g})\right)} \leq l_{max}$$

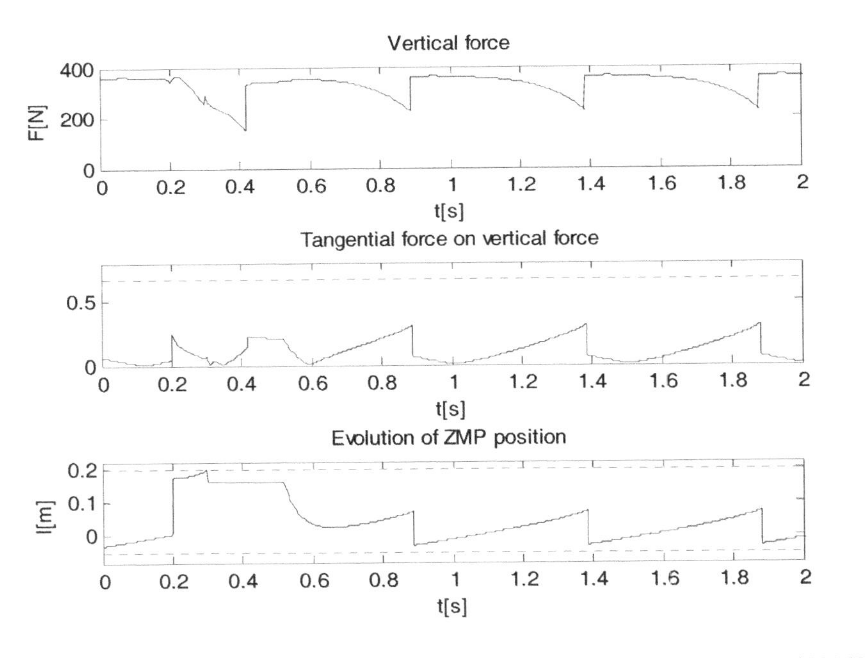

Figure 5.18. *The constraints on the ground reaction forces after a perturbation of 300 N. Control strategy [5.32] is used. Because of the perturbation, the robot's movement is faster than the desired movement and the ZMP moves as far to the forefront of the foot as is possible, in order to slow the movement down. In less than two steps, the robot's movements converge towards a cyclic movement. During the application of the perturbation, the ZMP position does not follow the desired position, as the perturbation is not known by the control*

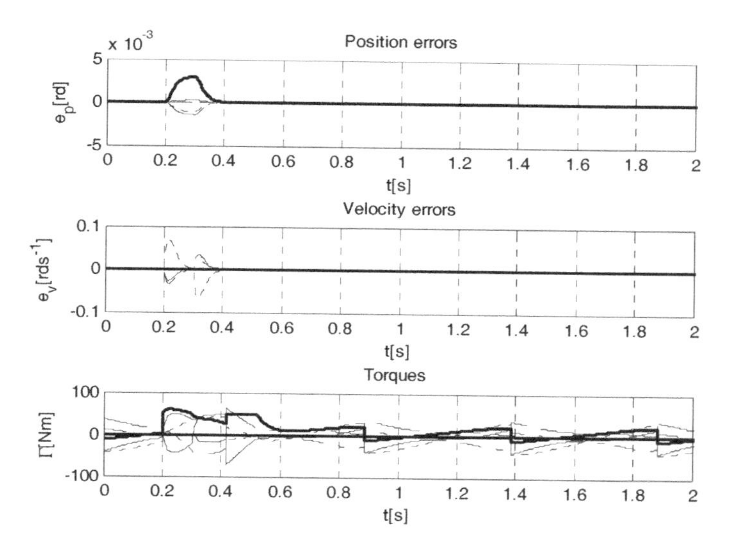

Figure 5.19. *There are tracking errors only during the application of the perturbation. At other times, the tracking errors are null and the reference is modified. After the perturbation, the higher torques are necessary but the useful torques quickly go back to the desired cyclic values*

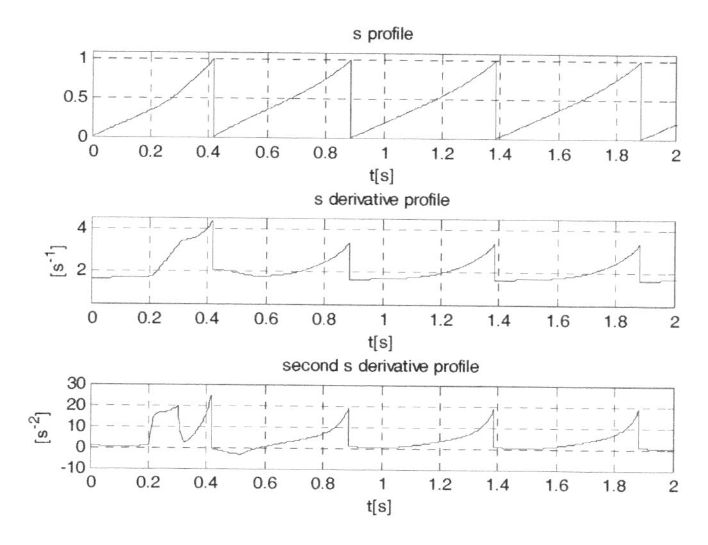

Figure 5.20. *Profile of the variable s and its derivatives in relation to time, as is illustrated in the above figures. For the successive steps, s varies from 0 to 1. During the perturbation, the* $\ddot{s}$ *acceleration is significantly modified and then* $\dot{s}$ *quickly converges towards a cyclic movement*

Due to its design, this control law means that the convergence occurs faster towards the desired cyclic movement than the preceding control.

The results which were obtained through simulation are illustrated by a response to a perturbation of 300 N (seen in Figures 5.18–5.20).

The maximum admissible perturbation (here 310 N) is linked to the chosen values of l_{min} and l_{max} (here l_{max} = 16 cm, l_{min} = –3 cm).

The greater the gap between the defined limits and the foot's effective size, the greater the admissible perturbation size. However, the profile bounds of the ZMP will therefore be smaller, and the movement will not converge as quickly towards a cyclic movement.

The principle of this control law is similar to control law [5.10]. However, the difference is that the reference is modified in this new version. This means that this approach is more robust when faced with perturbations. This constraint corresponds to $C_P \in \mathrm{Conv}(S)$ for a planar robot.

5.5. Taking an under-actuated phase into account

In section 5.4.2 we saw that we can impose a ZMP position and maintain an efficient control. This enables us to consider the control of a robot with point contact in the case of a 2D motion, or a linear contact in the case of a 3D motion.

We are also able to introduce rotation phases around the extremity of the toes during walking (second sub-phase of single support of walking as described in Chapter 2) [DJO 07].

We now present the case of a robot with no feet, as is the case for the planar *Rabbit* robot [WES 07]. The following laboratories took part in this project: Inria Sophia (Antipolis), Rhône (Alpes), IRCCyN (Nantes), LAG (Grenoble), LGIPM (Metz), LIRMM (Montpellier), LMS (Poitiers), LRP (Versailles), LSIIT (Strasbourg), LSS (Gif sur Yvette) and LVR (Bourges) [CHE 03]. The *Rabbit* robot can be found at LAG.

The control which was used for this robot respects the principles described in section 5.4.2, but was formulated differently. The reference trajectory has already been described in relation to the s parameter. For bipedal walking, it has been observed that there is a configuration variable which is monotonous. This variable can be the orientation axis with respect to the vertical axis of the virtual leg or the support ankle (see Figure 5.21).

The robot configuration can be represented by the q vector or by the vectors $[\alpha\, q_1\, q_2\, q_3\, q_4]^T$ or $[\theta\, q_1\, q_2\, q_3\, q_4]^T$.

The robot trajectory is therefore defined by the given datum $q(\alpha)$ or $q(\theta)$ with $q = [q_1\, q_2\, q_3\, q_4]^T$. This approach is limited because only the joint walking trajectories, with a monotonous evolution of θ or α, can be considered.

The advantage in comparison to an s parameterization is that the θ or α can be defined from the robot's configuration (or observation) measurements, whereas s is obtained from numerical integration.

The only controlled articulations are the actuated ones. This can be done via the intermediary of a computed torque control or via a simple control with proportional and derivatives actions.

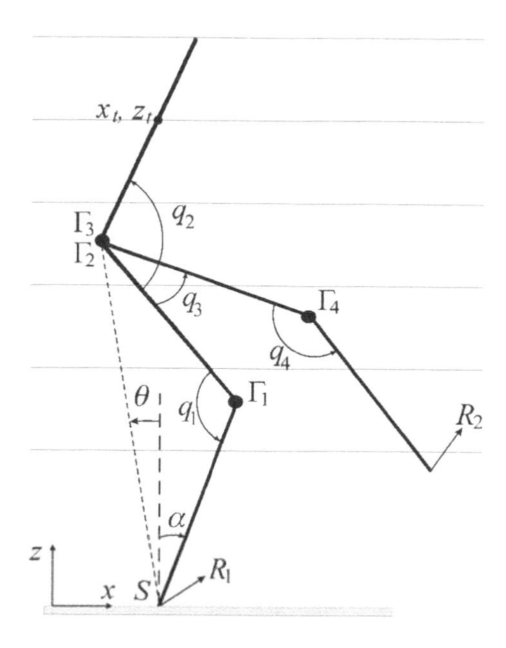

Figure 5.21. *Orientation of the stance leg*

An analysis of robot behavior can be carried out by using the method described in section 5.4.2. However, in the context of point contact, the ZMP position (or pressure center) is constant and known. It is merged with the extremity of the stance foot. The robot's equilibrium when rotating around the contact point (ZMP) is then written:

$$\dot{\sigma}_C = -mg(x_G - x_C) \tag{5.33}$$

This equation is obtained from equation [5.15] with $\dot{x}_C = 0$. The control enables us to ensure that the robot configuration is only a function of θ (or α). We obtain the system:

$$\begin{cases} \dot{\sigma} = -Mg\left(x_G(\theta) - x_C\right) \\ \dot{\theta} = \dfrac{\sigma}{I(\theta)}. \end{cases} \tag{5.34}$$

The combination of these two equations enables us to obtain:

$$\frac{d\zeta}{ds} = -mgI(\theta)(x_G(q^d(\theta)) - x_C) \quad \text{with} \quad \zeta = \frac{\sigma_C^2}{2} \tag{5.35}$$

The integration of this equation to the single support phase therefore gives:

$$\zeta(\theta) = \zeta(\theta_i) - \Phi(\theta),$$

$$\text{where} \quad \Phi(\theta) = mg\int_{\theta_i}^{\theta} I(\tau)(x_G(q^d(\tau)) - x_C)d\tau \tag{5.36}$$

where θ_i is the value of θ at the start of single support. The impact conditions are geometric conditions (both feet touch the ground). This ensures that, for any temporal profile which enables the step to accomplish itself, the step starts for a known $\theta = \theta_i$ and ends for a known $\theta = \theta_f$. Equation [5.20] enables us to determine, for step k, the value of $\dot{\theta}_k(\theta_f)$ at the end of a step in relation to the value $\dot{\theta}_k(\theta_i)$ at the start of a step:

$$\dot{\theta}_k(\theta_f) = \sqrt{\frac{I^2(\theta_i)\dot{\theta}_k^2(\theta_i) - 2\Phi(\theta_f)}{I^2(\theta_f)}} \tag{5.37}$$

In the case of the under-actuated *Rabbit* robot, the under-actuated and impact phases succeed each other. During the impact phase, the value of $\dot{\theta}_{k+1}(\theta_i)$ after impact is proportional to the value of $\dot{\theta}_k(\theta_f)$ before impact:

$$\dot{\theta}_{k+1}(\theta_i) = \delta_i \dot{\theta}_k(\theta_f) \quad [5.38]$$

The combination of these two equations results in the evolution $\dot{\theta}_k(\theta_f)$ from the current step to the following one. This gives:

$$\dot{\theta}_{k+1}(\theta_i) = \sqrt{\frac{I^2(\theta_i)\delta_i^2\dot{\theta}_k^2(\theta_i) - 2\delta_i^2\Phi(\theta_f)}{I^2(\theta_f)}} \quad [5.39]$$

This equation enables us to define the cyclic movement which corresponds to the cyclic trajectory $q^d(\theta)$:

$$\dot{\theta}(\theta_i) = \sqrt{\frac{-2\delta_i^2\Phi(\theta_f)}{I^2(\theta_f) - \delta_i^2 I^2(\theta_i)}} \quad [5.40]$$

This equation is only defined if the step can be obtained with monotonous θ, or:

$$-\frac{\delta_i^2 I^2(\theta_i)\Phi(\theta_f)}{I^2(\theta_f) - \delta_i^2 I^2(\theta_i)} > \max_{\theta_i \leq \theta \leq \theta_f} (\Phi(\theta)) \quad [5.41]$$

The control will therefore converge towards this cyclic movement if and only if:

$$\frac{\delta_i^2 I^2(\theta_i)}{I^2(\theta_f)} < 1 \quad [5.42]$$

The succession of a flat-footed phase (with a controlled ZMP position) with that of a rotational foot phase can also be treated. The combination of this study with the study mentioned in section 5.4.2 can be carried out to obtain a stability condition for the totality of step [DJO 07] for the type of movement seen in Figure 5.22.

Control with proportional and derivatives actions which uses the references expressed with respect to θ was used in Chapter 3 on three and five-link planar robots to illustrate the significant influence of mass distribution on the energetic efficiency of walking (see section 3.1). This control was also used experimentally on the *Rabbit* robot and gave very good results [WES 04]. This control is particularly robust. This robustness mainly comes from the fact that the stability conditions are inequalities. In the case of a model which is not perfectly known, the convergence towards a cyclic movement is not put into question, the obtained movement can be different from the prescribed cyclic movement.

The cyclic movement depends on the reference trajectory. The inclination of the trunk plays a particularly determinant role on the robot's forward velocity. The greater the trunk's forward inclination, the greater the forward position of the CoM, and the greater the cyclic velocity. It is possible to modify the reference trajectory at each step, either to improve velocity when converging towards the cyclic movement [AOU 03], or to regulate the forward velocity. This is done so that the robot's average velocity can be controlled and then follow the assigned forward velocity in [WES 03, WES 07].

This control approach was extended to the running gaits for the *Rabbit* robot [CHE 05b, MOR 06].

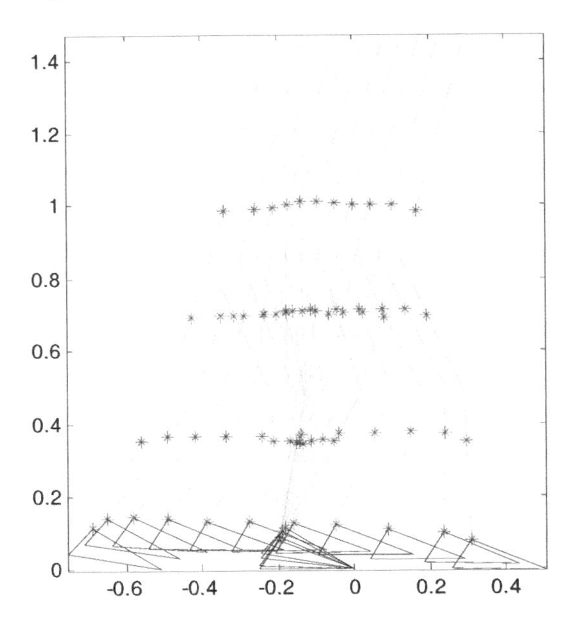

Figure 5.22. *Stick diagram corresponding to a reference trajectory composed of flat-footed stance phase and rotation around the extremity of the foot*

5.6. Taking the double support phase into account

During a walking gait, humans adopt double support phases which vary in duration depending on the forward displacement velocity. One of the main interests of these double support phases is that they improve walking stability. The objective here is therefore to study a bipedal robot's dynamic walk which is made up of double support phases of finite duration. A planar robot, with no feet, will lose two of its degrees of freedom (DoF) if we constrain one of its feet to stay on the ground (mechanically speaking). Consequently, an under-actuated or full-actuated biped during single support will then be over-actuated during double support. The number of motor torques is constant but the number of independent DoF decreases. Many solutions for the torques are therefore possible, with the condition of respecting the non-slide and non-take-off constraints. To illustrate the control problem during double support, a five-link robot with no feet is used as an example (Figure 5.21).

The reference movement for cyclic walking is calculated with the help of an optimization process, as seen in Chapter 4 (section 4.5.3.1). Let us re-state that each step is made up of finite single and double support phases. At the end of single support, the swing leg is supposed to touch the ground at zero velocity. There is therefore no impact.

In the preceding sections, we showed that it is very useful to define the controls from reference trajectories depending on the robot's state and not on time. This is the case particularly for the phases in which the robot is under-actuated, and to avoid specifying the impact instant. In this section we suppose that for a complete walking step, the references are expressed as functions of orientation α of the biped's supporting ankle.

During single support, the reference trajectories are defined for four actuated DoF $\delta_{i,SS}$ ($i = 1,\ldots,4$) by four-degree polynomials in α:

$$\delta_{i,SS}(\alpha) = a_{i0} + a_{i1}\alpha + a_{i2}\alpha^2 + a_{i3}\alpha^3 + a_{i4}\alpha^4 \qquad (i = 1,\ldots,4) \qquad [5.43]$$

During double support, the biped has only three DoF. The reference trajectories are defined for the relative variables $\delta_{i,DS} = \delta_{i,DS}(\alpha)$ ($i = 1, 2$).

Three-degree polynomials in α were chosen in order to coincide with single support:

$$\delta_{iDS}(\alpha) = a_{i0} + a_{i1}\alpha + a_{i1}\alpha^2 + a_{i1}\alpha^3 \qquad (i = 1,2) \qquad [5.44]$$

As the biped is over-actuated, the reference trajectory of variable α is defined as a time polynomial function:

$$\alpha_{DS}(t) = a_0 + a_1 t + a_2 t^2 + a_3 t^3 \qquad [5.45]$$

Consequently, the biped's movement is completely defined by the relative variables δ_{iDS} (i = 1, 2) and α. The fact that the relative variable's reference trajectory is a 3-degree polynomial enables us to define the initial, intermediary and final positions, as well as an initial velocity. The over-actuation enables us to closely follow reference trajectories [5.44] and [5.45], with the condition that the torque and ground reaction constraints (among others) are valid.

We will now look at control of α during the double support phase. The biped's movement solely depends on α (in the hypothesis that the reference trajectories are perfectly followed). During single support no action is possible on the support ankle orientation α because the biped is under-actuated. With the double support phase, it is possible to influence the dynamics of α to improve the convergence towards a cyclic walking movement.

An optimal time control was chosen during the double support phase. This consists of applying the maximum possible acceleration $\ddot{\alpha}$ for a convergence towards a cyclic movement α_c, and a continuation of this cyclic movement. The control law expression is based on the following acceleration expression:

$$\ddot{a} = \begin{cases} \ddot{a}_{\max}(a, \dot{a}) & \text{if } \dot{a}(a) - \dot{a}_c(a) < 0 \\ \ddot{a}_{\min}(a, \dot{a}) & \text{if } \dot{a}(a) - \dot{a}_c(a) > 0 \\ \ddot{a}_c(a, \dot{a}) & \text{if } \dot{a}(a) - \dot{a}_c(a) = 0 \end{cases} \qquad [5.46]$$

The variables $\ddot{\alpha}_{max}(\alpha, \dot{\alpha})$ and $\ddot{\alpha}_{min}(\alpha, \dot{\alpha})$ are each the maximum and minimum possible accelerations which are needed to fulfill the physical constraints of non-slide, non-take-off and torque constraints. The same simplified control principle was used in [GRI 94], where a cyclic walking movement with a double support phase was defined for a biped with telescopic legs. The authors chose a constant for the maximum acceleration so that the feet were maintained on the ground. In [MIO 05], these maximum and minimum velocities were determined at each instant through the use of a dynamic model. The physical constraints under consideration are:

$$\begin{cases} R_{iz} \geq R_{iz,min} & (i = 1,2) \\ -fR_{iz} \leq R_{ix} \leq fR_{iz} & (i = 1,2) \\ -\Gamma_{max} \leq \Gamma_j \leq \Gamma & (j = 1,...,4). \end{cases} \qquad [5.47]$$

The variable $R_{iz,min} > 0$ is the minimal value of the normal ground reaction component. Dimension f is the chosen ground friction coefficient (maximum index suppressed).

These constraints can be expressed with the help of α, $\dot{\alpha}$, $\ddot{\alpha}$ and torques τ by inserting equation [5.44] into dynamic model [2.38]. The following seven-dimension matrix equation is therefore obtained:

$$M_\alpha(\alpha)\ddot{\alpha} + H_\alpha(\alpha,\dot{\alpha}) = A^\tau \tau + J_1^T(\alpha) R_1 + J_2^T(\alpha) R_2 \qquad [5.48]$$

In this expression, the Lagrange multipliers used in Chapter 2 have been replaced by notations R_1 and R_2. This highlights the fact that they are the ground reaction forces on foot 1 and 2.

The Jacobian matrix linked to the contact constraints was presented in the form of two Jacobian matrices, and each corresponded to the contact constraints for each foot.

The eight unknown variables are the four torques τ_j and the two ground reaction components for each foot, R_{ix} and R_{iz} for seven equations only.

During double support, the biped is over-actuated. There is therefore an infinite number of torque solutions that we can express in relation to a parameter that can be R_{1x}, R_{1z}, R_{2x} or R_{2z}.

When we overview the forces and moments at G (the biped's center of gravity), we notice that the vertical components R_{1z} and R_{2z} of the ground reaction forces do not depend on torque distribution [MIO 02].

Among the possible parameters, only the horizontal components R_{1x} or R_{2x} remain. Through the parameterization of the τ_j torque solution with R_{2x}, we obtain the following type of equation:

$$\begin{bmatrix} R_{1x} & R_{1z} & R_{2z} & \Gamma^T \end{bmatrix}^T = B(\alpha)\ddot{\alpha} + C(\alpha,\dot{\alpha}) + D(\alpha) R_{2x} \qquad [5.49]$$

With the seven equations [5.49], it is possible to re-write the 14 inequalities of constraints [5.47] in the form:

$$E(\alpha)\ddot{\alpha}+F(\alpha,\dot{\alpha})+G(\alpha)R_{2x}+P\leq 0 \qquad [5.50]$$

The accelerations $\ddot{\alpha}_{max}(\alpha,\dot{\alpha})$ and $\ddot{\alpha}_{min}(\alpha,\dot{\alpha})$ are therefore obtained by solving two problems of *simplex* [DAN 63]:

$$\begin{cases} \ddot{\alpha}_{min}(\alpha,\dot{\alpha}) = \min\limits_{\ddot{\alpha},R_{2x}} \ddot{\alpha} \\ E(\alpha)\ddot{\alpha}+F(\alpha,\dot{\alpha})+G(\alpha)R_{2x}+P\leq 0 \\ \ddot{\alpha}_{max}(\alpha,\dot{\alpha}) = \max\limits_{\ddot{\alpha},R_{2x}} \ddot{\alpha} \\ E(\alpha)\ddot{\alpha}+F(\alpha,\dot{\alpha})+G(\alpha)R_{2x}+P\leq 0. \end{cases} \qquad [5.51]$$

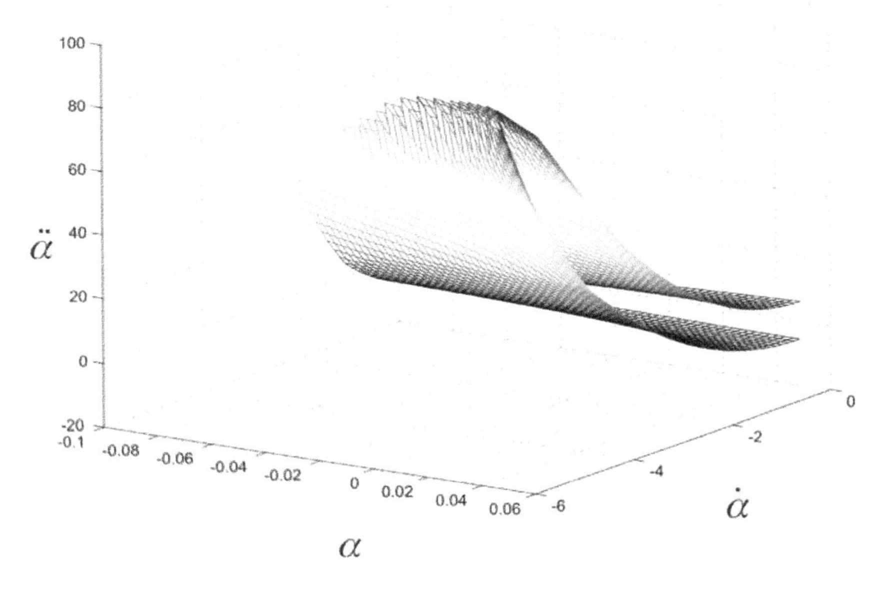

Figure 5.23. *Representation of $\ddot{\alpha}_{min}$ and $\ddot{\alpha}_{max}$ as functions of α and $\dot{\alpha}$*

Numeric tests of this control (for the given reference trajectories [5.43] and [5.44]) were carried out for a biped with no feet and with the same physical parameters as those given in Table 5.1 [MIO 05]. Figure 5.23 shows the constraints $\ddot{\alpha}_{min}$ and $\ddot{\alpha}_{max}$ which are functions of the variables α and $\dot{\alpha}$ only.

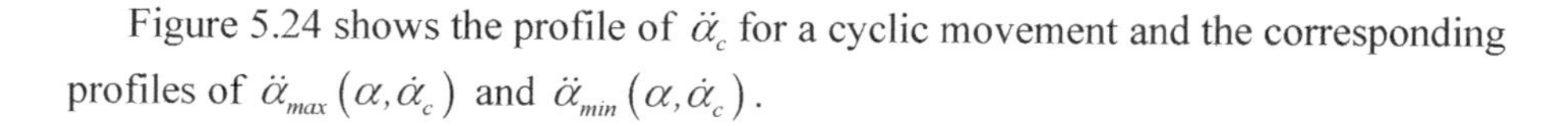

Figure 5.24 shows the profile of $\ddot{\alpha}_c$ for a cyclic movement and the corresponding profiles of $\ddot{\alpha}_{max}(\alpha, \dot{\alpha}_c)$ and $\ddot{\alpha}_{min}(\alpha, \dot{\alpha}_c)$.

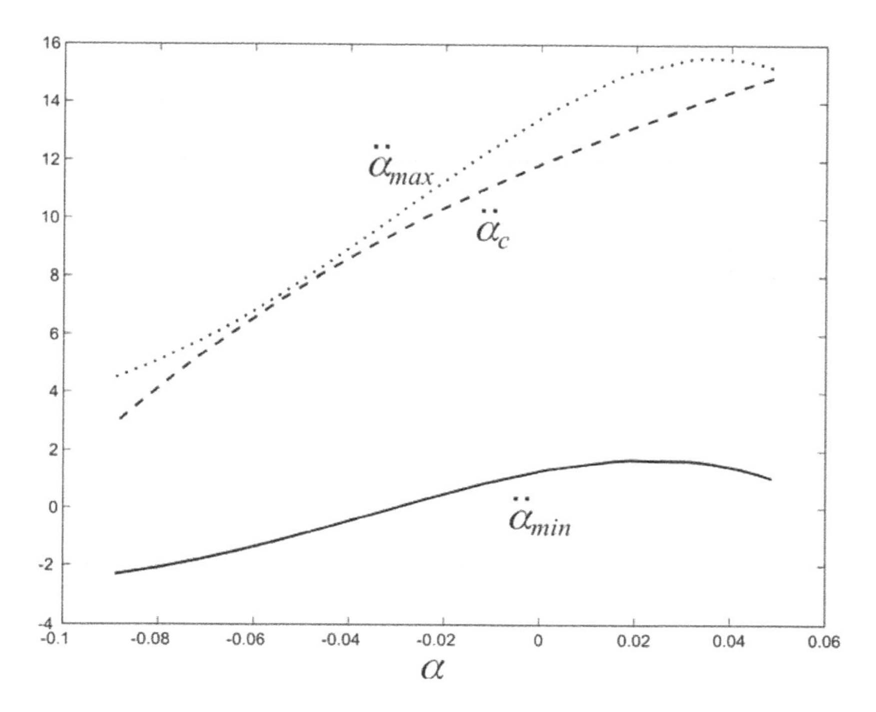

Figure 5.24. *Acceleration profile for the cyclic movement $\ddot{\alpha}_c$ with minimum and maximum possible accelerations $\ddot{\alpha}_{min}$ and $\ddot{\alpha}_{max}$ relative to α*

This approach enables us to improve the speed of the convergence towards a cyclic movement, as is illustrated in the Poincaré function which is relative to the initial velocity $\dot{\alpha}_{iDS}$ of the double support phase (Figure 5.25) for the cyclic movement under consideration.

There is a zero slope zone around the nominal walk. Within this zone, convergence is obtained in one step towards the nominal cycle. The attraction domain is limited by the biped's backwards fall and by the limitations of the actuating torques. During double support, the biped can start with a zero initial velocity $\dot{\alpha}_{iDS}$, at a stop, and then converge towards the nominal cycle in one step.

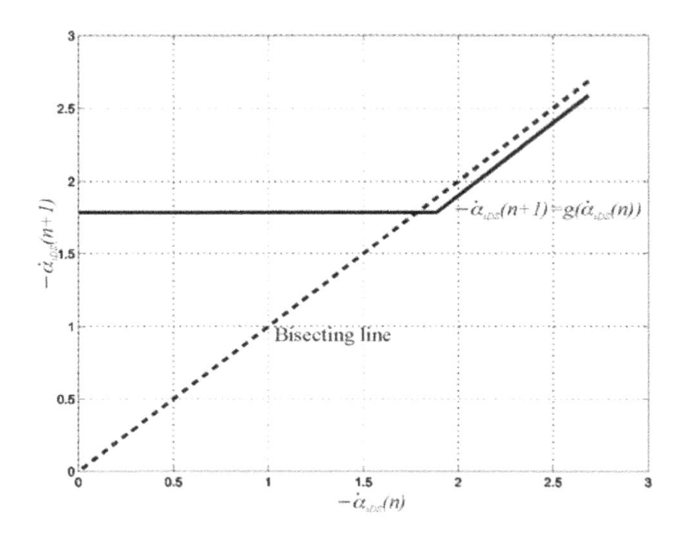

Figure 5.25. *Poincaré function drawn in relation to the initial velocity of single support*

5.7. Intuitive and neural network methods

5.7.1. *Intuitive methods*

Controls that are dedicated to robot walkers can sometimes be built according to a defined sequence of movements. These can move the robot's center of mass or provoke transitions between phases. An analogy with the locomotion of living beings, however, is not systematically sought. In this way, a dynamically stable walking gait called "Courbette", which does not exist in nature, was intuitively defined for *SemiQuad* [AOU 06].

This robot consists of a platform and two identical legs with knees and no feet. A half-step is made up of double support, single support on the rear leg and impact. The following half-step is made up of double support, single support on the front leg and ground impact (Figure 5.26). The different configuration lines which follow the walking phases are the result of the *SemiQuad's* simulated movement.

(a) Doublesupport (half step n) :
The projection of the plaform center
is in the middle of the leg tips

(b) Double support (half step n) :
The projection of the platform
center is closer to the back leg tip.

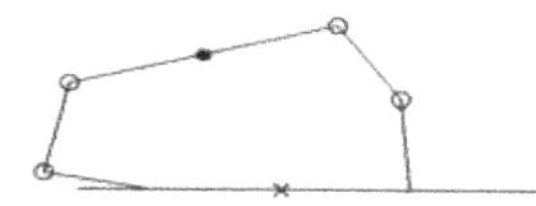

(c) Double support (half step n):
The front leg is unbent jsut before take off
before the single support.

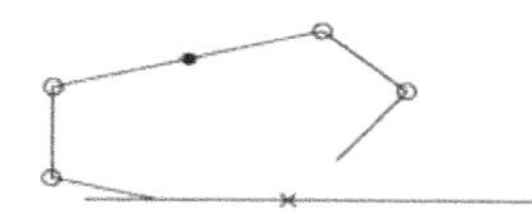

(d) Single support (half step n):
Just after jump of the front leg, the
front leg is bent

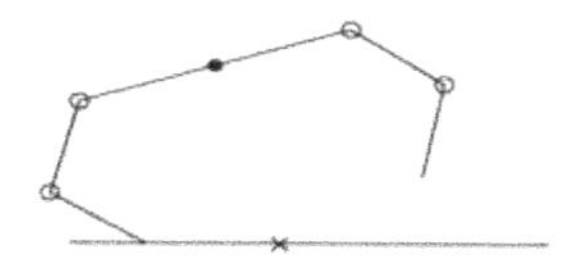

(e) Single support (half step n):
The distance between the leg tips is
larger than in the previous double
support phase.

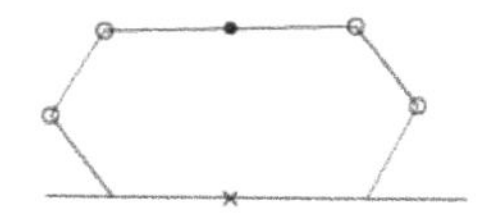

(f) Double support (end of half step n and start of
half step n+1): Just after landing with an impact
of the front leg. After half step n, the platform
center has moved forward.

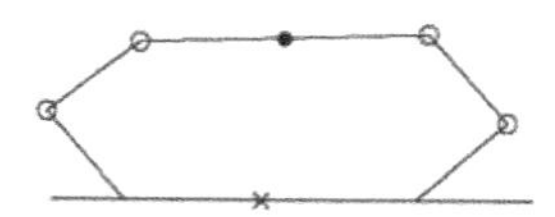

(g) Double support (half step n+1) :
The projection of the platform center
form is closer to the front leg tip.

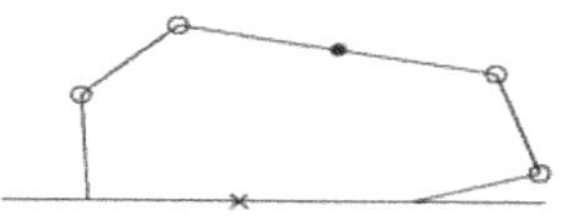

(h) Double support (half step n+1):
The back leg is unbent just before take off
before the next single support phase.

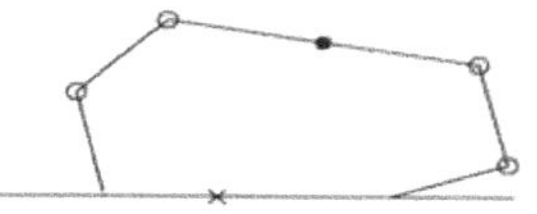

(i) Single support (half step n+1) :
Just after jump of the back leg,
the back leg is bent.

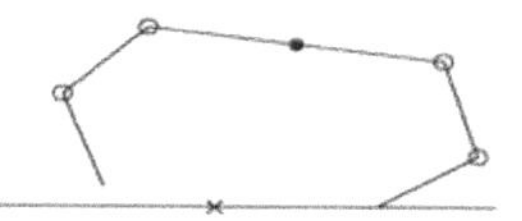

(j) Single support (half step n+1): The distance
between the leg tips becomes smaller than in
the previous double support phase.

Figure 5.26. *Transition configuration lines between the phases for the half-steps n and n + 1 during SemiQuad's walking gait*

The objective is to move the robot's center of mass during double support in order to trigger single support more easily. The actuated DoF are controlled during these single support phases, which vary the distance between the foot extremities. The movement was refined by comparing the experimental results with those obtained during the simulation.

A similar intuitive control principle was developed for the running gait of the quadruped called *Scout II*. It has four legs with no knees or feet [POU 05]. Each leg is made up of a lower and upper part, connected by a buffer. The whole makes up a prismatic elastic liaison. The desired movement is identical for the front and rear legs. Local correctors (with proportional and derivatives actions) control the motors which actuate the hips. Each corrector has a specific action depending on whether the leg is detected as being in the air, impacting or in single support. When the leg is in the air, the corrector's objective is to bring the hip's angular variable to a final desired value just before the impact phase. On impact, the maximum torque/maximum velocity template curve is taken into account for each motor, and a quasi-impulse control is applied to the stance leg.

The objective is to be able to support the robot's weight and the inertial forces. A control with proportional and derivative actions is then applied to each stance leg so that it can reach the limited angular value. With this type of intuitive control, the quadruped robot manages to perform a running gait which is dynamically stable.

This type of control was also successfully applied to the bipeds *Spring Turkey* and *Spring Flamingo* [PRA 98, PRA 01]. These bipedal robots have a trunk, two legs, knees and feet which can be actuated. *Spring Flamingo*'s mass is concentrated at its hips so that its leg mass can be considered as negligible. The motors produce the desired torques or forces which act at the robot's mass center. To explain this type of control strategy, let us assume that these robot walkers consist of a trunk, a single leg with a knee and a flat-footed contact (see Figure 5.27). Let us express the position of the hip as (x_h, z_h) and the orientation of the trunk in relation to the hip length l_1 and tibia length l_2 and angles θ_1, θ_2 and θ_3:

$$\begin{aligned} x_h &= -l_1 \sin\theta_1 - l_2 \sin(\theta_1 + \theta_2) \\ z_h &= l_1 \cos\theta_1 + l_2 \cos(\theta_1 + \theta_2) \\ \theta &= -\theta_1 - \theta_2 - \theta_3 \end{aligned} \qquad [5.52]$$

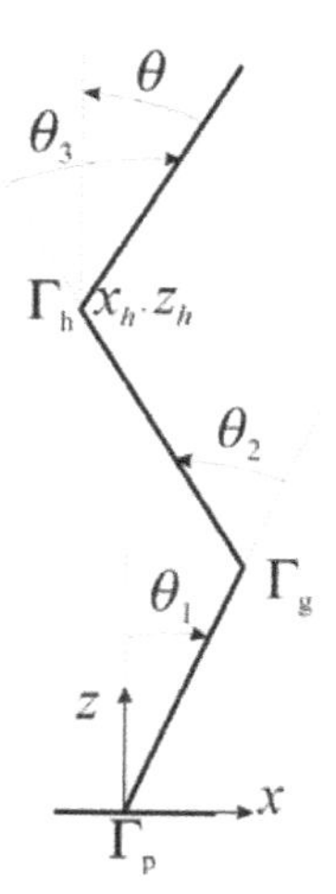

Figure 5.27. *Monopod with a trunk, hip, ankle and foot*

Starting from equation [5.52] we can describe the virtual displacement between the hip and the foot in relation to the virtual angular displacements $\delta\theta_1$, $\delta\theta_2$ and $\delta\theta_3$:

$$\begin{aligned}
\delta x_h &= -\left(l_1 \cos\theta_1 + l_2 \cos(\theta_1+\theta_2)\right)\delta\theta_1 - l_2 \cos(\theta_1+\theta_2)\delta\theta_2 \\
\delta z_h &= -\left(l_1 \sin\theta_1 + l_2 \sin(\theta_1+\theta_2)\right)\delta\theta_1 - l_2 \sin(\theta_1+\theta_2)\delta\theta_2 \\
\delta\theta &= -\delta\theta_1 - \delta\theta_2 - \delta\theta_3 .
\end{aligned} \qquad [5.53]$$

Using the theory of virtual studies, it is possible to define the relationship between the joint torques which act on the articulations of the hips, knee and foot as well as on the virtual efforts f_x, f_z and f_θ relative to x, z and θ:

$$\begin{pmatrix} \Gamma_p \\ \Gamma_g \\ \Gamma_h \end{pmatrix} = \begin{pmatrix} -l_1 \cos\theta_1 - l_2 \cos(\theta_1+\theta_2) & -l_1 \sin\theta_1 - l_2 \sin(\theta_1+\theta_2) & -1 \\ -l_2 \cos(\theta_1+\theta_2) & -l_2 \sin(\theta_1+\theta_2) & -1 \\ 0 & 0 & -1 \end{pmatrix} \begin{pmatrix} f_x \\ f_z \\ f_\theta \end{pmatrix} \qquad [5.54]$$

If we suppose that the foot is not actuated, this leads us to set $\Gamma_p = 0$ in equation [5.54]. In addition, to obtain a walking gait for a bipedal robot, we are led to favor the joint torques and forces which are applied to the vertical axis, rather than a force on the horizontal axis. This is why we can specify that the virtual force is f_z and the virtual moment is f_θ to find the virtual force f_x which corresponds to $\Gamma_p = 0$:

$$f_x = \frac{-1}{l_1 \cos\theta_1 + l_2 \cos(\theta_1+\theta_2)} \begin{pmatrix} l_1 \sin\theta_1 + l_2 \sin(\theta_1+\theta_2) & 1 \end{pmatrix} \begin{pmatrix} f_z \\ f_\theta \end{pmatrix} \qquad [5.55]$$

The actuated torques are deduced by substituting equation [5.55] into equation [5.54]:

$$\begin{pmatrix} \Gamma_g \\ \Gamma_h \end{pmatrix} = \begin{pmatrix} \dfrac{-l_1 l_2 \sin\theta_2}{l_1 \cos\theta_1 + l_2 \cos(\theta_1+\theta_2)} & \dfrac{-l_1 \cos\theta_1}{l_1 \cos\theta_1 + l_2 \cos(\theta_1+\theta_2)} \\ 0 & -1 \end{pmatrix} \begin{pmatrix} f_z \\ f_\theta \end{pmatrix} \qquad [5.56]$$

When the leg is straight, the matrix of equation [5.56] is singular. Pratt *et al.* [PRA 01] show that this principle of virtual actions can be applied to bipeds.

It is also possible to obtain a matrix relation between the torques which are applied to the hips and knees of legs 1 and 2 and the relative virtual moment and forces f_θ f_x, f_y applied during the double support phase:

$$\begin{pmatrix} \Gamma_{g_1} \\ \Gamma_{h_1} \\ \Gamma_{g_2} \\ \Gamma_{h_2} \end{pmatrix} = V(4\times3) \begin{pmatrix} f_x \\ f_z \\ f_\theta \end{pmatrix} \qquad [5.57]$$

The matrix $V(4\times3)$ is singular when there is no simultaneous flexion at the knees. For the other cases, the virtual forces f_x, f_z and the virtual moment f_θ are admissible. Using this virtual control model as a basis, an algorithm was created for the dynamically stable walks of *Spring Turkey* and *Spring Flamingo*. The objectives include:

– to maintain a certain trunk height and inclination during the single and double support phases;

– to correct the velocity perturbations during double support; and

– to guarantee the transferring leg's step length.

The transitions between single and double support are based on limited bipedal robot configuration choices. This control law was made for walking on non-horizontal terrain.

These different types of intuitive control lead to remarkable results when defining walking for experimental prototypes. The general advantage is that we do not have to use many sensors to define these control laws because they do not reflect the robot's whole state. In addition, these control laws are not based on predefined reference trajectories. The result is a certain robustness or dynamic stability. The main disadvantage, however, is that the definition of intuitive control requires many long and laborious trial and error tests. In addition, intuitive control characteristics are inherently linked to the robot walker under consideration, and generalizing them for a different robot structure is difficult.

5.7.2. *Neural network method*

Many animals control their movements through central neural networks. Within these networks the signals, which are received from sense captors, play an essential role in using the environment as a basis for movement coordination [REE 99]. These neural networks are called *Central Pattern Generators*.

A reflexive neural model was created to study the locomotion of a stick insect [CRU 81a, CRU 81b] and was then used to define the control of a bipedal robot [GEN 06]. This model uses information taken from one leg to generate actions for the other leg by using a reflexive mechanism. These two distinct methods: the CPG neural network method and the alternative method based on reflexive control have been referred to in literature about animal locomotion and robotics. These studies use the living world as an inspirational basis for legged coordination.

In this section, we will give a general overview of artificial neurons. We will also see how they can be used for two bipedal robot applications. The first case will use a reflexive mechanism and the second will use a CPG.

A biological neuron is a nervous system cell which treats and transmits information. It evaluates or sums up all the information which is transmitted by the other neurons at an entry point called the synapse. It then sends an electric signal to the synaptic terminals in order to transmit the analytical result to the following neuron.

For humans, the neural connective structure can modify itself. This is why we say that humans are capable of learning from plasticity [ANI 96].

In the 1940s, McCulloch and Pitts [MCC 43] created a biological neural model depicted in Figure 5.28.

The artificial neuron j has inputs u_i which are weighted by weight functions w_{ji} which model the excitation and inhibition models. It calculates the weighted sum of the received entries, to which it applies the activation function f. The weight entry bias b_j enables us to regulate the trigger threshold of the neuron.

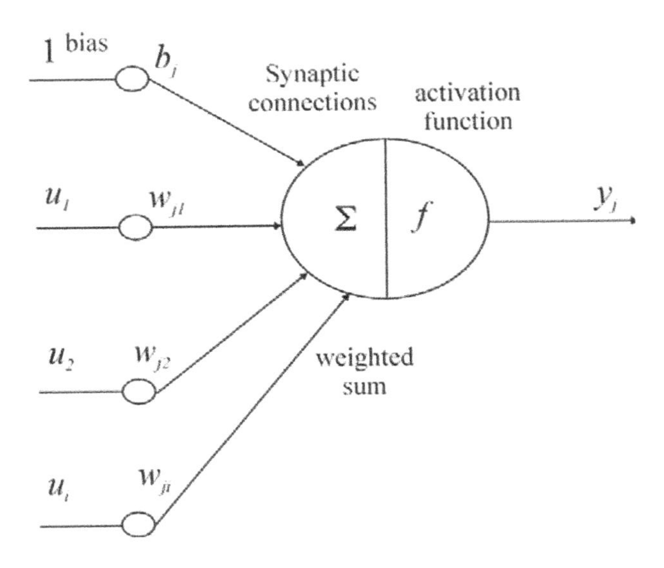

Figure 5.28. *Basic artificial neuron model*

Output y_j of McCulloch and Pitts's artificial neuron is an effective function of many variables:

$$y_j = f\left(\sum_{i=1}^{N} w_{ji} u_i\right) \qquad [5.58]$$

Function f is generally a sigmoid function, which varies from 0 to 1. It is differentiable and monotonous:

$$f(s) = \frac{1}{1+e^{-s}} \qquad [5.59]$$

Other artificial neural models have been created since. For example, Ekeberg's model [GAL 96] specifies the output variation y_j with a first order, nonlinear differential equation:

$$y_j = -\tau_j \frac{dy_j}{dt} + f\left(\sum_{i=1}^{N} w_{ji} u_i\right)$$

[5.60]

An artificial neuron is the elementary unit of an artificial neural network. Its function can be compared to a biological neuron. Within a network, the neurons work in parallel and are grouped together. Each neural group treats signals independently and transmits its result to the next group. There are therefore different network structures. They can be connected in layers, be recurrent, or be totally connected. The network's treatment capacity is stocked at the interconnections. This management function is based on an adaptation or learning process, which is a function for a series of reference models. The neural network structure is finally translated by the creation of network treatment processes which can take in many input variable types and then apply algorithms in order to arrive at a unique solution.

For the control of robot walkers, or to define reference trajectories for their DoF, methods based on neural networks can be used. These methods require only little information about the robot model.

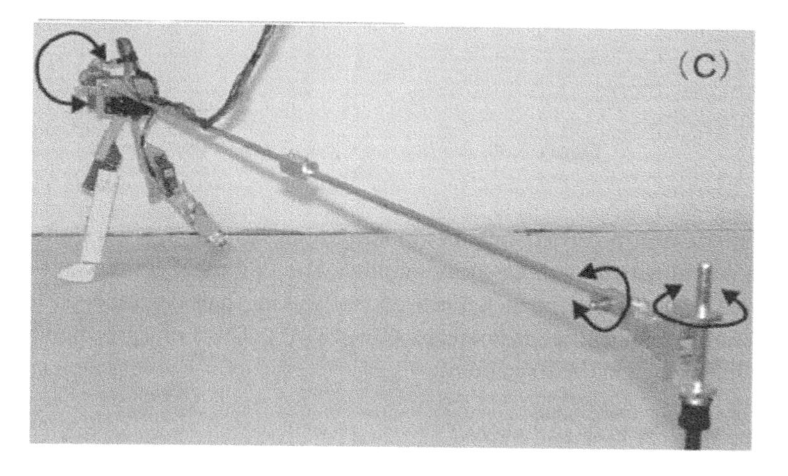

Figure 5.29. *RunBot robot [GEN 06]*

The control law based on the *reflexive* neural control schema was tested on a five-link robot (see Figure 5.29). It has a trunk and each leg has a knee and circular arc foot [GEN 06]. This biped is actuated by four motors situated at the hips and knees. The ground impact and instant where the joint position of the free leg reaches the hip's limited anterior angle (Figure 5.30) give rhythm to the nominal movement of each leg. In this way, the left leg's ground impact (solid line) locally triggers four reflex actions at the same instant: the left hip's flexion, the left knee's extension, the right hip's extension and the right knee's flexion. When the right hip reaches the limited anterior angle, only the right knee's extension is triggered. The reflexive control for the biped's walk has a hierarchical structure (Figure 5.31). The lower level is dedicated to the local reflex movements of the articulations which include the neural motors and sensors. The upper level is a neural network made up of hip receptors, neural impact sensors and artificial neurons which create the reflexes.

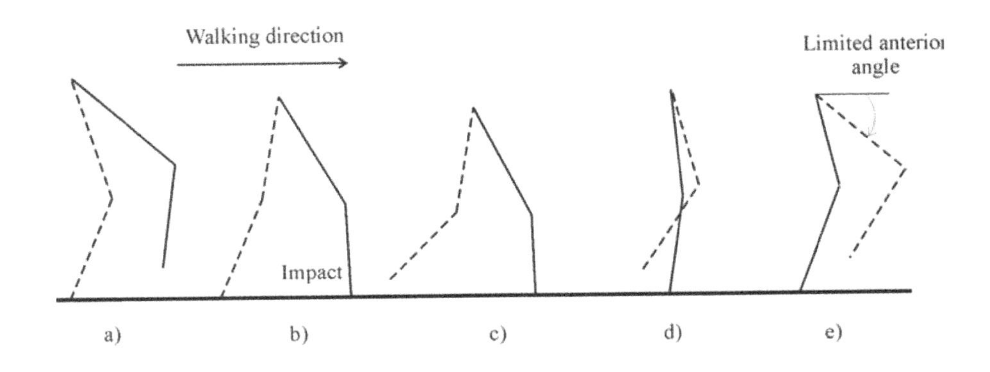

Figure 5.30. *RunBot robot's walk [GEN 06]*

A numeric study carried out on *RunBot* robot, which was controlled by the reflexive neural network, revealed its stability. The Poincaré section can be defined by the free leg's ground impact. Each cyclic walking gait corresponds to a fixed point X^* for the application of Poincaré $X^{k+1} = P(X^k)$. This Poincaré application can be made linear around the fixed point:

$$P\left(X^* + \delta X\right) \approx P\left(X^*\right) + J\delta X \qquad [5.61]$$

where $J = J\left(n \times n\right)$ is the Jacobian matrix of the partial derivatives of P in relation to the state vector X with a dimension n.

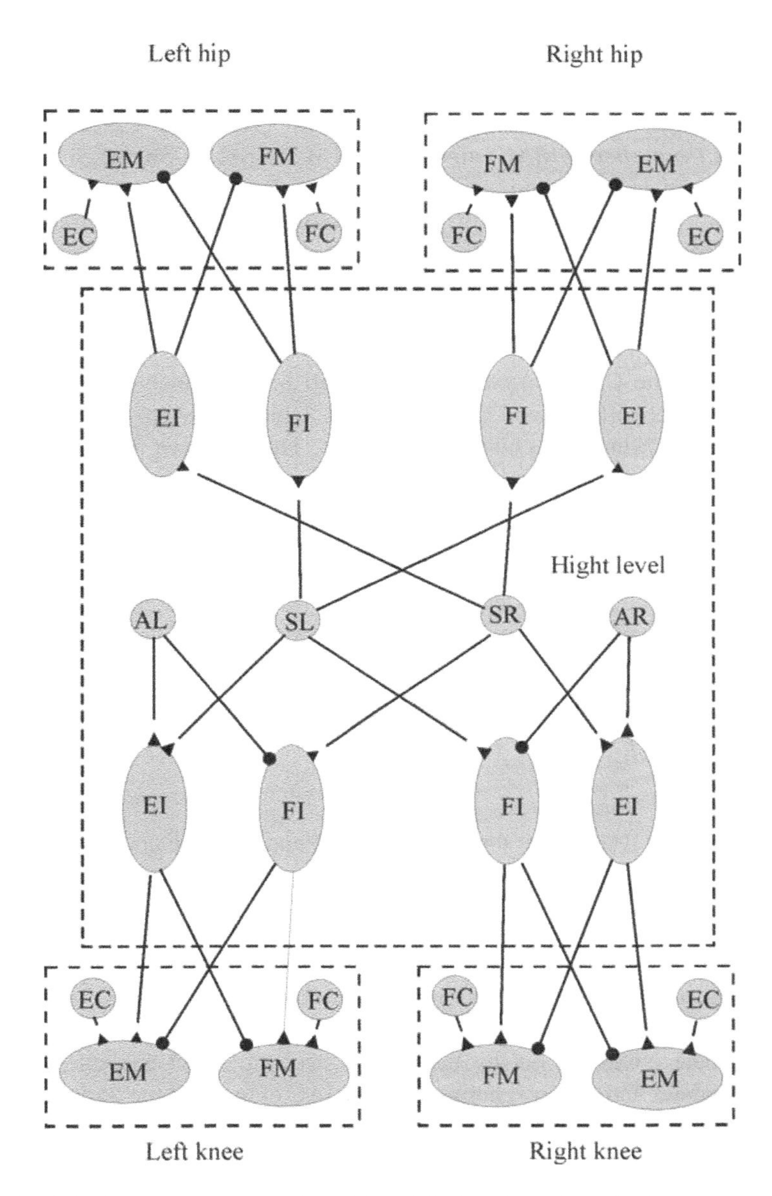

Figure 5.31. *Reflexive control neural model used for RunBot robot (AL, AR: receptors for the hips limited anterior; SL, SR: impact sensors for the left and right leg; EI, FI: extension neurons and flexion; EM, FM: motor extension neurons and flexion; EC, FC: extension sensor neurons and flexion) [GEN 06]*

Rosenblatt's Multilayered Perceptron (MLP) network [ROS 62] was used for bipedal control [HEN 94]. Its learning phase, however, which is based on a retro propagation algorithm, has a high calculation cost. The *Cerebellar Model Articulation Controller* (CMAC) algorithm [ALB 75a, ALB 75b] based its working principle on the human brain. It was applied to *Rabbit's* control [SAB 04] and required fewer calculations than the habitual multilayered neural networks. This was particularly true for the learning phase where, unlike the MLP network, there is no gradient calculation to be carried out. The model was also used for the control of a statically stable biped [BEN 97]. If the CMAC inputs are close to each other, the outputs will also be close to each other. However, if the network inputs are distanced, then the CMAC network outputs will be completely independent. The CMAC neural network is made up of N sensors, which are evenly distributed for C number of layers which give binary information. The receptive field's output has a value of 1 when the entry signal is included in its interval, otherwise it remains equal to 0.

The number of sensors N depends on the size of the receptive fields and on the number of layers C. The step length with a quantification q is characterized by the shift in the receptive fields at each layer. The CMAC weights w are updated during the learning phase, so that:

$$w(t_i) = w(t_{i-1}) + \frac{\beta e}{C}$$

where $w(t_i)$ and $w(t_{i-1})$ are the weights before and after the update at each step's time sample t_i. The variable e represents the error between the CMAC desired output Y^d and the CMAC calculated output. The term β is a parameter which is included in interval [0, 1]. If $\beta = 0$, there is no update. If $\beta = 1$, the weights are adjusted in such a way that they follow the desired output with precision. The effects of the preceding updates are therefore not taken into account. If $\beta = 0.5$, the neural network update is carried out by taking into account, in an equivalent way, the preceding updates and the error between the CMAC output and the desired output.

Parameter β enables us to adjust the memory effect during the learning process. An example of a CMAC network structures with 9 sensors, 2 layers and updated weights can be seen in Figure 5.32. The size of the receptive fields is equal to $2q$ (except for cells a_5 and a_9 where it is equal to q).

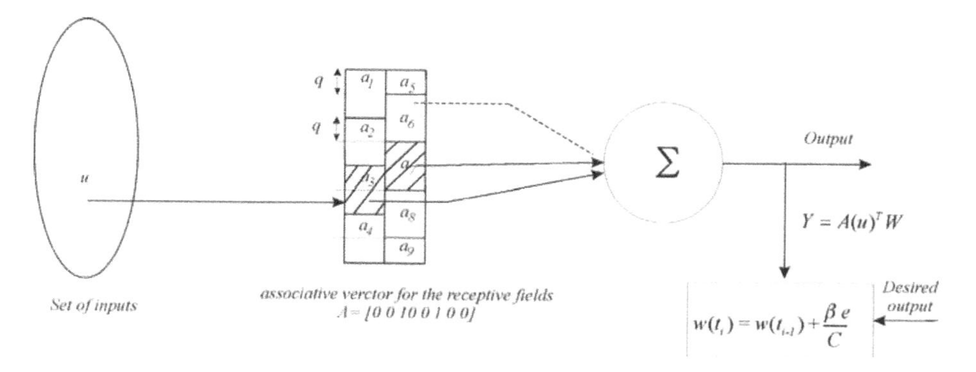

Figure 5.32. *Description of a CMAC network with 9 sensors distributed over 2 layers [SAB 04]*

For the *Rabbit* application, angle θ of the carrying leg was chosen as input u for the four networks which were associated with the two hip and knee articulations (see Figure 5.21).

The trajectories, which are learnt by the neural networks, are consequently not time functions but are part of the biped's geometric configuration.

The networks are initialized by intuitive trajectory. The step has an angle of 0.25° and a maximum receptive field size of 1.5°. We take into account the desired step length and the articulated position stops. Each network structure is based on 161 sensors which are distributed over 6 layers.

After the learning phase, during usage, the neural networks generate the desired positions and joint velocities for the trajectories which are injected into a corrector with proportional actions and derivatives.

The experimental results obtained by Sabourin for the *Rabbit* prototype show that this type of control is highly pertinent. Nevertheless, it is difficult to create stability criteria for bipedal walking when it is controlled because of external perturbations.

In addition, it is also difficult to pin-point the advantages of this type of control when we compare it to the PID control [WES 04], which had a reference trajectory which was also based on the monotonous configuration variable θ.

It is probable that neural controls will be very useful for future walking or running humanoid robots with many DoF, for whom the complete dynamic model would be difficult to apply.

5.8. Passive movements

Bipeds which could walk down slight slopes appeared as toys at the end of the 20th century. Their legs were straight and they moved laterally in an abrupt way to enable their feet to leave the ground. The behavioral analysis of these systems, however, is much more recent [MCG 90]. The interesting aspect of these systems is their low energy consumption. The energy given to the system comes from the variation in the potential energy linked to the slope. This compensates for the energy which is lost during impact.

The studies were carried out on constrained robots which moved along a sagittal plane [MCG 90]. The studies then progressed to robots which moved around in space, e.g. the robot depicted in Figure 5.33a [COL 05]. This robot has the characteristics: (a) the shape of its feet has been specially designed to favor stability in the frontal plane; (b) the foot dampers reduce the impact effects; and (c) the arms are important for lateral stability (to limit rolling and for changing the supporting leg).

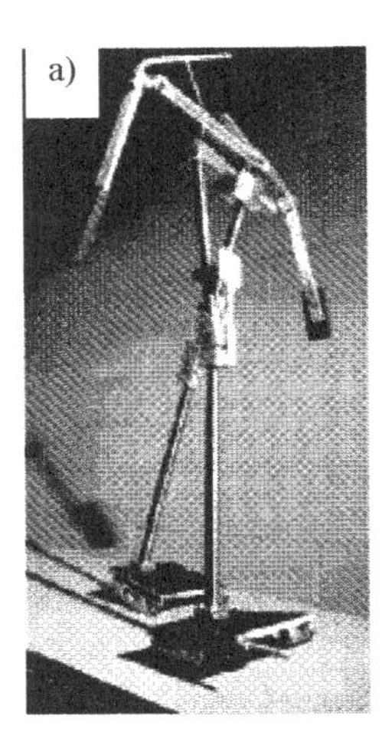

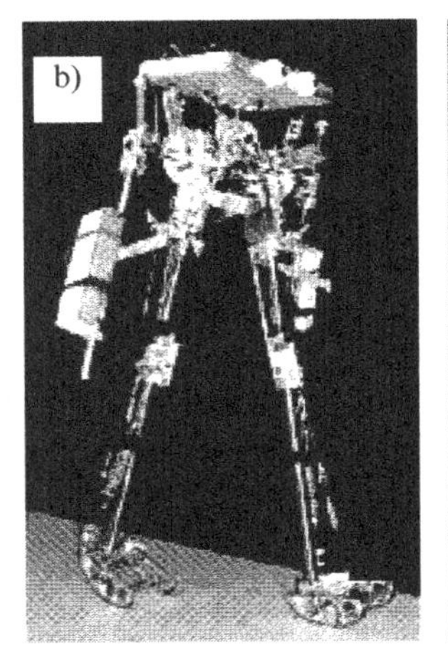

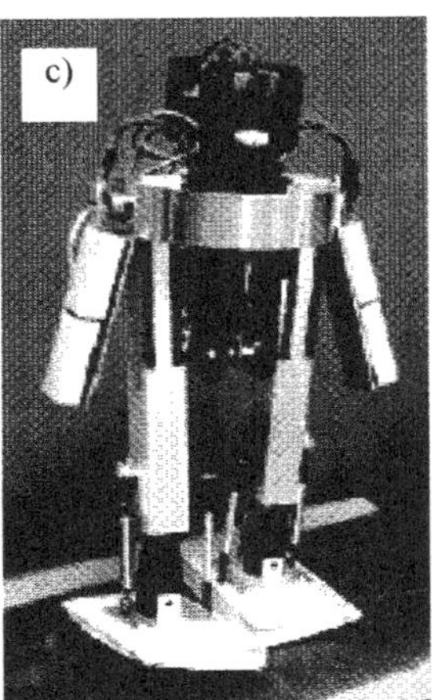

Figure 5.33. *(a) The passive bipedal robot, Cornell. This robot which has knees and arms is probably one of the passive robots which resembles humans the most; (b) the active bipedal robot Cornell (inspired from the passive robot) can move along flat ground and consumes very little energy (its energy efficiency is similar to that of a human); (c) the MIT biped is learning to walk, its movements based on passive walking [COL 05]*

For a robot with a trunk, it is not possible to obtain passive cyclic walking if the trunk is upright. Nevertheless, this approach can be extended to semi-passive gaits as the torques enable control of the relative trunk-thigh angle as the other articulations are passive [KHA 03].

Only knees with anti-counter-flexion position stops can be used for passive walking. If there are no position stops, the semi-passive approach can also be useful.

Research on passive cyclic movements can be carried out experimentally. It can also be carried out on robot models. For the latter, the principle is as follows. Starting from an appropriate initial condition (configuration and joint velocities), a cyclic movement is determined for a robot on a slope (with no torques). The loss of kinetic energy during impact compensates fully for the variation in potential energy due to the slope. Researching initial conditions which lead to a passive movement can be carried out numerically by using a linear or nonlinear dynamic model as a starting point. After a step, the robot must return to its initial state.

When a dynamic model is supposed linear, the step duration is known. The linear hybrid system becomes:

$$\begin{aligned} &\dot{x} = Ax \text{ if } t < T \\ &x(T^+) = \Delta x(T^-) \end{aligned} \qquad [5.62]$$

Via integration after a step, the state is $x(T^+) = \Delta\, e^{AT}\, x(T^+)$. There is a cyclic movement if matrix Δe^{AT} has a unitary eigenvalue. The corresponding initial conditions are determined by the associated eigenvector. The periodic cycle is stable if the other eigenvalues are inferior to 1 in norm. If the initial conditions are to be found in the cycle's surrounding area, then there is a convergence towards this cycle.

We can avoid using the linear model by using the Poincaré section and the tools presented in section 5.2. The system's evolutionary profile can be calculated from the hybrid model [5.1] with control $u = 0$.

In general, the choice of using the Poincaré section is such that we consider the robot's state just before impact (we therefore have $\phi(x) = 0$). The state which corresponds to the cyclic movement is the fixed Poincaré application point. The J_P eigenvalues enable us to know if the cycle is stable.

The main advantages of these passive movements include the following:

– the "automatic" generation of a joint cycle, corresponding to the robot's forward displacement. These movements are visually similar to human movement;

– on a slope, these movements are energetically efficient because the torques are zero. The robot morphology is adapted to make these movements possible. The foot shape is studied in particular (point contact or spherical soles). If the robot has knees, a mechanical lock is put in place for when the leg is straight, in order to avoid counter-flexions. An upright trunk (where the trunk's mass center is above the hips) is incompatible with a purely passive movement;

– by using these gaits as an inspirational starting point, controls can be developed to obtain non-passive movements on flat ground. For compass type robots, controls based on the system's energy are appropriate [ESP 94]. For robots which are completely actuated, joint torques can compensate for the effect of gravity [SPO 03];

– the disadvantage of this type of approach is that it is locally stable but not very robust: the limit cycle's attraction domain is small. Control can help us to widen this attraction basin while maintaining the limit cycle's beneficial properties (low energy cost). In order to do this, the control laws mentioned in section 5.5 can be used with a reference trajectory which results from a passive approach [WES 06]. Intuitive approaches, which are dedicated to this type of approach, can also be used [WIS 05].

During displacement, a passive planar robot will easily fall when faced with a disruption. Creating an attraction basin for passive walking can enable us to account for this result and help to remedy it.

As this attraction basin is created from a numerical problem, it has a high calculation cost. This is why studies have been based on the simplest robot walker [SCH 01]: a compass robot (see Figure 5.34a). The robot's initial state at the start of a step is determined by state ($\theta, \dot{\theta}$).

Figure 5.34b shows that a backwards fall occurs in zone B if the biped's initial kinetic energy is not sufficient for the robot to pass over the vertical support point which corresponds to the maximum potential energy.

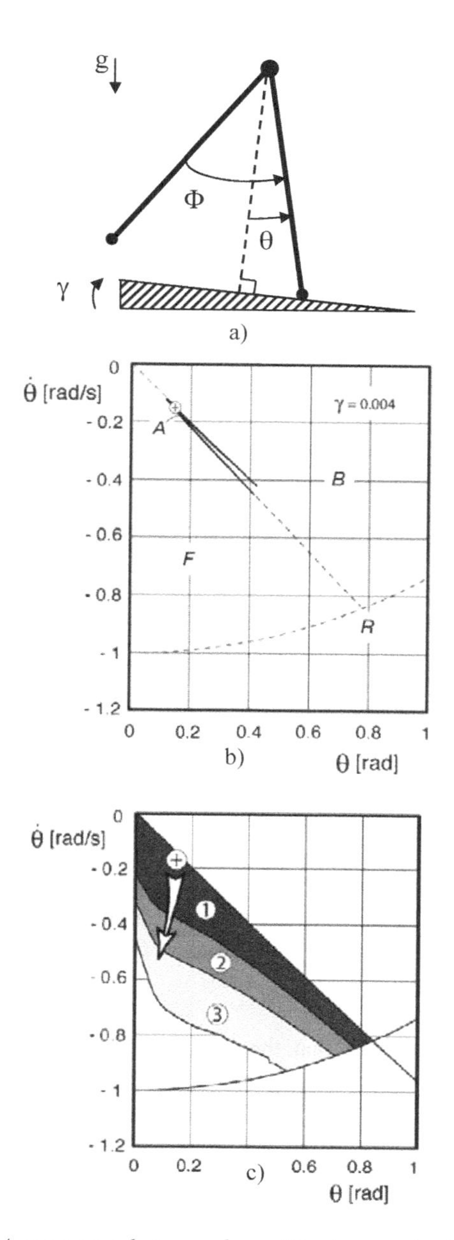

Figure 5.34. *(a) A compass robot on a slope; (b) attraction basin for the passive walking gait on a 0.004 rad slope (the attraction basin for the cyclic walk defined for $\theta = 0.1534$, $\dot{\theta} = -0.1561$ is indicated by +); (c) the attraction basin with proportional and derivatives actions [5.63], k = 25 leads to zone (1), k = 50 leads to zone (2) and k = 100 leads to zone (3) [WIS 05]*

Figure 5.34c shows that the proportional and derivative control (with torque between hips) does not stop the fall occurring in zone B. Zone R corresponds to the zone where an over-actuated velocity results in the supporting leg's take-off (via the centrifugal effect).

Zone F corresponds to a zone where the robot falls forward because the swing leg is not correctly positioned during impact. For wheels with no rim, however, Coleman demonstrated that a forward fall is not possible if the distance between the radii is sufficiently small [COL 97]. For a 0.004 rad slope, a 0.3 rad inter-radii angle is an appropriate value, and corresponds to the initial compass angle of 0.15 rad. Resultantly, the authors deduced that control, which quickly brings the swing leg past and in front of the supporting leg (so that the angle between the two legs is 0.3 rad), enables them to create a more robust robot walk. Angle control is implanted between the two legs in the form of:

$$\Gamma = -k_p(\Phi - \Phi_d) - k_v\dot{\Phi} \qquad [5.63]$$

where Φ_d is the desired angle for Φ ($\Phi_d = 0.3$ rad), k_p is positive and $k_v = 2\sqrt{k_p}$ to have a critical damping coefficient. The influence of k_p on the attraction basin is shown in Figure 5.34c. This type of control was successfully implanted in the Mike robot [WIS 05].

5.9. Conclusion

One of the difficulties of walking is unilateral contacts. The control must avoid a robot fall, which can occur due to external disruptions. This is mainly due to neglecting the constraints relating to contact (non-slide, non-take-off and non-rotation of the stance foot).

This chapter demonstrated that there are various solutions to these problems. One useful solution is online reference modification. The Poincaré stability analysis is a convenient and efficient tool for the evaluation of control performance during cyclic walking. When it is supposed that reference trajectories are perfectly followed, the resultant stability analysis leads to formal conditions. In this simplified model, reference trajectories were defined for the controlled variables in relation to a configuration variable or to a monotonous parameter. The results are interesting because we can predict, explain and improve a robot walker's control stability, or modify the reference trajectory itself online.

It is also possible to obtain good individual results for dynamically stable robot walkers by adopting either human or animal walking invariants. Controls may also be based on intuitive (or physically coherent) criteria, or on neural networks. The neural network approach, which is not explicitly based on the robot's dynamic model, leads to walking gaits which can reveal themselves to be experimentally, but unexplainably, robust. By using passive walking as a starting point, it is also possible to build energetically efficient controls. Intuitive adaptations of these controls can make them more robust.

5.10. Bibliography

[ALB 75a] ALBUS J.S., "A new approach to manipulator control: the cereballer model articulation controller (CMAC)", *Journal of Dynamic Systems, Measurement and Control*, vol. 97, no. 3, p. 200–227, 1975.

[ALB 75b] ALBUS J.S., "Data storage in the cerebellar model articulation controller (CMAC)", *Journal of Dynamic Systems, Measurement and Control*, vol. 97, no. 3, p. 228–233, 1975.

[ANI 96] ANIL K.J., JIANCHANG M., "Artificial neural networks: a tutorial", *Computer*, vol. 29, no. 3, p. 31–44, 1996.

[AOU 03] AOUSTIN Y., FORMAL'SKY A.M., "Control design for a biped reference trajectory based on driven angles as functions of the undriven angle", *Journal of Computer and System Sciences International*, vol. 42, no. 4, p. 159–176, 2003.

[AOU 06] AOUSTIN Y., CHEVALLEREAU C., FORMAL'SKY A., "Numerical and experimental study of a virtual quadrupedal walking robot – Semiquad", *Multibody System Dynamics*, vol. 16, p. 1–20, 2006.

[AZE 02] AZEVEDO C., POIGNET P., "Commande prédictive pour la marche d'un robot bipède sous-actionné", *Actes de la Conférence Internationale Francophone d'Automatique*, 2002.

[BEN 97] BENBRAHIM H., FRANKLIN J.A., "Biped dynamic walking using reinforcement learning", *Robotics and Autonomous Systems*, vol. 22, p. 283–302, 1997.

[BHAT 98] BHAT S.P., BERNSTEIN D.S., "Continuous finite-time stabilization of the translational and rotational double integrator", *IEEE Transaction on Automatic Control*, vol. 43, no. 5, p. 678–682, 1998.

[CHE 03] CHEVALLEREAU C., LORIA A., "Commande pour la marche et la course d'un robot bipède", *Actes des Journées Robea*, Toulouse, 2003.

[CHE 05a] CHEMORI A., Quelques contributions à la commande non linéaire des robots marcheurs bipèdes sous-actionnés, PhD Thesis, Institut national polytechnique de Grenoble, 2005.

[CHE 05b] CHEVALLEREAU C., WESTERVELT E.R., GRIZZLE J.W., "Asymptotically stable running for a five-link, four-actuator, planar bipedal robot", *International Journal of Robotics Research*, vol. 24, no. 6, p. 431–464, 2005.

[COL 97] COLEMAN M.J., CHATTERJEE A., RUINA A., "Motions of a rimless spoked wheel: a simple 3d system with impacts", *Dynamic Stability Systems*, vol. 12, no. 3, p. 139–160, 1997.

[COL 05] COLLINS S., RUINA A., TEDRAKE R, WISSE M., "Efficient bipedal robots based on passive-dynamic walkers", *Science Magazine*, vol. 307, no. 19, p. 1082–1085, 2005.

[CRU 81a] CRUSE H., "Is the position of the femur-tibia joint under feedback control in the walking stick insect: I. force measurements", *Journal of Experimental Biology*, vol. 92, p. 67–85, 1981.

[CRU 81b] CRUSE H., PFLÜGER H.J., "Is the position of the femur-tibia joint under feedback control in the walking stick insect? I. electrophysiological recordings", *Journal of Experimental Biology*, vol. 92, p. 97–107, 1981.

[DAN 63] DANTZIG G., *Linear Programming and Extensions*, Princeton University Press, Princeton, New Jersey, 1963.

[DJO 07] DJOUDI D., Contribution à la commande d'un robot bipède, PhD Thesis, Ecole centrale de Nantes and University of Nantes, 2007.

[ELA 99] EL ALI B., Contribution à la commande du center de masse d'un robot bipède PhD Thesis, Institut national polytechnique de Grenoble, 1999.

[ESP 94] ESPIAU B., GOSWAMI A., "Compass gait revisited", *IFAC Symposium on Robot Control*, p. 839–846, 1994.

[GAL 96] GALLAGHER J., BEER R., ESPENSCHIELD K., QUINN R., "Application of evolved locomotion controllers to a hexapod robot", *Robotics and Autonomous Systems*, vol. 19, p. 95–103, 1996.

[GEN 06] GENG T., PORR B., WÖRGÖTTER F., "Coupling of neural computation with physical computation for stable dynamic biped walking control", *Neural Computation*, vol. 18(5), 1156–1196, 2006.

[GOS 97] GOSWAMI A., ESPIAU B., KERAMANE A., "Limit cycles in a passive compass gait biped and passivity-mimicking control laws", *Autonomous Robots*, vol. 4, no. 3, p. 273–286, 1997.

[GRI 94] GRISHIN A.A., FORMAL'SKY A.M., LENSKY A.V., ZHITOMIRSKY S.V., "Dynamic walking of a vehicle with two telescopic legs controlled by two drives", *The International Journal of Robotics Research*, vol. 13, no. 2, p. 137–147, 1994.

[GRI 01] GRIZZLE J.W., ABBA G., PLESTAN F., "Asymptotically stable walking for biped robots: Analysis *via* systems with impulse effects", *IEEE Transaction on Automatic Control*, vol. 46, no. 1, p. 51–64, 2001.

[GUC 85] GUCKENHEIMER J., HOLMES P., *Nonlinear Oscillations, Dynamical Systems, and Bifurcations of Vector Fields*, Springer-Verlag, New York, 1985.

[HEN 94] HENAFF P., Mises en œuvre de commandes neuronales par rétropropagation indirecte: applications à la robotique mobile, PhD Thesis, Pierre and Marie Curie University, 1994.

[HIR 98] HIRAI K., HIROSE M., HAIKAWA Y., TAKENAKA T., "The development of Honda humanoid robot", *Proceedings of the international conference on robotics and automation*, p. 1321–1326, Leuven, Belgium, 1998.

[HUR 94] HURMUZLU Y., BASTOGAN C., "On the measurement of dynamic stability of human locomotion", *Journal of Biomechanical Engineering*, vol. 116, no. 1, p. 30–36, 1994.

[KHA 03] KHRAIEF N., M'SIRDI N.K., SPONG M.W., "An almost passive walking of a kneeless biped robot with torso", *Proceedings of the European Control Conference*, 2003.

[MCC 43] MCCULLOCH W.S., PITTS W.H., "A logical calculus of the ideas immanent in neural net", *Bulletin of Mathematical Biophysics*, vol. 5, no. 115–133, p. 16–39, 1994.

[MCG 90] MCGEER. T., "Passive dynamic walking", *International Journal of Robotics Research*, vol. 9, no. 2, p. 62–82, 1990.

[MIO 02] MIOSSEC S., AOUSTIN Y., "Mouvement de marche compose de simple et double supports pour un robot bipède planaire sans pieds", *Actes de la Conférence Internationale Francophone d'Automatique*, 2002.

[MIO 05] MIOSSEC S., AOUSTIN Y., "A simplified stability study for a biped walk with underactuated and overactuated phases", *The International Journal of Robotics Research*, vol. 24, no. 7, p. 537–551, 2005.

[MOR 06] MORRIS B., WESTERVELT E.R., CHEVALLEREAU C., BUCHE G., GRIZZLE J.W., "Achieving bipedal running with Rabbit: Six steps toward infinity", in M. DIEHL, K. MOMBAUR (ed.), *Fast Motions in Biomechanics and Robotics*, Springer-Verlag, Heidelberg, p. 277–297, 2006.

[PLE 03] PLESTAN F., LAGHROUCHE S., "A new sliding mode approach for Rabbit's walking", *Proceedings of the 6th International Conference on Climbing and Walking Robots Clawar'03*, p. 181–188, Catania, Sicily, 2003.

[POI 04] POINCARÉ H., *Œuvres complètes*, J. Gabay, Paris, 2004.

[POU 05] POULAKAKIS I., SMITH J.A., BUEHLER M., "Modeling and Experiments of Untethered Quadruped Running with a Bounding Gait: The Scout II Robot", *International Journal of Robotics Research*, vol. 24, no. 4, p. 239–256, 2005.

[PRA 98] PRATT J., PRATT G., "Intuitive control of a planar bipedal walking robot", *Proceedings of the International Conference in Robotics and Automation*, p. 2014–2021, 1998.

[PRA 01] PRATT J., CHEW C.H., TORRES A., DILWORTH P., PRATT G., "Virtual Model Control: An intuitive Approach for Bipedal Locomotion", *The International Journal of Robotics Research*, vol. 20, no. 2, p. 129–143, 2001.

[REE 99] REEVE R., Generating walking behaviors in legged robots, Unpublished PhD Thesis, University of Edinburgh, 1999.

[ROS 62] ROSENBATT F., *Principles of Neurodynamics*, Spartan Books, USA, 1962.

[SAB 04] SABOURIN C., Approche bio-inspirée pour le contrôle de la marche dynamique d'un bipède sous-actionné: validation expérimentale sur le robot Rabbit, PhD Thesis, University of Orleans, 2004.

[SCH 01] SCHWAB A.L., WISSE M., "Basin of attraction of the simplest walking model", *International Conference on Noise and Vibration*, 2001.

[SLO 91] SLOTINE J.-J., LI W., *Applied Nonlinear Control*, Prentice Hall, Englewood Cliffs, New Jersey, 1991.

[SON 88] SONG S., WALDRON K., *Machines That Walk: The Adaptive Suspension Vehicle*, MIT Press, Cambridge, Massachusetts, 1988.

[SPO 03] SPONG M.W., BULLO F., "Controlled symmetries and passive walking", *IEEE Transactions on Automatic Control*, vol. 50, no. 7, p. 1025–1031, 2005.

[TZA 03] TZAFESTAS S., RAIBERT F.M., TZAFESTAS C., "Robust sliding-mode control applied to a 5-link biped robot", *Journal of Intelligent and Robotics Systems*, vol. 15, no. 1, 1996.

[WES 03] WESTERVELT E.R., GRIZZLE J.W., CANUDAS DE WIT C., "Switching and PI control for planar biped walkers", *IEEE Transactions on Automatic Control*, vol. 48, no. 2, p. 308–312, 2003.

[WES 04] WESTERVELT E.R., BUCHE G., GRIZZLE J.W., "Experimental validation of a framework for the design of controllers that induce stable walking in planar bipeds", *The International Journal of Robotics Research*, vol. 24, no. 6, p. 559–582, 2004.

[WES 06] WESTERVELT E.R., MORRIS B., FARRELL K.D., "Sample-based HZD control for robustness and slope invariance of planar passive bipedal gaits", *The IEEE 14th Mediterranean Conference on Control Automation*, 2006.

[WES 07] WESTERVELT E.R., GRIZZLE J.W., CHEVALLEREAU C., CHOI J.-H., MORRIS B., *Feedback Control of Dynamic Bipedal Robot Locomotion*, Taylor and Francis/CRC Press, Boca Raton, Florida, 2007.

[WIS 05] WISSE M., SCHWAB A.L., VAN DER LINDE R.Q., VAN DER HELM F.C.T., "How to keep from falling forward: elementary swing leg action for passive dynamic walkers", *IEEE Transactions on Robotics*, vol. 21, no. 3, p. 393–401, 2005.

Index

M, N, O

P, R

S, T

U, W, Z

Printed and bound by CPI Group (UK) Ltd, Croydon, CR0 4YY

12/01/2025

14624099-0001